Metal Forming Plasticity

International Union of Theoretical
and Applied Mechanics

Metal Forming Plasticity

Symposium Tutzing/Germany
August 28 – September 3, 1978

Editor **H. Lippmann**
in Cooperation with
W. Johnson, H. Kudo, G. S. Pisarenko
W. Szczepinski

With 188 Figures

Springer-Verlag
Berlin Heidelberg New York 1979

Dr. rer. nat. HORST LIPPMANN
o. Professor am Lehrstuhl A für Mechanik
und Leiter des Staatlichen Materialprüfamtes für den Maschinenbau
der Technischen Universität München

ISBN-13:978-3-642-81357-3 e-ISBN-13:978-3-642-81355-9
DOI: 10.1007/978-3-642-81355-9

Softcover reprint of the hardcover 1st edition 1979

2060/3020/543210

Preface

The International Union of Theoretical and Applied Mechanics (IUTAM) which is the head organisation of most of the existing national and international societies of mechanics, decided to sponsor a Symposium on

METAL FORMING PLASTICITY.

It was held near Munich (Federal Republic of Germany) between August 28 and September 3, 1978, in the "Evangelische Academy" in the Castle of Tutzing which is situated in a park at Lake Starnberg overlooking the Alps.

The subjects of the Symposium were basic aspects of the theoretical and experimental mechanics of metal forming processes rather than technological details, or plasticity as such. Thus the spectrum of the Conference extended from necessary physical background, through experimental, analytical, or numerical methods, to applications to specific technological deformation processes such as rolling, deep drawing, extrusion, etc.

The following persons were by the IUTAM-bureau appointed to membership of the "Scientific Committee" which was responsible for the nomination of participants as well as for the form of the scientific program:

W. Johnson (U.K.), H. Kudo (Japan),
H. Lippmann (F.R.G, chairman),
G.S. Pisarenko (USSR), and W. Szczepiński (Poland).

The technical organisation was in the hands of a "Local Organizing Committee" formed by

F. Fischer, K. Heckel, G. Kuhn,
H. Lippmann (chairman), K. Magnus,
V. Mannl, G. Sonntag,
- all of them from Munich -
and K. Lange (Stuttgart), O. Pawelski (Düsseldorf).

This committee was supported by two secretaries, i.e. Mrs. U. Boldt and Mrs. M. Lippmann, as well as by 11 members of the scientific staff of the Technological University of Munich Institute of Mechanics. I am greatly indebted, and express my sincere thanks to all those who did their very best in helping me to prepare or to organize the Meeting in such a way that it became successful. Among them Bill Johnson, Günter Kuhn, Volker Mannl and my wife Martina had to share the major load.

Including the members of the committees and the 11 staff members mentioned above, 75 invited scientists belonging to the following countries participated: Australia (2), Austria (2), Bulgaria (2), Canada (2), France (3), FRG (19 + 11), Italy (1), Japan (8), Norway (1), Poland (5), South-Corea (1), Turkey (1), U.K. (8), USA (8), USSR (1). Approximately 25 scientific lectures were delivered in all and are published subsequently in this volume; also because of the wide selection of papers offered for the Conference in the first place, members of the Scientific Committee agreed to refrain from presenting own contributions. The following topics characterize the scientific progress achieved:

(1) New directions of research are perceived to have shifted from being preoccupied with the loads necessary to perform a metal forming operation to,

(1a) the flow of material, and the properties of the product generated by it,

(1b) residual stresses,

(1c) limits and defects of deformation and

(1d) optimization of processing.

(2) Interrelations between the flow of material, and the design of tools, esp. with regard to optimization or friction.

(3) Improvement, adaptation, and the extension of the analytical or numerical methods.

(4) Emphasis on the real behaviour of material rather than to a simplified, idealized one.

(5) Improvement, adaptation, and extension of experimental techniques, also with a view to the mathematical description.

The preliminary consequences of an investigation initiated and conducted by the Japanese colleagues H. Kudo and S. Matsubara, for comparing 10 Finite-Element Computer-Programmes (FEM) which had been developed independently at different institutes across the world, stimulated participants to several discussions until deep in the night, and to a final official round-table discussion which lasted several hours on Sunday morning. It turned out that there were grave differences in the results. It was agreed that similar comparisons concerning different applications of FEM, would be of the utmost use, and that a report on the present state of the project should additionally be included into these Proceedings.

During the Symposium, an extensive personal exchange of experience was made possible because sufficient time was available for each lecture (i.e. 30 or 45 min. as desired), and at least 15 min. for (vivid) discussion. Besides, there were the official and non-official accompanying discussions partly mentioned above. The success of the Conference was enhanced by the exclusive atmosphere of the Evangelische Akademie in Tutzing in which most of the participants were mutually accommodated and taken care of - or possibly also because of the bad weather! Last not least, it is my duty and my pleasure to gratefully acknowledge the financial and technical support given to the Symposium by the following institutions or industrial companies:

Deutsche Forschungsgemeinschaft (DFG),
D-5300 Bonn-Bad Godesberg

Bayerisches Staatsministerium für Unterricht und
Kultus, D-8000 München 1

International Union of Theoretical and Applied
Mechanics (IUTAM)

Siemens AG, D-8000 München 2

AEG-Telefunken, Energie- und Industrietechnik,
D-6000 Frankfurt / Main 71

L. Schuler GmbH, D-7320 Göppingen

Deutsche Lufthansa AG, D-6000 Frankfurt / Main

MAN, D-8000 München 50

BHS-Bayerische Berg-, Hütten- und Salzwerke AG,
D-8000 München.

Thanks are also due to the Springer-Verlag for the excellent co-operation regarding the publication of this volume.

München, December 1978

H. Lippmann

Participants

(Authors are identified by a circle)

AL-KHATTAT, I., Dept. of Mechanical Engineering, Stanford University, Stanford, Cal. 94305, USA

ANDRESEN, K., Mechanikzentrum TU Braunschweig, Pockelsstraße 14, D-3300 Braunschweig, FRG

AVITZUR, B., Dept. of Metallurgy and Materials Science, Lehigh University, Whitaker Laboratory 5, Bethlehem, Pa. 18015, USA

BARGMANN, H., Montanuniversität, A-8700 Leoben, Österreich

°BALTOV, A., Bulgarian Academy of Sciences, Inst. of Mathematics and Mechanics, P.O. Box 373, Sofia 13, Bulgaria

°BEDNARSKI, T., Inst. of Metal Forming Technology, Warsaw Technical University, Narbutta 85, 02-524 Warsaw, Poland

BEHRENS, A., Hochschule der Bundeswehr Hamburg, Inst. für Kontruktionstechnik, Fachgebiet Fertigungstechnik, Holstenhofweg 85, D-2000 Hamburg 70, FRG

BLANKENAGEL, H.J., in Firma MBB Apparate, Postfach 801149, D-8000 München 80, FRG

°BLAZYNSKI, T.Z., The University of Leeds, Dept. of Mechanical Engineering, Leeds LS2 - 9JT, U.K.

°BONTCHEVA, N., Bulgarian Academy of Sciences, Inst. of Mathematics and Mechanics, P.O. Box 373, Sofia 13, Bulgaria

BRAMLEY, A.N., Dept. of Mechanical Engineering, University of Leeds, Leeds LS2 - 9JT, U.K.

°BRATT, J.F., Faculty of Mechanical Engineering, Norwegian Institute of Technology, Trondheim, Norge

°BRUHNS, O., Ruhr-Universität Bochum, Institut für Mechanik, D-4630 Bochum 1, FRG

BUCHHOLZ, F.-G., Gesamthochschule Paderborn, Fachbereich 10, Pohlweg 47-49, D-4790 Paderborn, FRG

°COLLINS, I.F., Dept. of Mathematics, University of Manchester Inst. of Science and Technology, P.O. Box 88, Manchester M60 1QD, U.K.

°CRISTESCU, N., Facultatea de Matematica, str. Academiei, 14, Bukuresti 1, Romania

°DIETRICH, L., Inst. of Fundamental Technological Research, Polish Academy of Sciences, ul.Swietokrzyska 21, 00-049 Warsaw, Poland

EL-MAGD, E., Lehrgebiet für Werkstoffkunde, Rhein.-Westf. TH Aachen, Augustinerbach 4, D-5100 Aachen, FRG

FISCHER, F., Lehrstuhl für Verformungskunde und Gießereiwesen, TU München, Tengstr. 11/I, D-8000 München 40, FRG

GATTO, F., Istituto Sperimentale dei Metalli Leggeri, Via G. Fauser 4, Novara, Italy

GUNASEKERA, J.S., Dept. of Mechanical Engineering, Monash University, Clayton,Victoria, Australia 3168

HAWKYARD, J.B., Dept. of Mechanical Engineering, U.M.I.S.T., Sackville Street, Manchester M60 1QD, U.K.

HECKEL, K., FB Luft- und Raumfahrttechnik, Hochschule der Bundeswehr München, Fliegerhorst, D-8014 Neubiberg, FRG

HOPKINS, H.G., The University of Manchester Inst. of Science and Technology, P.O. Box 88, Manchester M60 1QD, U.K.

IKEGAMI, K., Tokyo Institute of Technology, Res. Lab. of Precision Machinery and Electronics, Nagatsuta, Midori-Ku, Yokohama 227, Japan

INOUE, T., Dept. of Mechanical Engineering, Kyoto University, Kyoto, Japan

JOHNSON, W., University Engineering Dept., Trumpington Street, Cambridge CB2 1PZ, U.K.

°KAFTANOǦLU, B., Dept. of Mechanical Engineering, Middle East Techn. University, Ankara, Turkey

°KLIE, W., Institut für Mechanik, TU Hannover, Appelstr. 24B, D-3000 Hannover, FRG

°KOBAYASHI, S., Mechanical Engineering Dept., Mechanical Design Division, University of California, Berkeley, Cal. 94720, USA

°KUDO, H., Yokohama National University, Dept. of Mechanical Engineering, Ohka, Minami-Ku, 2-31-1 Yokohama 232, Japan

KUHN, G., Institut für Mechanik, TU München, Arcisstr. 21, D-8000 München 2, FRG

°LEE, E.H., Dept. of Mechanical Engineering, Stanford University, Stanford, Cal. 94305, USA

LIPPMANN, H., Institut für Mechanik, TU München, Arcisstr. 21, D-8000 München 2, FRG

LUBORDA, V., Dept. of Mechanical Engineering, Stanford University, Stanford, Cal. 94305, USA

MAGNUS, K., Lehrstuhl B für Mechanik, TU München, Arcisstr. 21, D-8000 München 2, FRG

°MALLETT, R.L., Division of Applied Mechanics, Stanford University, Stanford, Cal. 94305, USA

°MALININ, N.N., Moscow High Technical School, 2-ya Baumanskaya ul. 5, Moskow, USSR

MAMALIS, A.G., University Engineering Dept., Trumpington Street, Cambridge CB2 1PZ, U.K.

MANNL, V., Institut für Mechanik, TU München, Arcisstr. 21, D-8000 München 2, FRG

MARECZEK, G., Abt. Numerische Methoden in der Umformtechnik, TU Hannover, D-3000 Hannover, FRG

NAJAR, J., Polish Academy of Sciences, Inst. of Fundamental Techn. Research, ul. Swietokrzyska 21, 00-049 Warsaw, Poland

°NEALE, K.W., University of Sherbrooke, Sherbrooke, Québec, Canada

°NEVEZ, P. LE, Laboratoire de Mécanique et Technologie (E.N.S.E.T.), 61, Av. du Président Wilson, F-94230 Cachan, France

°OXLEY, P.L.B., School of Mechanical and Industrial Engineering, P.O. Box 1, Kensington, New South Wales, Australia

°PAWELSKI, O., Max-Planck-Institut f. Eisenforschung, Max-Planck-Str. 1, D-4000 Düsseldorf 1, FRG

°PETRYK, H., Inst. of Fundamental Technological Research, Polish Academy of Sciences, ul. Swietokrzyska 21, 00-049 Warsaw, Poland

PISARENKO, G.S., Inst. for Problems of Strength, Academy of Sciences of the Ukr.SSR, ul. Timiryazevskaya 2, 252014 Kiev 14, UkSSR

PREDELEANU, M., 68, Bl.St. Michel, Paris (6^e), France

°RASP, W., Max-Planck-Institut f. Eisenforschung, Max-Planck-Str. 1, D-4000 Düsseldorf 1, FRG

°RICHMOND, O., U.S. Steel Corporation Research Laboratory, Monroeville, Pa. 15146, USA

ROLL, K., Institut für Umformtechnik, Universität Stuttgart, Holzgartenstr. 17, D-7000 Stuttgart, FRG

°SAITŌ, K., Dept. of Mechanical Engineering, Osaka Prefectural University, 804 Mozuume 4-chome, Sakai 591, Japan

SAITŌ, Y., Dept. of Metallurgical Engineering, Osaka University (SUITA), Oaza-Yamadaue, Suita 565, Japan

SAKURAI, T., Japan Advanced Numerical Analysis, Inc., 1-1-71, Nakameguro, Meguro-Ku, Tokyo, Japan

°SCHEY, J.A., Faculty of Engineering, Dept. of Mechanical Engineering, Waterloo, Ontario N2L3G1, Canada

°SHIMA, S., Dept. of Mechanical Engineering, Faculty of Engineering, Kyoto University, Yoshida-Honmachi, Sakyo-ku, Kyoto 606, Japan

°SLATER, R.A.C., Dept. of Mechanical Engineering, City University, St. John Street, London EC IV 4 PB, U.K.

SONNTAG, G., Lehrstuhl C für Mechanik, TU München, Arcisstr. 21, D-8000 München 2, FRG

STECK, E., Lehrstuhl B für Mechanik, TU Braunschweig, Pockelsstraße 14, D-3300 Braunschweig, FRG

STÜWE, H.-P., Institut für Metallphysik der Montanistischen Hochschule, A-8700 Leoben, Österreich

SZCZEPIŃSKI, W., Polish Academy of Sciences, Inst. of Fundamental Techn. Research, ul. Swietokrzyska 21, 00-049 Warsaw, Poland

TANAKA, K., Dept. of Aeronautical Engineering, University of Osaka Prefecture, Mazu-Umemachi, Sakai, Osaka 591, Japan

WIDERA, G.E.O., College of Engineering, University of Illinois at Chicago Circle, Dept. of Materials Eng., P.O. Box 4348, Chicago 60680, USA

°YAMADA, Y., Res. Institute of Industrial Science, The University of Tokyo, 22-1, Roppongi 7-chome, Minato-ku, Tokyo 106, Japan

°YANG, D.-Y., Dept. of Mechanical Science, The Korea Advanced Institute of Science, P.O. Box 150, Chongyangni, Seoul, Korea

ZARKA, H., Ecole Polytechnique, Laboratoire de Mécanique des Solides, F-91120 Palaiseau, France

ZIMMERMANN, R., Richard-Wagner-Str. 15, D-4923 Extertal 1, FRG

FURTHER LOCAL PARTICIPANTS

from the Institut für Mechanik, Technische Universität (TU), Arcisstr. 21, D-8000 München 2, FRG

ALP, O.	MEYER, H.
BARTUSSEK, W.	NEUREITER, W.
BURGERT, W.	NIERING, E.
DREXLER, W.	OTTO, P.
FUCHS, J.	WINTER, W.
HEIGL, H.	

Contents

The consecution of papers in this volume corresponds to the sequence of presentation during the Symposium.

On the Prediction of Necking in Anisotropic Sheets

J.L. BASSANI
Dept. of Mechanical Engineering
Massachusetts Inst. of Technology
Cambridge, Mass. USA 02139

J.W. HUTCHINSON
Division of Applied Sciences
Harvard University
Cambridge, Mass. USA 02138

K.W. NEALE
Faculté des sciences appliquées
Université de Sherbrooke
Sherbrooke, Qué. CANADA J1K 2R1

Summary

Thin sheets with transversely isotropic material properties are studied, and we examine the effects of anisotropy on necking as predicted by various constitutive laws [1-3] which have been proposed for such metals. Both flow theory and deformation theory versions of the respective constitutive laws are considered. The results for both plasticity theories indicate that the necking strains and forming limit curves are strongly dependent on the *shape* of the anisotropic yield surface.

Introduction

A major difficulty which arises in the analysis of necking failures in thin sheets concerns the choice of an appropriate constitutive law to be incorporated in the analysis, since the predicted critical strains are often strongly dependent on the form of constitutive equation employed. For example, the use of the simplest finite-strain flow theory of plasticity leads to some questionable results for isotropic sheet materials [4]. In particular, for sheets subjected to in-plane biaxial stretching (where both principal strains are positive) this simple flow theory produces unrealistic "forming limit curves" and predicts that the critical strain for necking is exceptionally sensitive to small geometric imperfections. Furthermore, it indicates that a local necking mode of bifurcation cannot occur in a perfect sheet. On the other hand, the predicted trends obtained using a finite-strain deformation theory appear to be in much better agreement with reported experimental data. Even when the respective plasticity theories are extended to include the influence of material strain-rate-dependence, deformation theory seems to give a better qualitative agreement with the trends of experiments than does simple flow theory [5].

A realistic analysis of necking in sheet metals must eventually account for the effects of anisotropy. In this study we restrict attention to sheets with *transversely isotropic* material properties. Constitutive laws based on the yield criteria proposed by Hill [1,2] and Bassani [3] for such metals are used to determine the influence of anisotropy on necking. Since the detailed shapes of the yield surfaces depend on the parameters in the yield functions, it becomes possible to examine whether slight changes in the *shape* of the assumed anisotropic yield surface are important. We shall see that this is the case, and that the predicted necking or "limit" strains are very sensitive to the shape of the yield surface.

Yield Functions for Transversely Isotropic Sheets

To date, the most widely known constitutive law for transversely isotropic materials is based on a yield criterion due to Hill [1], who proposed the following yield function for plane stress

$$(\sigma_1+\sigma_2)^2 + (1+2R)\,(\sigma_1-\sigma_2)^2 = 2(1+R)\sigma_u^2 \tag{1}$$

Here, σ_1 and σ_2 are the principal stresses in the plane of the sheet ($\sigma_3=0$), the 3-axis is the axis of transverse isotropy, σ_u is the yield stress in uniaxial tension and R is the corresponding ratio of in-plane transverse plastic strain-increment to the thickness plastic strain-increment. As a consequence of (1), the ratio of yield stress in equi-biaxial tension ($\sigma_1=\sigma_2=\sigma_b$) to yield stress in uniaxial tension becomes

$$\frac{\sigma_b}{\sigma_u} = (\frac{1+R}{2})^{1/2} \quad (2)$$

implying that the σ_b/σ_u ratio lies on the same side of unity as does R. This trend is not always confirmed experimentally and leads to what has been termed as "anomalous behaviour" by Woodthorpe and Pearce [6].

Hill [2] has since modified his original criterion (1) and suggested the following three-parameter yield function

$$|\sigma_1+\sigma_2|^n + (1+2R)|\sigma_1-\sigma_2|^n = 2(1+R)\sigma_u^n \quad (3)$$

for which the σ_b/σ_u ratio now becomes

$$\frac{\sigma_b}{\sigma_u} = \frac{1}{2}[2(1+R)]^{1/n} \quad (4)$$

The "anomalous behaviour" can clearly be incorporated here, and this theory obviously reduces to Hill's original function (1) for n=2. Parmar and Mellor [7] have recently employed Hill's new yield criterion (3) together with the flow theory of plasticity to predict the limit strains occurring in in-plane stretching. They report that, for the test results given in [8] for aluminum, an improved correlation between theory and experiment is obtained when Hill's original criterion (1) is replaced by (3).

In a recent study [3], Bassani has compared several families of yield functions for polycrystals, including (3), with those calculated from a Bishop-Hill procedure. Ideal textures giving rise to transversely isotropic properties that could be expected for typical sheet metals were considered. Hill's new criterion (3) did not prove to be entirely adequate for predicting details of the yield loci obtained from the Bishop-Hill calculations. The following four-parameter family of yield functions

$$|\sigma_1+\sigma_2|^n + \frac{n}{m}(1+2R)\sigma_u^{n-m}|\sigma_1-\sigma_2|^m = [1+\frac{n}{m}(1+2R)]\sigma_u^n \quad (5)$$

was then introduced by Bassani in [3], and it was found to be sufficiently flexible to accurately approximate yielding for a wide range of transversely isotropic materials. Convexity of the yield function requires that n,m $\geq$ 1, and for n=m (5) reduces to (3). Bassani's criterion (5) gives the following relationship for the equi-biaxial to uni-axial yield stress ratio

$$\frac{\sigma_b}{\sigma_u} = \frac{1}{2}[1+\frac{n}{m}(1+2R)]^{1/n} \quad (6)$$

which again can account for the anomalous behaviour. A typical comparison between a yield locus calculated by the Bishop-Hill procedure (solid curve) and the phenomenological yield loci obtained from (3) and (5) (broken lines) is shown in Fig. 1.* For this case R=2.7 and the ratio σ_b/σ_u=1.02.

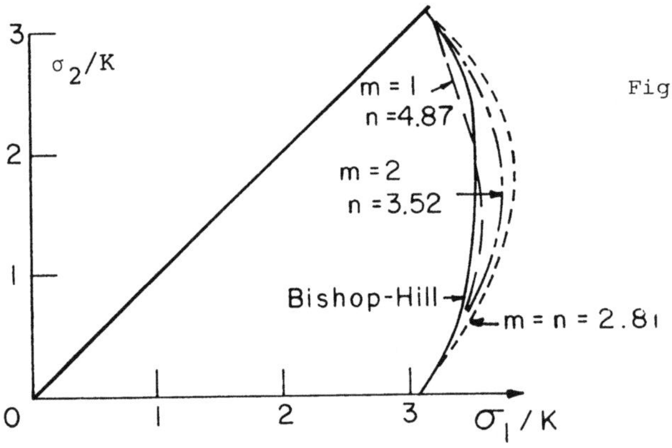

Fig. 1: Yield loci for [321] ideal texture, R=2.7

In the subsequent analysis the material is assumed to be rigid-plastic. Bassani's yield function is employed, with $\sigma_1 \geq \sigma_2$, so the expression for effective stress σ_e is given by the implicit relation

$$\sigma_e[1+\frac{n}{m}(1+2R)]^{1/n} = [(\sigma_1+\sigma_2)^n + \frac{n}{m}(1+2R)\sigma_e^{n-m}(\sigma_1-\sigma_2)^m]^{1/n} \quad (7)$$

---

* This figure is taken from [3] (Fig. 7).

The effective stress is homogeneous of degree one in the stress components. Using σ_e as the flow potential leads to the following flow theory constitutive law

$$d\varepsilon_i = d\lambda \frac{\partial \sigma_e}{\partial \sigma_i} \qquad i=1,2 \tag{8}$$

From the condition that the plastic work satisfies $dW = \sigma_i d\varepsilon_i = \sigma_e d\varepsilon_e$, where $d\varepsilon_e$ is the effective strain-increment, it follows that $d\lambda = d\varepsilon_e$. The constitutive equation (8) then takes the form

$$\frac{d\varepsilon_1}{(\sigma_1+\sigma_2)^{n-1}+(1+2R)(\sigma_1-\sigma_2)^{m-1}\sigma_e^{\,n-m}}$$

$$= \frac{d\varepsilon_2}{(\sigma_1+\sigma_2)^{n-1}-(1+2R)(\sigma_1-\sigma_2)^{m-1}\sigma_e^{\,n-m}} = \frac{-d\varepsilon_3}{2(\sigma_1+\sigma_2)^{n-1}} \tag{9}$$

$$= \frac{d\varepsilon_e}{[1+\frac{n}{m}(1+2R)]\,\sigma_e^{\,n-1}+(1-\frac{n}{m})(1+2R)(\sigma_1-\sigma_2)^m\sigma_e^{\,n-m-1}}$$

Since σ_1 and σ_2 are principal stresses, the corresponding finite-strain deformation theory constitutive law [4] is obtained by simply replacing $d\varepsilon_i$, $d\varepsilon_e$ in (9) by their total (logarithmic) values ε_i and ε_e.

M-K Analysis for Imperfect Sheets

The present analysis, for determining limit strains in biaxially-stretched sheets, is along the lines of that introduced by Marciniak and Kuczynski ("M-K") [9]. We consider a sheet (Fig. 2) subjected to proportional straining $(\varepsilon_1,\varepsilon_2)$ or stressing (σ_1,σ_2) along its edges such that

$$\frac{\varepsilon_2}{\varepsilon_1} = \rho = \text{const} \quad , \quad \frac{\sigma_2}{\sigma_1} = \alpha = \text{const} \tag{10}$$

where the fixed values of strain ratio ρ will be taken from $\rho=0$ (plane strain) to $\rho=1$ (equi-biaxial tension). The sheet initially has a nonuniformity in the form of a narrow band perpendicular to the major principal strain ε_1. The thickness at the minimum section in the band is denoted by $h^B(t)$, with an initial value $h^B(0)$; whereas the region outside the band, referred to as the "uniform" section, has thickness $h(t)$ with an initial value $h(0)$. The initial geometric nonuniformity is defined as

$$\xi = \frac{h(0)-h^B(0)}{h(0)} \geqslant 0 \tag{11}$$

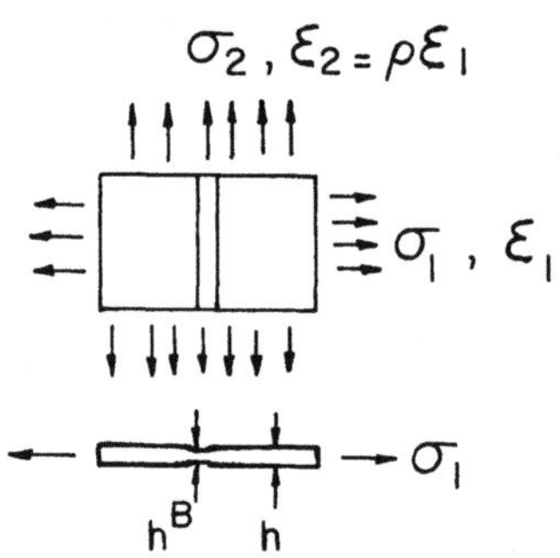

Fig. 2: Sheet geometry and loading

Throughout the analysis a superscript B will denote quantities along the minimum band section, and the absence of this symbol will refer to quantities in the uniform region. As discussed in [10], it is tacitly assumed in this M-K type analysis that the width-to-thickness ratio of the band is large.

According to the M-K simplification, the stress state over each cross-section is considered to be uniform; that is, the stress and strain components are quantities which are averaged through the thickness. The equilibrium condition across the band is simply

$$\sigma_1 h = \sigma_1^B h^B \tag{12}$$

We assume the relation between effective stress σ_e and the effective strain ε_e, obtained from a uniaxial true stress-natural strain curve, to be of the form

$$\sigma_e = K\varepsilon_e^N \tag{13}$$

This, together with (12) and the definitions

$$\varepsilon_3 = \ln \frac{h}{h(0)} \quad , \quad \varepsilon_3^B = \ln \frac{h^B}{h^B(0)} \tag{14}$$

leads to the following expression

$$\frac{\sigma_1}{\sigma_e} = (1-\xi) \frac{\sigma_1^B}{\sigma_e^B} \left(\frac{\varepsilon_e^B}{\varepsilon_e}\right)^N \exp(\varepsilon_3^B - \varepsilon_3) \tag{15}$$

The object of the subsequent analysis is to express all stress and strain quantities in (15) in terms of ε_e and ε_e^B. From the relationship between ε_e and ε_e^B we can calculate the development of the band and, in particular, determine the limit (maximum) strains attained in the uniform region of the sheet.

In the uniform section, where (10) holds, the expression (7) becomes

$$[(1+\alpha)\psi]^n + \frac{n}{m}(1+2R)\ [(1-\alpha)\psi]^m = 1 + \frac{n}{m}(1+2R) \tag{16}$$

in which $\psi = \sigma_1/\sigma_e$ = const. The constitutive law (9), or its deformation theory equivalent, then gives

$$\rho = \frac{d\varepsilon_2}{d\varepsilon_1} = \frac{\varepsilon_2}{\varepsilon_1} = \frac{[(1+\alpha)\psi]^{n-1} - (1+2R)\ [(1-\alpha)\psi]^{m-1}}{[(1+\alpha)\psi]^{n-1} + (1+2R)\ [(1-\alpha)\psi]^{m-1}} \tag{17}$$

So given ρ we can solve (16) and (17) for α and ψ. From (9), (10) and the definition for ψ it follows that

$$\frac{d\varepsilon_2}{d\varepsilon_e} = \frac{\varepsilon_2}{\varepsilon_e} = \frac{[(1+\alpha)\psi]^{n-1} - (1+2R)\ [(1-\alpha)\psi]^{m-1}}{[1+\frac{n}{m}(1+2R)] + (1-\frac{n}{m})(1+2R)\ [(1-\alpha)\psi]^m} \equiv b \tag{18}$$

and

$$\frac{d\varepsilon_3}{d\varepsilon_e} = \frac{\varepsilon_3}{\varepsilon_e} = \frac{-2\,[(1+\alpha)\psi]^{n-1}}{[1+\frac{n}{m}(1+2R)] + (1-\frac{n}{m})(1+2R)\ [(1-\alpha)\psi]^m} \equiv -c \tag{19}$$

Again, both flow theory and deformation theory lead to the above results since proportional loading (b,c = const.) occurs in the uniform region.

For straining within the band, we introduce the following variables analogous to α and ψ

$$\alpha^B = \frac{\sigma_2^B}{\sigma_1^B} \quad , \quad \psi^B = \frac{\sigma_1^B}{\sigma_e^B} \tag{20}$$

which, in general, do not remain constant during deformation. Equation (7) becomes [c.f. (16)]

$$[(1+\alpha^B)\psi^B]^n + \frac{n}{m}(1+2R)[(1-\alpha^B)\psi^B]^m = 1+\frac{n}{m}(1+2R) \tag{21}$$

and using (19), (20) and the relations $\psi = \sigma_1/\sigma_e$, the equilibrium condition (15) can be written as

$$\frac{\psi}{\psi^B} = (1-\xi)\left(\frac{\varepsilon_e^B}{\varepsilon_e}\right)^N \exp(c\varepsilon_e + \varepsilon_3^B) \tag{22}$$

For flow theory, the expressions analogous to (18) and (19) are

$$\frac{d\varepsilon_2^B}{d\varepsilon_e^B} = \frac{[(1+\alpha^B)\psi^B]^{n-1} - (1+2R)[(1-\alpha^B)\psi^B]^{m-1}}{[1+\frac{n}{m}(1+2R)] + (1-\frac{n}{m})(1+2R)[(1-\alpha^B)\psi^B]^m} \tag{23}$$

and

$$\frac{d\varepsilon_3^B}{d\varepsilon_e^B} = \frac{-2[(1+\alpha^B)\psi^B]^{n-1}}{[1+\frac{n}{m}(1+2R)] + (1-\frac{n}{m})(1+2R)[(1-\alpha^B)\psi^B]^m} \tag{24}$$

The corresponding relations for deformation theory have $\varepsilon_2^B/\varepsilon_e^B$ and $\varepsilon_3^B/\varepsilon_e^B$ in place of $d\varepsilon_2^B/d\varepsilon_e^B$ and $d\varepsilon_3^B/d\varepsilon_e^B$ on the left-sides of (23) and (24).

When the compatibility requirement $d\varepsilon_2^B = d\varepsilon_2$ (or $\varepsilon_2^B = \varepsilon_2$) is used together with (18) and (23), the following expression is obtained

$$\frac{d\varepsilon_e^B}{d\varepsilon_e} = b\,\frac{[1+\frac{n}{m}(1+2R)]+(1-\frac{n}{m})(1+2R)[(1-\alpha^B)\psi^B]^m}{[(1+\alpha^B)\psi^B]^{n-1}-(1+2R)[(1-\alpha^B)\psi^B]^{m-1}} \tag{25}$$

This relation is valid for the flow theory case and, as previously, the corresponding deformation theory expression consists of substituting $\varepsilon_e^B/\varepsilon_e$ for $d\varepsilon_e^B/d\varepsilon_e$ in (25). For n=m, i.e., with Hill's new criterion (3), equations (21) to (25) reduce to those derived by Parmar and Mellor in [7]. They also reduce to the expressions given in [8] when n=m=2.

Flow theory results were obtained from a straightforward incremental solution of (24) and (25). This determined the band deformation ε_e^B in terms of the prescribed uniform deformation ε_e. In the numerical solution, the strain ratio ρ was initially specified, and (16) and (17) were solved for α and ψ using a Newton-Raphson technique. A similar iterative procedure was employed at each increment to solve (21) and (22) for α^B and ψ^B. On the other hand, the results of ε_e^B for given ε_e according to deformation theory were obtained from a direct numerical solution of (21), (22) together with the deformation theory relations corresponding to (24) and (25). In all solutions the maximum value of strain attained in the uniform part of the sheet was determined for prescribed values of ρ. We refer to this limiting value of uniform strain, where $d\varepsilon_e/d\varepsilon_e^B=0$, as the "limit" strain or critical strain for localized necking as it represents the state where the deformation becomes concentrated in the band while the remainder of the sheet begins to unload.

Results and Discussion

The previous analysis was applied to generate numerical results for initial strain ratios varying from $\rho=0$ (plane strain) to $\rho=1$ (equi-biaxial tension). Theoretical limit strains are plotted in Figs. 3-6 as "forming limit curves", i.e., curves which illustrate the dependence of the limit strain ε_1^* on imposed strain ratio. In these figures, the solid and dashed curves refer to the deformation theory and flow theory predictions, respectively. Results

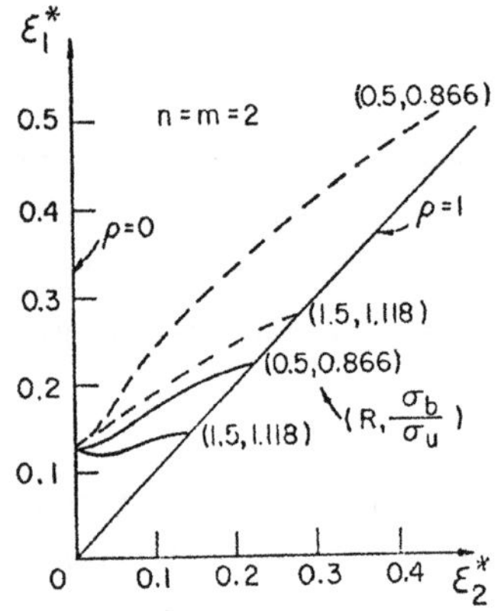

Fig. 3

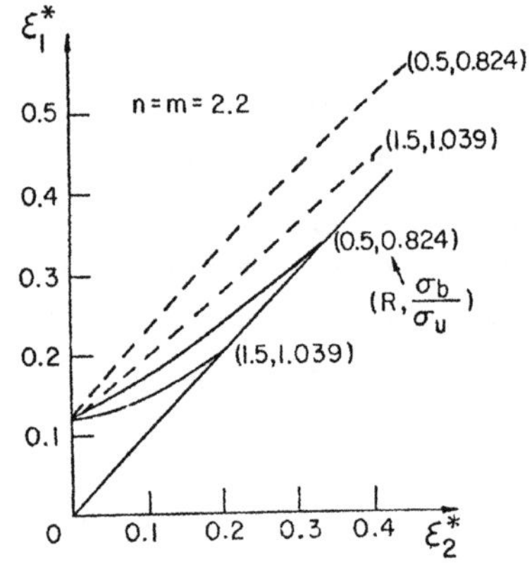

Fig. 4

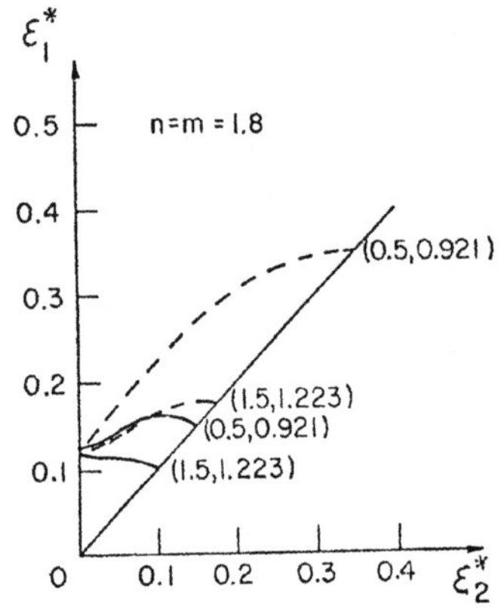

Fig. 5

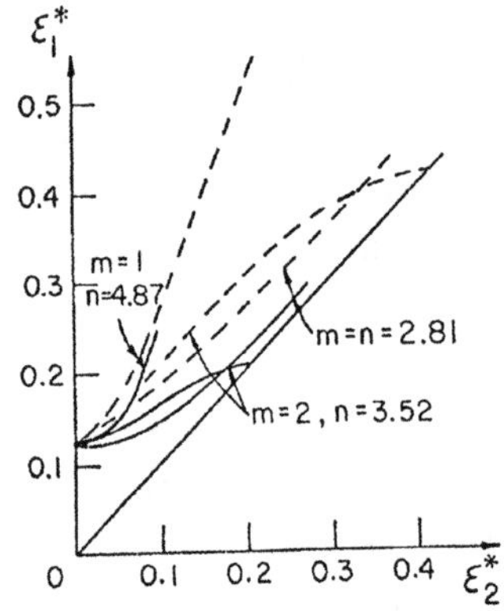

Fig. 6

Figs. 3-6:
Forming limit curves for N=0.20 and ξ=.02 (solid lines refer to deformation theory and dashed curves to flow theory).

are shown for a strain-hardening exponent $N=0.20$ and an initial imperfection $\xi=0.02$. The numbers in parentheses next to each curve in Figs. 3-5 denote the R-values and associated σ_b/σ_u ratios, respectively.

The results in Fig. 3 are obtained with Hill's original yield function (1), i.e., $n=m=2$, and the flow theory curves here are representative of previously given M-K analyses [7,8]. These curves rise much more steeply with increasing ρ than the corresponding deformation theory curves, especially as the R-value is decreased. Experimentally observed forming limit curves (e.g. [8]) do not appear to rise as rapidly as flow theory would predict, but seem to be more in line with the trends of deformation theory. It was apparently this discrepancy between the flow theory predictions and test data that prompted Parmar and Mellor [7] to seek a better correlation using Hill's new criterion (3) in their flow theory analysis.

In Figs. 4 and 5, results determined using Hill's new criterion (3) are shown with $n=m=2.2$ and 1.8, respectively. The shapes of the forming limit curves here are seen to be extremely sensitive to small changes in the $n=m$ values. In particular, as we approach the equi-biaxial stretching condition ($\rho=1$) the limit strain for given R is reduced drastically as the σ_b/σ_u ratio is increased. The flow theory curve for $n=m=1.8$ (Fig. 5) and $R=0.5$ is identical to that given in [7] and, as mentioned previously, this trend is in better agreement with experiments than the corresponding curve in Fig. 3. Nevertheless, the experimental forming limit curve [8] for aluminum still tends to be much flatter than the flow theory prediction for $n=m=1.8$, and bears more resemblance to the theoretical deformation theory curve.

Bassani's yield function (5) was employed to compute the results shown in Fig. 6. An R-value = 2.7 and a fixed ratio $\sigma_b/\sigma_u=1.02$ were assumed for these calculations; and m,n were chosen to produce the phenomenological yield surfaces of Fig. 1. In this way one could examine the effect of the *shape* of the yield surface on the predicted trends. It is clear from Fig. 6 that the forming limit curves are strongly dependent on the parameters m and n, that is on the details of the yield locus. This observation holds for both plasticity theories, despite the fact that the flow theory curves generally lie well above the deformation theory curves and predict much higher limit strains as $\rho \to 1$.

What seems to be particularly important is the shape of the yield surface near the equi-biaxial stress state. In Fig. 1 the yield locus for m=2, n=3.52 appears to best fit the results of the Bishop-Hill calculation at the equi-biaxial state ($\sigma_1=\sigma_2$), and it is for this case alone in Fig. 6 that we obtain "reasonable" forming limit curves. Although the choice m=1, n=4.87 gives perhaps the best overall approximation to the Bishop-Hill locus, it produces a poor fit near $\sigma_1=\sigma_2$ and leads to unrealistic forming limit curves. Furthermore, computational difficulties arise with m=1 because of the corner which exists in the yield surface at the equi-biaxial state. The forming limit curves in Fig. 6 for n=m=2.81 also seem somewhat unrealistic.

We finally note that the additional calculations were carried out for other values of m, n, R and σ_b/σ_u considered by Bassani in [3]. These cases examined corresponded to Figs. 6 and 9 of [3], where again m=2 gives a better approximation to the Bishop-Hill yield curve near the equi-biaxial state than m=1 or m=n. The trends and conclusions here are essentially the same as discussed above; namely, (i) flow theory generally predicts much higher limit strains than deformation theory as $\rho \to 1$, and (ii) the theoretical forming limit curves for both plasticity theories are strongly dependent on the detailed shapes of the anisotropic yield surfaces, especially on their shapes in the vicinity of the equi-biaxial stress state.

Acknowledgments

The work of J.L.B. and J.W.H. was supported in part by the National Science Foundation (Grant ENG 76-04019) and by the Division of Applied Sciences, Harvard University. The work of K.W.N. was supported by the National Research Council of Canada (Grant A-8584).

References

1 Hill, R.: The Mathematical Theory of Plasticity. Oxford University Press, London (1978).

2 Hill, R.: private communication (1976).

3 Bassani, J.L.: Yield characterization of metals with transversely isotropic plastic properties. Int. J. Mech. Sci. 19 (1977) 651-660.

4 Hutchinson, J.W.; Neale, K.W.: Sheet necking - II Time-independent behaviour. Proc. GM Symp. on Mechanics of Sheet Metal Forming. Plenum Press, New York (1978).

5 Hutchinson, J.W.; Neale, K.W.: Sheet necking - III Strain-rate effects. ibid.

6 Woodthorpe, J.; Pearce, R.: The anomalous behaviour of aluminum sheet under balanced biaxial tension. Int. J. Mech. Sci. 12 (1970) 341-347.

7 Parmar, A.; Mellor, P.B.: Predictions of limit strains in sheet metal using a more general yield criterion. Int. J. Mech. Sci. 20 (1970) 385-391.

8 Marciniak, Z.; Kuczynski, K.; Pokora, T.: Influence of the plastic properties of a material on the forming limit diagram for sheet metal in tension. Int. J. Mech. Sci. 15 (1973) 789-805.

9 Marciniak, Z.; Kuczynski, K.: Limit strains in the processes of stretch-forming sheet metal. Int. J. Mech. Sci. 9 (1976) 609-620.

10 Hutchinson, J.W.; Neale, K.W.; Needleman, A.: Sheet necking-I Validity of plane stress assumptions of the long-wavelength approximation. Proc. GM Symp. on Mechanics of Sheet Metal Forming. Plenum Press, New York, (1978).

On the Theoretical and Experimental Description of the Mechanical Behaviour of Metals during Metal Forming Processes

A.BALTOV, N.BONTCHEVA, R.KAZANDJIEV
Inst. of Mechanics and Biomechanics
Bulg.Acad.Sci., Sofia, Bulgaria
I.RADOVANOV
Centr.Labor.Phys.Chem.Mechanics
Bulg.Acad.Sci., Sofia, Bulgaria
ST.VODENICHAROV
Inst. of Metal Research
Bulg.Acad.Sci., Sofia, Bulgaria

Summary

A theoretical and experimental method for describing the mechanical behaviour of metals after metal forming processes is proposed. The experimental method is based on Szczepinski's macroexperimental method and on Vassilev's X-ray method. The relation between the residual microstresses and the plastic deformation is established.

Introduction

It is known, that during metal forming processes, due to plastic deformation, residual microstresses appear in metal bodies and deformational anisotropy often takes place. Separate experimental methods exist for determining the plastic deformational anisotropy [1] and the residual microstresses [2], dealing with different mechanical models [3 - 6]. The problem set here is to propose a theoretical model, which enables, using both methods, to determine relations between plastic deformation and residual microstresses.

Theoretical model

The model is based on the following assumptions:
1. A material, which is micrononhomogeneous and macrohomogeneous is considered. σ'_{ij} and ε'_{ij} $(i,j=1,2,3)$ are the stress and the strain tensors on microlevel. A characteristic small volume V exists, in which the residual microstresses are selfequilibrated. The following average macrotensors are defined in V:

$$\bar{\sigma}_{ij}=\frac{1}{V}\int_V \sigma'_{ij}dV, \qquad \bar{\varepsilon}_{ij}=\frac{1}{V}\int_V \varepsilon'_{ij}dV \tag{1}$$

$\bar{\varepsilon}_{ij}$ is the tensor, which we measure in mechanical experiments. The deviations of the micromeasures from the average macromeasures are:

$$\sigma^*_{ij}=\sigma'_{ij}-\sigma_{ij}, \qquad \varepsilon^*_{ij}=\varepsilon'_{ij}-\varepsilon_{ij} \tag{2}$$

The following relations are valid for them:

$$\int_V \sigma^*_{ij}dV=0, \qquad \int_V \varepsilon^*_{ij}dV=0 \tag{3}$$

2. The volume V is assumed consisting of two parts - V_a and V_b such, that the microstresses σ'_{ij} are nonnegative everywhere in V_a and negative in V_b. This may be caused by structural reasons. According to Vassilev [2], the domain (a) belongs to the phase of the crystal grains and the domain (b) - to the phase of the boundaries between the grains. The relations (3) then take the form:

$$\int_{V_a} \sigma^*_{ij}dV= -\int_{V_b} \sigma^*_{ij}dV, \qquad \int_{V_a} \varepsilon^*_{ij}dV= -\int_{V_b} \varepsilon^*_{ij}dV \tag{4}$$

We define the phase-average stresses and strains:

$$\sigma_{ij(\alpha)} = \frac{1}{V_\alpha}\int_{V_\alpha} \sigma^*_{ij} dV, \qquad \varepsilon_{ij(\alpha)} = \frac{1}{V_\alpha}\int_{V_\alpha} \varepsilon^*_{ij} dV \tag{5}$$

$$(\alpha = a, b)$$

By means of (5) the relations (4) yield to:

$$\upsilon_a \sigma_{ij(a)} + \upsilon_b \sigma_{ij(b)} = 0 \tag{6}$$

$$\upsilon_a \varepsilon_{ij(a)} + \upsilon_b \varepsilon_{ij(b)} = 0$$

where

$$\upsilon_a = \frac{V_a}{V}, \quad \upsilon_b = \frac{V_b}{V}, \quad \upsilon_a + \upsilon_b = 1 \tag{7}$$

3. We introduce according to Novozhilov [4], the micro-stress tensor σ^μ_{ij}, which is a macromeasure of the stresses on microlevel. The change of the specific strain energy W is:

$$\delta W = \frac{1}{V}\int_V \sigma'_{ij} \delta\varepsilon'_{ij} dV = \delta\overline{W} + \delta W^* \tag{8}$$

where

$$\delta\overline{W} = \overline{\sigma}_{ij}\delta\overline{\varepsilon}_{ij}, \qquad \delta W^* = \frac{1}{V}\int_V \sigma^*_{ij}\delta\varepsilon^*_{ij} dV$$

We assume that δW^* may be expressed in the following way:

$$\delta W^* = \sigma^\mu_{ij}\delta\overline{\varepsilon}_{ij} \tag{9}$$

and obtain

$$\delta W = \sigma_{ij}\delta\overline{\varepsilon}_{ij}, \text{ where } \sigma_{ij} = \overline{\sigma}_{ij} + \sigma^\mu_{ij} \tag{10}$$

The stress tensor σ_{ij} is the macromeasure of the stress state of the body. The average microstress tensor $\overline{\sigma}_{ij}$ corresponds to the active stress tensor in Kadashevich-Novozhilov's theory [5].

4. Voigt model is assumed to be valid on microlevel for each phase:

$$\varepsilon^*_{ij}=\varepsilon_{ij(\alpha)} \text{ in } V_\alpha, \quad (\alpha=a,b) \tag{11}$$

We introduce the deviations from the average phase stresses:

$$\sigma^*_{ij(\alpha)}=\sigma^*_{ij}-\sigma_{ij(\alpha)} \text{ in } V_\alpha, \quad (\alpha=a,b) \tag{12}$$

The following relations are valid for them:

$$\int_{V_\alpha}\sigma^*_{ij(\alpha)}\, dV=0, \quad (\alpha=a,b) \tag{13}$$

Taking into account (11), (12) and (13), the expression for δW^* in (8) reduces to:

$$\delta W^*=\upsilon_a\sigma_{ij(a)}\delta\varepsilon_{ij(a)}+\upsilon_b\sigma_{ij(b)}\delta\varepsilon_{ij(b)}=\eta_a\delta W_a \tag{14}$$

where

$$\delta W_a=\sigma_{ij(a)}\delta\varepsilon_{ij(a)}, \quad \eta_\alpha=\frac{\upsilon_a}{\upsilon_b}=\frac{V_a}{V_b} \tag{15}$$

Integrating (14) along the deformation path from state (o) up to state (1), we obtain:

$$W^*=\int_{(o)}^{(1)}\eta_a\delta W_a \tag{16}$$

5. The strain tensor on microlevel consists of an elastic and a plastic part:

$$\varepsilon'_{ij}=\varepsilon'^{e}_{ij}+\varepsilon'^{p}_{ij} \tag{17}$$

Then the following relations follow directly:

$$\varepsilon_{ij(\alpha)}=\varepsilon^e_{ij(\alpha)}+\varepsilon^p_{ij(\alpha)}, \quad \varepsilon_{ij}=\varepsilon^e_{ij}+\varepsilon^p_{ij}, \quad (\alpha=a,b) \tag{18}$$

The expression (16) then takes the form:

$$W^* = \int_{(o)}^{(1)} \eta_a \delta W_a^e + \int_{(o)}^{(1)} \eta_a \delta W_a^p \tag{19}$$

where

$$\delta W_a^e = \sigma_{ij(a)} \delta \varepsilon_{ij(a)}^e, \quad \delta W_a^p = \sigma_{ij(a)} \delta \varepsilon_{ij(a)}^p$$

The elastic parts of the strains are connected with the stresses by means of Hook's law:

$$\sigma_{ij(a)} = E_{ijkl}^a \varepsilon_{kl(a)}^e, \quad \sigma_{ij} = E_{ijkl} \varepsilon_{kl}^e \tag{20}$$

where E_{ijkl}^a is the elastic moduli tensor on microlevel for phase (a) and E_{ijkl} is the elastic moduli tensor on macrolevel. We shall consider the case of elastic isotropy. The incremental plastic strain on macrolevel is obtained by means of the flow law:

$$\delta \varepsilon_{ij}^p = \delta \lambda \frac{\partial F}{\partial \sigma_{ij}}, \quad \varepsilon_{jj}^{!p} = 0, \quad \varepsilon_{jj(a)}^p = 0, \quad \varepsilon_{jj}^p = 0 \tag{21}$$

where plastic incompressibillity is assumed. The positive factor $\delta\lambda$ depends on the assumed hardening law and F=0 is the assumed yield condition. We assume Baltov-Sawczuk's yield condition for anisotropic hardening [6]:

$$F \equiv \frac{1}{2} N_{ijkl} s_{ij} s_{kl} - \tau_p^2, \quad (i,j,k,l=1,2,3) \tag{22}$$

where

$$\bar{s}_{ij} = \bar{\sigma}_{ij} - \frac{1}{3}\delta_{ij}\bar{\sigma}_{kk}, \quad s_{ij}^\mu = \sigma_{ij} - \frac{1}{3}\delta_{ij}\sigma_{kk}^\mu,$$

$$N_{ijkl} = I_{ijkl} + A s_{ij}^\mu s_{kl}^\mu, \quad A = \text{const}, \quad s_{ij}^\mu = C\varepsilon_{ij}^p,$$

$$I_{ijkl} = \frac{1}{2}(\delta_{il}\delta_{jk} + \delta_{ik}\delta_{jl}) - \frac{1}{3}\delta_{ij}\delta_{kl}, \tag{23}$$

$$C = const, \quad \tau_p = \tau_p(\gamma_o^p), \quad \gamma_o^p = \sqrt{\frac{1}{2}\varepsilon_{ij}^p \varepsilon_{ij}^p}$$

After integrating (9) along the deformation path from state (o) up to state (1) and taking into account (18) and (23) we obtain:

$$W^* = C(\gamma_o^p)^2 \tag{24}$$

The cases of isotropic or kinematic hardening may be obtained as particular cases from the above relations.

6. The ratio η_a between the domains of both phases changes during the deformation process, so that the function

$$\eta_a = \eta_a(\gamma_o^p) \tag{25}$$

exists. This change, which is due to the plastic deformation, may be interpreted as a stress redestribution on microlevel, or as a breaking of the crystal grains, caused by the appearence of dislocation lines.

If a closed loading cycle is realized and plastic deformation is available, residual elastic and plastic strains ($\varepsilon_{ij(\alpha)}^{e(r)}$, $\varepsilon_{ij(\alpha)}^{p}$, α=a,b) will occure after unloading. Taking into account Hook's law on microlevel and equation (25), the expression (19) for W* yields:

$$W^* = \eta_a(\gamma_o^p) W_a^{e(r)} + \int_{(o)}^{(1)} \eta_a \delta W_a^p = \eta_a^r W_a^{e(r)} \tag{26}$$

where the elastic residual strain energy for phase (a) is:

$$W_a^{e(r)} = \frac{1}{2}E_{ijkl}^a \varepsilon_{ij(a)}^{e(r)} \varepsilon_{kl(a)}^{e(r)}, \qquad (i,j,k,l=1,2,3)$$

$$\eta_a^r = \eta_a + \frac{1}{W_a^{e(r)}} \int_{(o)}^{(1)} \eta_a \delta W_a^p \qquad (27)$$

7. The residual elastic strains for the considered metals with crystal structure are due to plastic deformation, which causes rearrangement of the crystals. It is known that unloading on microlevel is not possible to be obtained by means of unloading on macrolevel. Deformations of the crystal lattice remain and restricted elastic deformation exists. This means, that the residual elastic strain energy of phase (a) may be treated as a function of the plastic strain:

$$W_a^{e(r)} = f(\gamma_o^p) \qquad (28)$$

Comparing (24) with (26) and taking into accoüt (28), we obtain:

$$\eta_a^r = \frac{C(\gamma_o^p)^2}{f(\gamma_o^p)} \qquad (29)$$

This expression gives an indirect information about the changes of the domains of both phases, due to plastic deformation. If we assume for $\delta\varepsilon_{ij(a)}^p$ a flow law on microlevel and treat $(27)_2$ as an integral equation, we may determine the function $\eta_a(\gamma_o^p)$.
The obtained relations make possible the connection of the plastic strains with the residual elastic microstrains.

Experimental method

O n e d i m e n s i o n a l t e n s i l e e x p e r i m e n t s. By means of onedimensional tensile experiments all mechanical macrocharacteristics of the material are obtained: Young's modul E, Poisson's coeffitient ν, the initial yield point and the hardening law $\tau_p = \tau_p(\gamma_o^p)$.

T w o - d i m e n s i o n a l   e x p e r i m e n t s. According to Szczepinski's method [1], thin plates are extended uniaxially, so that different residual plastic strains ε_1^p are realized in them. Specimens in directions making different angles α with the axis of the preliminary tension are cut out from these plates. Each of the specimens is loaded by uniaxial tension and the yield point σ_{Ip}^{α} is obtained. The yield points correspond to points, lying on a plane curve [1], belonging to the yield surface in the threedimensional stress space $(\sigma_{11}, \sigma_{22}, \sigma_{12})$:

$$\sigma_{11}=\sigma_{Ip}^{\alpha}\cos^2\alpha, \quad \sigma_{22}=\sigma_{Ip}^{\alpha}\sin^2\alpha,$$
$$\sigma_{12}=\sigma_{Ip}^{\alpha}\sin\alpha\cos\alpha \tag{30}$$

Thus different curves, lying on yield surfaces are experimentally obtained for each preliminary plastic strain ε_1^p. Those surfaces are approximated with ellypsoides [7] according to the assumed yield condition (22):

$$F\equiv R_1(\sigma_{11}-a_1)^2+R_2\sigma_{22}^2-R_1\sigma_{22}(\sigma_{11}-a_1)+$$
$$\sigma_{12}^2-\tau_p^2=0 \tag{31}$$

where

$$R_1=\frac{1}{3}+\frac{1}{2}AC^2(\varepsilon_1^p)^2, \quad R_2=\frac{1}{4}(R_1+1), \quad a_1=\frac{3}{2}C\varepsilon_1^p,$$
$$\tau_p=\tau_p(\gamma_o^p), \quad \gamma_o^p=\frac{\sqrt{3}}{2}\varepsilon_1^p \tag{32}$$

In that case the plane curves are ellipses.
Thus the parameters of anisotropic hardening, the constants A and C, are obtained by means of macroexperiments.

X-ray stress measurements.

X-ray stress measurements are made on the surface of the specimens which had been used in the macroexperiments. These measurements estimate the change in the spacing of the lattice planes of the crystals, by means of the change in the reflecting angle. This enables to calculate the average elastic microstrains in the direction of the normal n to the reflecting plane. According to the assumed two-phase microstructure, we are able to measure components of the average alastic microstrain tensor, belonging to phase (a), the phase of the crystal grains [2].

Specimens of the material under consideration, in which no initial technologycal residual stresses and no preliminary plastic strain exist, are loaded by uniaxial tensile stress $\sigma_1=\sigma_{1(a)}$. At different levels of the stress σ_1 X-ray measurements are made and the average elastic microstrains $\varepsilon^{(a)}_{o,\psi}$ in directions of the normals n, closing various angels ψ with the normal to the specimen surface, are obtained. Using the linearity of the relation between $\varepsilon^{(a)}_{o,\psi}$ and σ_1:

$$\varepsilon^{(a)}_{o,\psi}=\frac{1}{2}S_2\sigma_1\sin^2\psi+S_1\sigma_1$$
$$S_1=-\frac{\nu_a}{E_a},\quad S_2=2\frac{1+\nu_a}{E_a}, \tag{33}$$

we may calculate the values of the elastic characteristics of phase (a) on microlevel, E_a and ν_a.

The macroexperiments must now be connected with the X-ray microexperiments. The specimens from the two-dimensional macroexperiments, cut out from the plates in the direction $\alpha=0$, in which a preliminary plastic strain has been realized, are now investigated by the

X-ray method. Thus we obtain the residual average elastic microstrains of phase (a), $\varepsilon_{o,\psi}^{(a)r}$. They are connected with the residual microstresses $\sigma_{1(a)}^{r}$ and $\sigma_{2(a)}^{r}$ by the relation:

$$\varepsilon_{o,\psi}^{(a)r} = m \sin^2\psi + a \tag{34}$$

where

$$m=\frac{1}{2} S_2\sigma_{1(a)}^{r}, \quad a= S_1(\sigma_{1(a)}^{r}+\sigma_{2(a)}^{r}) \tag{35}$$

Taking measurements at different values of the angle ψ and using the linearity in (34) we obtain the values of m and a. Then we calculate by means of (35) the values of the main residual microstresses $\sigma_{1(a)}^{r}$ and $\sigma_{2(a)}^{r}$. The main residual average elastic microstrains of phase (a) are connected with them by Hook's law:

$$\varepsilon_{1(a)}^{e(r)} = \frac{1}{E_a}(\sigma_{1(a)}^{r} - \nu_a\sigma_{2(a)}^{r}) \tag{36}$$

$$\varepsilon_{2(a)}^{e(r)} = \frac{1}{E_a}(\sigma_{2(a)}^{r} - \nu_a\sigma_{1(a)}^{r})$$

The expression for the specific residual elastic strain energy has then the form:

$$W_a^{e(r)} = \frac{1}{2}(\sigma_{1(a)}^{r}\varepsilon_{1(a)}^{e(r)} + \sigma_{2(a)}^{r}\varepsilon_{2(a)}^{e(r)}) \tag{37}$$

Different values of $W_a^{e(r)}$ are experimentally obtained, corresponding to different values of the effective strain γ_o^p. This makes possible an approximation of those experimental points by means of an analytical function $f(\gamma_o^p)$.

By means of the constant C, obtained by the macroexperiments and the function f, obtained by the microexperiments, the Function η_a^r in (29) is found. Thus, knowing by X-ray measurements the residual microstrains, we may estimate the intensity of the preliminary plastic strain which the body has undergone. Or knowing the preliminary plastic deformation, we may give an average estimation of the residual microstresses.

E x a m p l e. Experiments were made, using electrolytic copper, thermally treated at 800°C, for two hours. The results of the onedimensional tensile experiments are shown on fig.1. Plates of the size 100x200x3 were used for the two-dimensional experiments. The residual plastic strains ε_1^p realized in them were 2·5, 5, 10 and 20%. Specimens with a measuring lenght 50 mm and width 15 mm were cut out from the plates in directions making angles α 0°, 30°, 45°, 60° and 90° with the tension axis. Their yield points σ_{Ip}^{α} were determined corresponding to a 0·02% plastic strain. The constants A and C in the equation of the ellypsoide (31), approximating the expiremental points, were obtained by means of the method of least squares. The value C=213·30N/mm² was obtained. The ellypses, lying on the yield surfaces (31), as well as the experimental points are shown on fig.2, plotted in the planes $\sigma_{11}-\sigma_{22}$ and $\sigma_{12}-p$, where

$$p=\sigma_{Ip}^{\alpha}\sqrt{1-\sin^2\alpha\cos^2\alpha}\cos\xi\cos\left(\frac{3}{4}\pi-\phi\right)$$

$$\mathrm{tg}\,\xi=\frac{\mathrm{tg}\,\alpha}{1+\mathrm{tg}^4\alpha}\qquad\qquad \mathrm{tg}\,\phi=\mathrm{tg}^2\alpha \tag{38}$$

X-ray measurements were made by means of TuR M-62 with a difractometer HZG-3. A Ni-filtered CuK_α-ray was used. The surfaces of the specimens were chemically treated. An equipment was constructed, allowing the simultaneous realization of macro- and microexperiments. Fig.3 shows the experimentally obtained relation (36) for various ε_1^p. The values of $W_a^{e(r)}$, corresponding to differenf γ_o^p are represented by points in the plane $W_a^{e(r)}-\gamma_o^p$ on

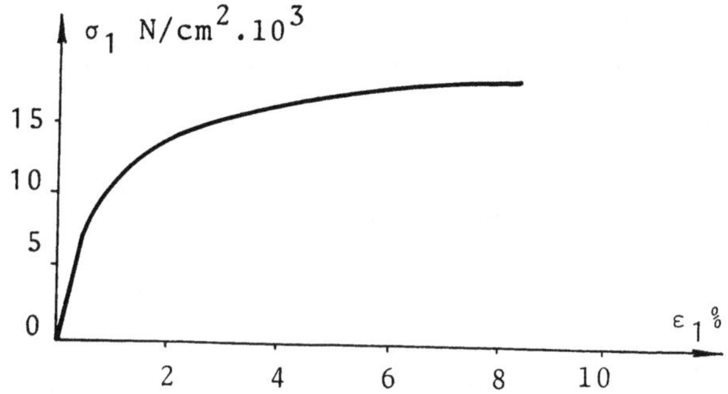

fig. 1

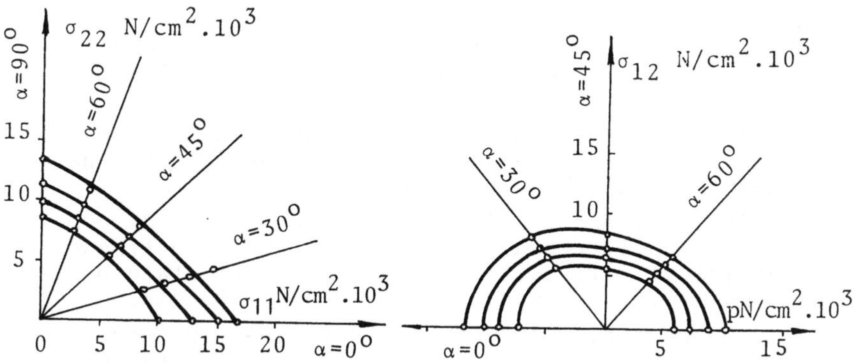

fig. 2

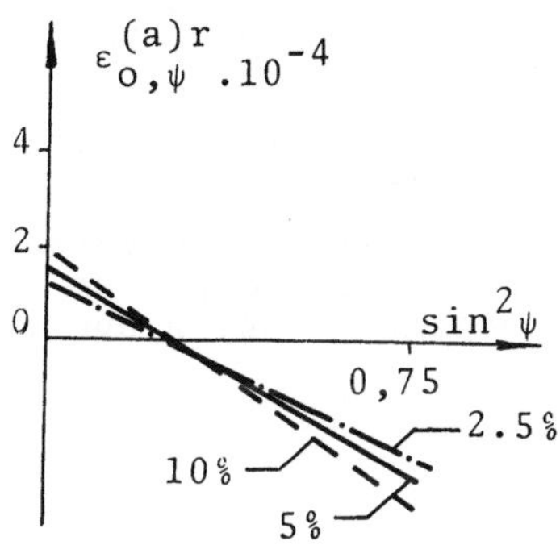

fig. 3

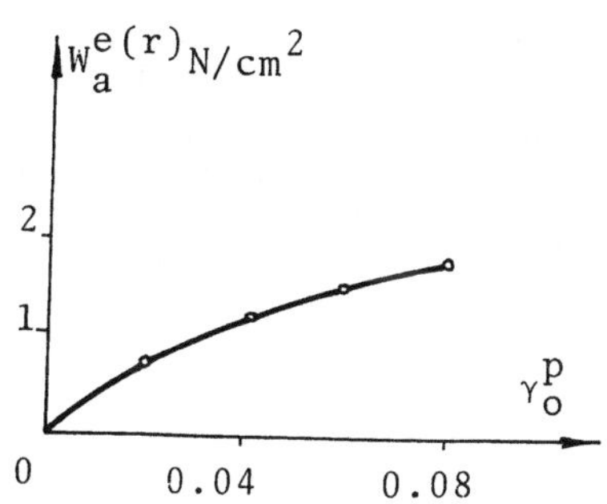

fig. 4

fig.4. An analytical approximation with a second-order curve was made, based on the least squares method and the function

$$f = -184{\cdot}2(\gamma_o^p)^2 + 36{\cdot}43\gamma_o^p \tag{39}$$

was found. The expression for η_a^r is then:

$$\eta_a^r = \frac{21330}{-184{\cdot}2(\gamma_o^p)^2 + 36{\cdot}43\gamma_o^p} \tag{40}$$

References

1 Szczepiński, W.: On the effect of plastic deformation on yield condition. Arch.Mech.Stos. 15 (1963) 275-296.

2 Васильов, Д. М.: О микронапряжениях, возникающих в поликристаллических образцах при пластическом деформировании. ЖТФ 28 (1958) 2527-2541.

3 Kafka, V.: Theory of slow elastic-plastic deformation of polycristalline metals with microstresses as latent variables descriptive of the state of the material. Arch.Mech.Stos. 24 (1972) 403-418.

4 Новожилов, В. В.: О сложном нагружении и перспективах феноменологического подхода к исследованию микронапряжений. ПММ 28 (1964) 5-9

5 Кадашевич. Ю. И.; Новожилов, А. А.: Теория пластичности учитывающая остаточные микронапряжения. ПММ 22 (1958) 78-84

6 Baltov, A.; Sawczuk, A.: A rule of anisotropic hardening. Acta Mechanica, 1 (1965) 81-92.

7 Балтов, А.; Воденичаров, Ст.: Методика за изследване влиянието на историята на скоростта на деформация върху пластичните свойства на метали. Теор. и прил. механика 10 (1979)

Spin-Forging of Sheet Metal Cones Having Various Cone Angles from 70/30 Brass and Commercially Pure Aluminium

R.A.C. SLATER

Department of Mechanical Engineering
The City University, London, U.K.

SUMMARY

Using mandrels having different cone angles and a roller having a specific profile, cones were produced by spin-forging from 70/30 brass sheet. It is shown that the three component forces exerted at the workpiece-roller interface vary inversely as the mandrel speed and directly as the feed rate.
An optimum mandrel speed exists for a given cone angle and constant feed rate. This optimum mandrel speed varies directly proportional to the cone semi-angle and increases with feed rate.
Experimental values of the dimensionless tangential component force as a function of fractional reduction in wall thickness are compared with values predicted by approximate upper- bound estimates.
Cones were similarly produced from commercially pure aluminium sheet to examine the effects of mandrel speed, feed rate and cone angle on the external surface finish of the formed cones.
The mean value of the centre-line average, R_a, which varied from the nose to the rim, attained a minimum at a particular mandrel speed for a given cone angle and constant feed rate.

INTRODUCTION

The rapid increase in the cost of raw materials and the increasing awareness of their limited availability has caused manufacturing industry to critically examine forming techniques with the aim of making a more economical use of available materials. This situation has stimulated the development of metal forming processes which do not involve metal removal to manufacture the desired product. One of these newer methods is that referred to here as *spin-forging* which is a derivative of the old established sheet metal forming process known as spinning.

Spin-forging is used for the manufacture of hollow components, containers, shells and casings having rotational symmetry to a high degree of dimensional accuracy and surface finish. Important applications in the aerospace, automobile and other manufacturing industries include, for example, satellite nose cones, rocket components, truck wheel rims and various parts for nuclear reactors. The concepts of spinning and spin-forging have been discussed previously at length, for example, in the technical literature [1] , [2] , [3]. Only spin-forging, therefore, will be briefly referred to here.
It is only during the last three decades that automation of the spinning and spin-forging processes has been developed. More recently, N.C. and C.N.C. systems have been introduced to transfer skills from the operator to machine which is expected to increase the speed, accuracy and consistency with which the forming cycle is effected. There is therefore the need to establish reliable data with particular respect to the effect of the process variables on the important forming parameters involved, for example, during spin-forging of sheet metal cones. This information is required for the design of the control system and the roller assembly where predictions of the force, energy and power requirements are essential and for the formulation of computer programs to control the forming cycle. The spin-forging investigations undertaken in the Manufacturing Processes Laboratory at The City University, London during the past four years have therefore been directed with these aims in mind.
During the spinning of sheet metal, the shape of the circular blank only is changed and its area remains unchanged. Implicit in this definition is that the initial wall thickness, t_0, remains unchanged. However, the spin-forging operation is characterised by the plastically deforming zone being localised to a very small region of the workpiece in contact with a forming roller. The final thickness, t_f, becomes less than the initial thickness, t_0, of the circular blank and is dictated by the sine of the cone semi-angle, α , such that in t r u e s p i n - f o r g i n g,

$$t_f = t_0 \sin \alpha \qquad (1)$$

The fractional reduction in wall thickness, $R = (t_0 - t_f)/t_0$.

Therefore $$R = 1 - (t_f/t_o) = 1 - \sin\alpha \qquad (2)$$
Also, in spin-forging there is not an appreciable change in the radial position of each element from its initial position in the circular blank. The maximum diameter of the formed cone thus remains approximately the same as the initial blank diameter as illustrated in figure 1.

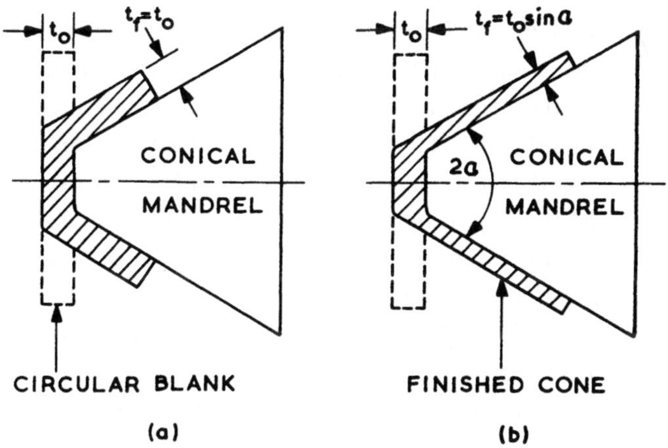

Figure 1. Illustrating the forming of a sheet metal cone by (a) the spinning process and (b) spin-forging

The simplest solution for the tangential component, F_t, of the resultant force exerted at the workpiece-roller interface during the spin-forging of a cone is given by Paulton and Colding [4] which is based on the concept of the energy dissipated during homogeneous deformation. Another elementary theory is suggested by Kalpakcioglu [5] who analysed the spin-forging of cones by assuming simple shear deformation under the roller. Avitzur and Yang [6] have derived an equation for predicting the tangential component force by also assuming shear deformation and determining the strain rate in the deforming material under the roller. Their equation is of a complex nature and requires the aid of a computer for its solution. The energy dissipation due to redundant deformation is, however, neglected which is likely to be significant particularly at lower values of feed rate. The forming forces involved in spin-forging of cones have also been calculated by Kobayashi, Hall and Thomsen [7],[8].

The analysis by these authors is similar to that given by Kalpakcioglu except that their total effective strain includes the strain due to bending of the element under the roller.
However, the effect of the process variables on the component roller forces appear not to be well appreciated as indicated by the lack of published literature in this field. Joorabchian [9], and Slater and Joorabchian [10], have therefore reported an experimental study of the spin-forging of commercially pure aluminium sheet metal cones which was directed towards obtaining a better understanding of the process to enable the appropriate process variables for a specific application to be selected. Furthermore, no attempt appeared to have been made to apply the upper-bound theorem to obtain an estimate of the forming parameters by assuming, for example, either plane strain or axisymmetric deformation. A tentative upper-bound estimate of the dimensionless tangential component force was consequently presented by Slater and Joorabchian [11] assuming plane strain deformation generally following the method proposed by Johnson [12].
A more systematic approach for obtaining a good upper-bound estimate for forming parameters was given by Kudo [13], [14]. He introduced the concept of the "unit rectangular deforming region" and by considering various types of kinematically admissible velocity fields in the deforming regions, he suggested that an improved upper-bound solution may be obtained by considering "a rigid triangular velocity field". According to Kudo this is because this velocity field always involves a lower rate of energy dissipation than the parallel velocity field.
This type of velocity field has also been proposed by Green [15] for plane strain drawing and ironing which may be applied to an axisymmetric flow through a conical die assuming the deformation pattern to be similar.
The principal objectives in presenting this paper, which is concerned with recent investigations undertaken at The City University, London, may be summarised as:

(a) to present some of the results of an experimental study concerned with the effect of the process variables on the components of the resultant force exerted at the workpiece-roller interface and on the external surface finish of the formed cones during the spin-forging of 70/30 brass and commercially pure aluminium sheet metal cones respectively, and

(b) to give approximate upper-bound estimates for the dimensionless tangential component force assuming plane strain and axisymmetric deformation and compare the theoretical predictions with the experimental results.

BRIEF DESCRIPTION OF THE EXPERIMENTAL EQUIPMENT

Part 1. Spin-Forging of 70/30 Brass Sheet Metal Cones

For the spin-forging of sheet metal cones from 70/30 brass and commercially pure aluminium, a robust centre lathe having continuously variable spindle speed between 13-2000 rev/min and equipped with a hydraulic copying attachment was employed. Forming of the cones was effected using a roller having a specific profile. This experimental equipment has been previously described in some detail [10] and the photograph of figure 2 presents a close-up view of a partially formed sheet metal cone, showing the roller, conical mandrel and corresponding template.

Figure 2. Close-up view of a partially formed sheet metal cone, showing the roller, conical mandrel and corresponding template

During the present investigations, four conical mandrel heads were used having semi-angles, α , of 30, 35, 40 and 45°. A mandrel adapter was machined from mild steel having a threaded axial hole to enable the selected conical head to be secured to the adapter which, in turn, was clamped into a four jaw chuck on the lathe. With each mandrel head having a particular cone semi-angle, α , a corresponding triangular template having the same included angle, α , was used in conjunction with the hydraulic copying attachment on the lathe. This arrangement ensured that the roller traversed parallel to the conical mandrel head.
A three component force dynamometer was used to measure the components of the resultant force exerted at the workpiece-roller interface. This dynamometer and its calibration have also been previously described in some detail [10].

Part 2 External Surface Finish Assessment of Commercially Pure Aluminium Sheet Metal Cones

A stylus tracer instrument known as "Talysurf 4" comprising a pen recorder, an electronic averaging unit and a gear box was used to determine the surface roughness of the external surface of the commercially pure aluminium sheet metal cones.
This instrument provided information in two forms:

(a) A trace representing the surface profile with the vertical height of the irregularities magnified by from x500 to x100 000 and the horizontal spacing magnified by either x20 or x100.

(b) A meter indication of the centre-line average, R_a, which is the numerical assessment of the surface profile in accordance with B.S.1134/1961.

TENTATIVE UPPER-BOUND ESTIMATES FOR THE DIMENSIONLESS TANGENTIAL COMPONENT FORCE DURING SPIN-FORGING OF SHEET METAL CONES

(1) Assuming plane strain deformation

Little attention has been given in recent years to the analysis of the spin-forging process for producing, for example, sheet metal

cones and it appeared that no attempt had been made to apply the upper-bound theorem to obtain an estimate of the forming parameters by assuming either plane strain or axisymmetric deformation.
It is clear, however, that in any such analysis a number of assumptions are necessary with regard to the mode of deformation and the behaviour of the deforming material in the interests of mathematical tractability. This is because the spin-forging of sheet metal cones is essentially an unsteady state process since the positions of the elements in the plastically deforming region change as the cone is formed and the diameter increases. At the same time, for a constant mandrel speed, the velocity of elements increase from the nose to the rim of the cone. The effects of strain-hardening and temperature are also difficult to assess.
Nevertheless, Slater and Joorabchian [11] found experimentally that the deviation from steady state conditions, as characterised by variation in the magnitudes of the components of the resultant force exerted at the workpiece-roller interface, was not very significant during the spin-forging of commercially pure aluminium sheet metal cones.
Despite the many limitations, an approximate upper-bound estimate for the dimensionless tangential component force was presented assuming plane strain deformation and involving a kinematically admissible velocity field comprising straight lines of velocity discontinuity to form a single triangle on a median plane.
The dimensionless tangential component force, $F_t/\bar{\sigma}t_0f$, is then given by

$$F_t/\bar{\sigma}t_0f = \left\{R/2\sin\beta\sin\phi\sin(\beta+\phi)\right\}\left[\sin^2\phi/(1-R) + \sin^2\beta - \sin2\phi\sin(\beta+\phi) + (1-R)\sin^2(\beta+\phi) + \left\{R/(1-R)\sin^2\phi\sin\beta\right\}\right] \quad (3)$$

The tangential component force is F_t, f is the feed per revolution, t_0, the initial thickness of the blank, R, the fractional reduction in wall thickness, $\bar{\sigma}$, the average flow stress, β, the angle of inclination of the roller land to the cone surface and, ϕ, is the angle of inclination of the exit line to the cone surface.
The optimum value of the angle, ϕ, which gives a least upper-bound estimate was determined for various values of, R, by computer.

The main assumptions made in deriving equation (3) are that: (a) steady state flow prevails, (b) the workpiece material is homogeneous and isotropic, (c) no elastic deformation occurs and consequently there are no volume changes, (d) a value of mean flow stress may be introduced to make an allowance for strain hardening, (e) the Tresca yield criterion is applicable, (f) thermal effects are insignificant, (g) the workpiece-mandrel interface is perfectly smooth and (h) the workpiece-roller interface is perfectly rough.

Other straight line patterns of tangential velocity discontinuities can be considered depending on the value of the fractional reduction in wall thickness R. However, no attempt was made to refine the analysis because this would certainly lead to an even greater complexity of the solution which may not have been justified in view of the many assumptions already introduced.

(2) Assuming axisymmetric deformation with straight lines of tangential velocity discontinuity on a median plane

The assumed velocity field comprises straight lines of tangential velocity discontinuity on a median plane where the surfaces of velocity discontinuity in the deforming region are conical. An upper-bound estimate for the dimensionless tangential component force may then be obtained following the method of Kudo and introducing Green's assumption in plane strain drawing that the angle $\theta = \phi + \beta$ is also valid in the axisymmetric upper-bound approach considered here since the proposed velocity field on a median plane is similar to that originally suggested by Green.

The following expression for the dimensionless tangential component force may then be obtained:

$$F_t/\bar{\sigma}t_0 f = (R/2)\Big[\Big\{\sin\phi\sin\beta/(1-R)\sin^3(\beta+\phi)\Big\} + \Big\{\sin\beta/\sin\phi\sin(\beta+\phi)\Big\} + \Big\{R\sin\phi/(1-R)\sin\beta\sin(\beta+\phi)\Big\} + 2\Big\{(1-R)(1+\ln R)\Big\}\Big] \tag{4}$$

the derivation of which is given in Appendix 1.

The various assumptions made in deriving equation (3) are also applicable here and the optimum value of ϕ to produce a least upper-bound estimate was determined by the same method as before.

RESULTS AND DISCUSSION

Part 1. Spin-Forging of 70/30 Brass Sheet Metal Cones

Effect of mandrel speed on the three component forces at constant feed rate for a given cone semi-angle

The relationships between the tangential, radial and axial components of the resultant force exerted at the workpiece-roller interface and the mandrel speed for a constant saddle feed rate of 4.1 cm/min are shown in figure 3. The results showing the relationship between the tangential component force and mandrel speed for constant saddle feed rates of 4.1, 6.4 and 10.2 cm/min are presented in figure 4. During these investigations the mandrel used had a cone semi-angle of 35^{0} and the directions of the component forces are referred to the roller axis.

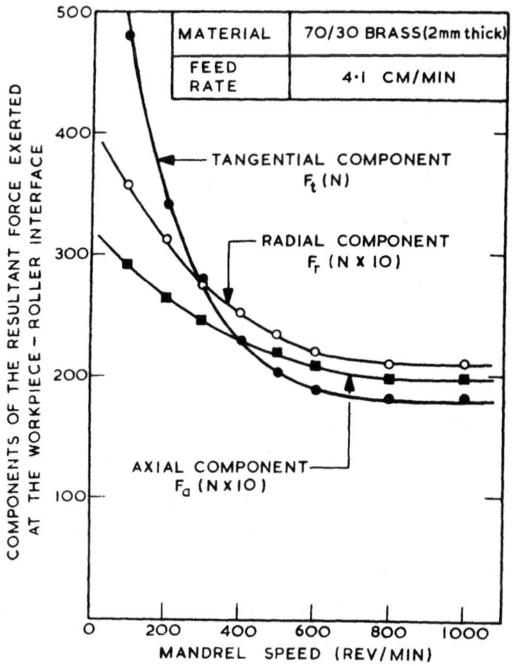

Figure 3. Relation between the tangential, radial and axial component forces and the mandrel speed at a constant feed rate (mandrel cone semi-angle = 35^{0})

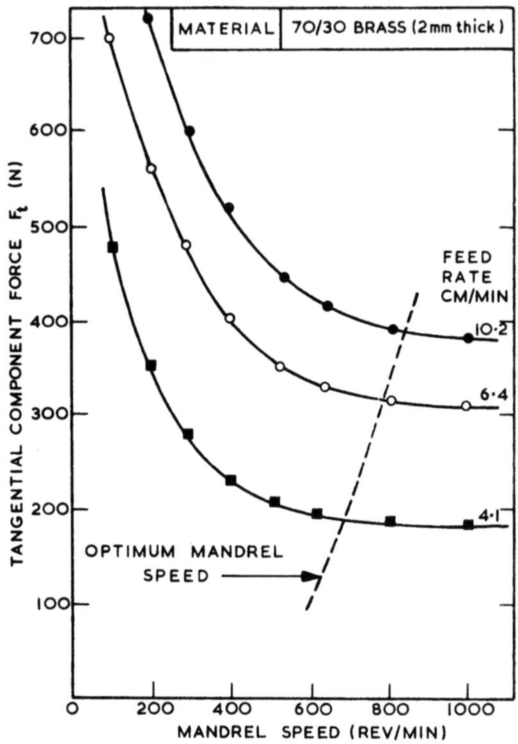

Figure 4. Relation between the tangential component force and the mandrel speed for varying feed rate showing that an optimum mandrel speed exists (mandrel cone semi-angle = 35^0)

Referring to figure 3 it will be seen that the three component forces decrease in magnitude as the mandrel speed increases becoming asymptotic to the mandrel speed axis. It may also be noted that the tangential component force is affected to the greatest extent and the axial component least by variation in the mandrel speed.

When a constant feed rate of 4.1 cm/min was employed, the decrease in magnitude of the three component forces was associated with an increase in the mandrel speed until a mandrel speed of approximately 650 rev/min was attained. Further increase in mandrel speed has insignificant effect on the magnitude of the component forces. By referring to figure 4 it can be seen that a similar situation exists for the other two feed rates of 6.4 and 10.2 cm/min except that the corresponding mandrel speeds are approximately 750 and 800 rev/min respectively.

The results therefore indicate that there exists an o p t i m u m mandrel speed for a given cone angle and workpiece material corresponding to a constant feed rate, that is, there is a least value of mandrel speed corresponding to a minimum value of the tangential component of the roller force. However, when interpreting these results it must be remembered that optimum mandrel speeds are only valid for t r u e spin-forging. Any deviation from this condition will affect the optimum value of mandrel speed as well as the magnitudes of the three components of the resultant force exerted at the workpiece-roller interface.

Effect of feed rate on the three component forces at constant mandrel speed

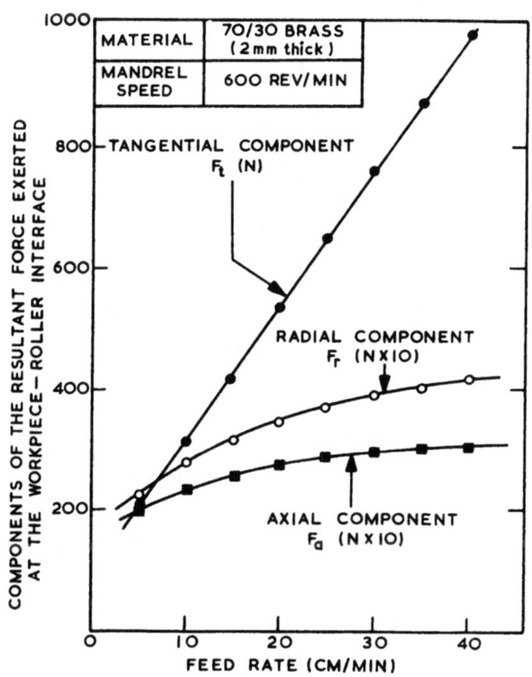

Figure 5. Effect of feed rate on the tangential, radial and axial component forces at a constant mandrel speed (mandrel cone semi-angle = 35°)

Figure 5 shows the effect of feed rate on the tangential, radial and axial components of the resultant force exerted at the workpiece-roller interface. These results show, in particular, that the tangential component varies linearly with the feed rate whilst the radial and axial components do not. It will also be noted that, as the feed rate increases, the radial and axial components do not increase significantly and appear to approach maximum values at some feed rate in excess of 40 cm/min.

Relation between the tangential component force and mandrel speed using mandrels having various cone semi-angles

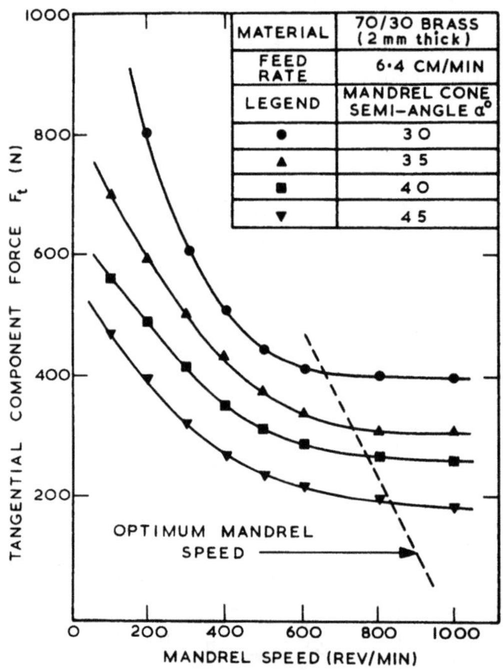

Figure 6. Relation between the tangential component force and mandrel speed for various cone semi-angles at constant feed rate

In figure 6, the relationship between the tangential component force and the mandrel speed is presented for cone semi-angles of 30, 35, 40 and 45^0 when the feed rate was maintained constant at 6.4 cm/min. Results obtained showing the relationship between the optimum mandrel speed and the mandrel cone semi-angle, α , for a constant feed rate of 6.4 cm/min were deduced from figure 6. Together with similar results corresponding to constant feed rates of 4.1 and 8.3 cm/min these are presented in figure 7.

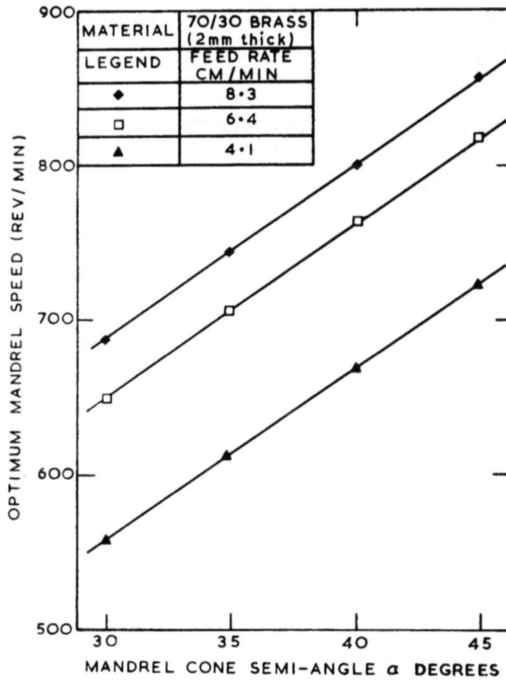

Figure 7. Relation between the optimum mandrel speed and cone semi-angle for various feed rates

Figure 7 indicates that the optimum mandrel speed varies directly proportional to the cone semi-angle, α , for each of the feed rates considered but is also affected by variation in the feed rate. It is therefore important when selecting an optimum mandrel speed to consider the effect of the cone semi-angle in addition to the feed rate.

Relation between the tangential component force and feed rate using mandrels having various cone semi-angles

This relationship is presented in figure 8 for the four mandrel cone semi-angles of 30, 35, 40 and 45^{0} for a constant mandrel speed of 600 rev/min. In each case, the tangential component force is shown to vary directly proportional to the feed rate. However, as the mandrel cone semi-angle, α , increases from 30 to 45^0 the relative effect becomes less. From figure 8, it is possible to select the highest feed rate for a given cone semi-angle, α , corresponding to the power available at the mandrel in order to perform the most efficient spin-forging operation.

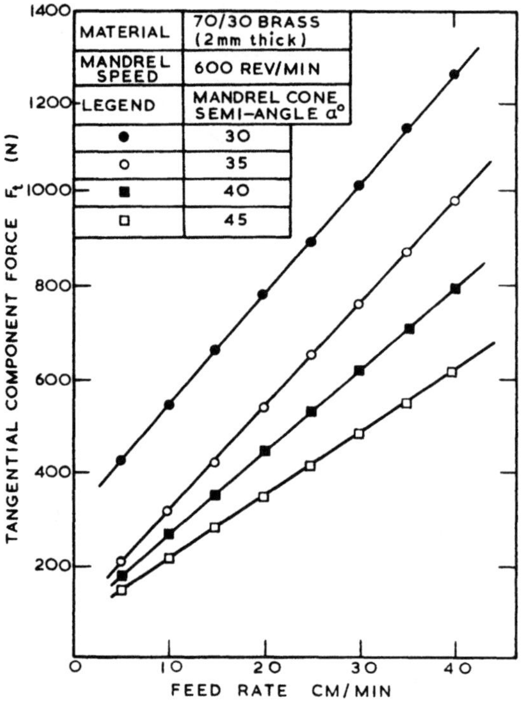

Figure 8. Relation between the tangential component force and feed rate for various cone semi-angles at constant mandrel speed

The dependence of the three component forces on the fractional reduction in wall thickness during spin-forging with a constant cone semi-angle and mandrel speed

Figure 9 illustrates the relationship existing between the tangential, radial and axial components of the resultant force exerted at the workpiece-roller interface and the fractional reduction in wall thickness of the formed conical workpiece. An important feature of these curves is that inflexions are present in the relationships for the tangential and radial component forces which occur at a fractional reduction in wall thickness of approximately 0.43. This corresponds to a fractional reduction in wall thickness, R, for t r u e s p i n - f o r g i n g given by $R = 1-\sin\alpha = 1-0.5736 \approx 0.43$ when the cone semi-angle, α , = 35^0.

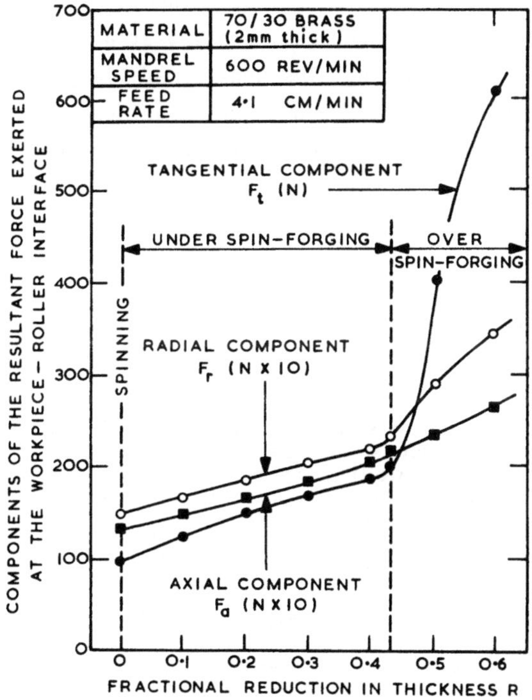

Figure 9. Relation between the component forces and fractional reduction in wall thickness (mandrel cone semi-angle = 35^0)

When R< 0.43, the undeformed flange tended to move forward and made contact with the roller. The rubbing action of the flange on the roller significantly affected the magnitude of the component forces, wrinkling of the flange in the circumferential direction could occur and the external surface finish of the formed cone was unsatisfactory. Also, the maximum diameter of the formed cone was found to be less than the diameter of the original circular blank. This condition is referred to here as u n d e r s p i n - f o r g i n g.

During t r u e s p i n - f o r g i n g, which for the case when

$\alpha = 35^0$, R = 0.43, the flange remained straight at an angle of 90^0 to the axis of rotation. The external surface finish of the cones was satisfactory and the rim diameter of the formed cones was equal to the diameter of the original circular blank.

When the wall thickness of the formed cone was less than that dictated by the sine-law of equation (1), the magnitudes of the tangential and radial component forces increased at a rate greater than that during under spin-forging. The undeformed flange tended to move backwards away from the roller and cracks developed on the external surface at the root of the flange resulting in fracture in some cases as the fractional reduction in wall thickness approached a value of approximately 0.65. This condition is referred to as o v e r s p i n - f o r g i n g .

Approximate upper-bound estimates for the dimensionless tangential component force during the spin-forging of cones

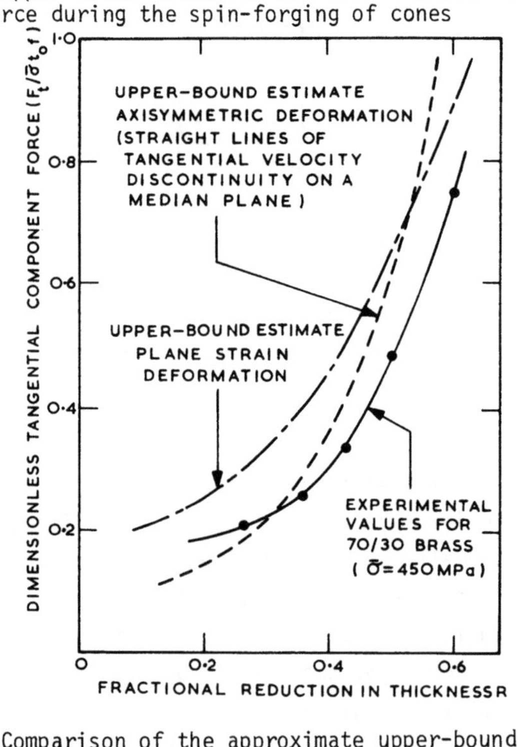

Figure 10. Comparison of the approximate upper-bound estimates with experimental values of the dimensionless tangential component force for 70/30 brass

The values of the dimensionless tangential component force during the spin-forging of cones as predicted by approximate upper-bound estimates assuming (a) plane strain deformation and (b) axisymmetric deformation with straight lines of tangential velocity discontinuity on a median plane for varying fractional reduction in wall thickness are produced in figure 10. For purposes of comparison the experimental values obtained for 70/30 brass sheet metal cones are also included.

It is believed that the most important observation to be made when referring to figure 10 is that the curves for the two approximate upper-bound estimates and the experimental results have a similar characteristic shape, that is, all three are c o n c a v e u p w a r d s . When either simple shear deformation or homogeneous deformation is assumed the corresponding curve exhibits a characteristic shape which is c o n v e x u p w a r d s [10] .

Despite the various assumptions made in deriving the expressions giving the approximate upper-bound estimates for the dimensionless tangential component force, acceptable correlation with the experimental results is achieved. However, it should be borne in mind that discrepancies between the upper-bound estimates and the experimental values could arise because a value of average flow stress $\bar{\sigma}$ = 450 MPa for the deforming material was used to determine the experimental values.

This value of average flow stress was derived from the compressive true stress - compressive natural strain curve obtained for the 70/30 brass sheet material by means of axisymmetric compression tests for the range of maximum natural strain involved. However, the average flow stress would actually vary with the value of fractional reduction in wall thickness.

Part 2. External Surface Finish Assessment of Commercially Pure Aluminium Sheet Metal Cones

Effect of mandrel speed on the external surface roughness for a constant mandrel cone semi-angle and various saddle feed rates

It was found that the value of the centre-line average, R_a, varied from the nose to the rim of a formed cone such that, in general, a least value was attained at some section between the nose and rim. The relationships between the mean value of centre-line average, R_a, and mandrel speed at saddle feed rates of 6.4, 8.3 and 10.2 cm/min, for true spin-forging, using a mandrel having a cone semi-angle of 30^0 are presented in figure 11.

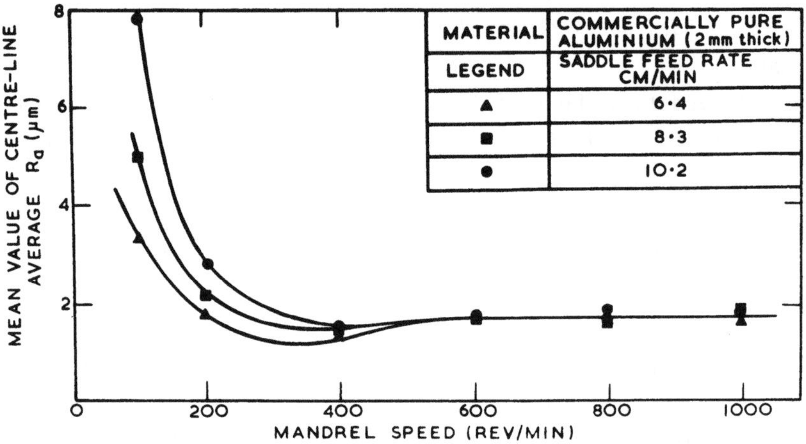

Figure 11. Relation between surface roughness and mandrel speed at various saddle feed rates (mandrel cone semi-angle = 30^0)

It will be noted that these curves have the same characteristic shape for the different saddle feed rates and that the mean R_a value remained sensibly constant at 1.7 μm for mandrel speeds between 600 and 1000 rev/min. It thus emerges that an optimum mandrel speed exists, that is, there is a least value of mandrel speed corresponding to a minimum value of the mean centre-line average, R_a, for a given cone semi-angle, α, and constant saddle feed rate. This optimum mandrel speed is approximately 600 rev/min. It therefore bears similarity with the optimum mandrel speed associated with the minimum value of the tangential component force during the spin-forging of commercially pure aluminium sheet metal cones using a mandrel having a cone semi-angle of 30^0 [10].

Relation between surface roughness and mandrel speed for various mandrel cone semi-angles at constant saddle feed rate

Figure 12 shows the relationship between the surface roughness and mandrel speed using mandrels having cone semi-angles, α, of 30, 40 and 45° with the saddle feed rate remaining constant at 8.3 cm/min. Again, it can be seen that the mean value of centre-line average, R_a, remains sensibly constant for mandrel speeds between 600 and 1000 rev/min for a given mandrel cone semi-angle. However, the smaller the cone semi-angle the higher is the R_a value obtained although the increase is not very significant.

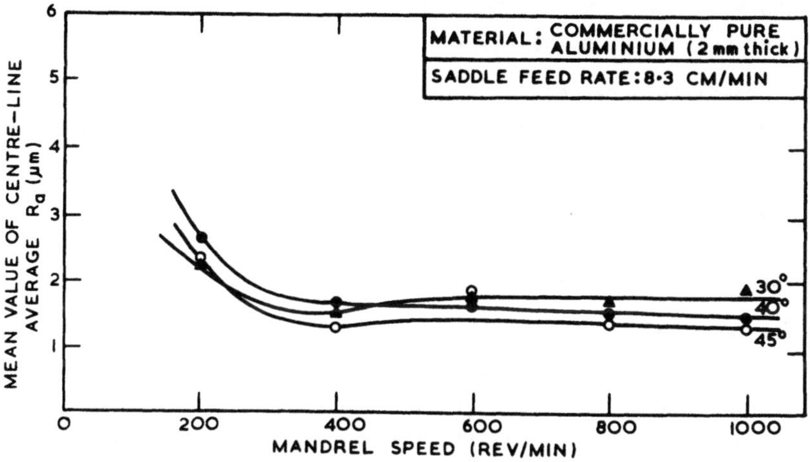

Figure 12. Relation between surface roughness and mandrel speed for various mandrel cone semi-angles at constant saddle feed rate

CONCLUSIONS

(1) During the spin-forging of cones, the tangential, radial and axial component roller forces vary inversely as the mandrel speed and an o p t i m u m mandrel speed exists for a given cone angle and workpiece material corresponding to a constant feed rate.

(2) The optimum mandrel speed varies directly proportional to the cone semi-angle and increases with feed rate.

(3) At constant mandrel speed, the tangential component force varies directly proportional to the feed rate for various values of mandrel cone semi-angle but the effect is reduced as the cone semi-angle increases.

(4) Inflexions have been obtained in the relationships which exist

between the tangential and radial component roller forces and the fractional reduction in wall thickness of formed cones. These inflexions occur at a fractional reduction in wall thickness, $R = 1 - \sin\alpha \approx 0.43$ which corresponds with that for the condition of true spin-forging when $\alpha = 35^{o}$. Spin-forging operations may thus be defined in terms of three regimes: under spin-forging when $R < 1 - \sin\alpha$, true spin-forging when $R = 1 - \sin\alpha$ and over spin-forging when $R > 1 - \sin\alpha$.

(5) Approximate upper-bound estimates assuming (a) plane strain deformation and (b) axisymmetric deformation for the dimensionless tangential component roller force by considering a single triangular velocity field on a median plane appear to provide acceptable correlation with the experimental values during the spin-forging of 70/30 brass sheet metal cones.

(6) The mean value of the centre-line average, R_a, which varied from the nose to the rim, attained a minimum at a particular mandrel speed for a given cone angle and constant feed rate during the spin-forging of commercially pure aluminium sheet metal cones.

ACKNOWLEDGEMENTS

The author offers his thanks to Professor J.C.Levy, Head of Department of Mechanical Engineering, The City University, for making it possible for the investigations to be undertaken.
The technical assistance provided by Mr.H.V.Marsh of the Manufacturing Processes Laboratory and the staff of the University Central Workshop in preparing the specimens for spin-forging and instrumentation is sincerely appreciated.
Thanks are also due to Mr.E.Willett who was responsible for the photographic reproductions.

REFERENCES

1 Astrop,A.W.: Modern spinning and flow-forming. Machinery and Production Engineering (23 June 1976) 578-581.

2 Packham,C.L.: Metal spinning and shear and flow-forming. Metallurgia and Metal Forming (June 1976) 168-170.

3 Packham,C.L.: Metal spinning and shear and flow-forming. Metallurgia and Metal Forming, Part 2: Spinning, (July 1976) 203-206; Part 3: Shear Forming, (August 1976) 250-252; Part 4: Flow-forming (September 1976) 281-285.

4 Paulton,R.A.;Colding,B.N.: Two new industrial processes for plastic deformation. Proc. Conf. Technol. Eng. Manuf., Instn. mech. Engrs. (London) 54-62.

5 Kalpakcioglu,S.: On the mechanics of shear spinning. Trans. Amer. Soc.mech. Engrs.,Ser.B,J.Engng. for Ind. 83 (1961) 125-130.

6 Avitzur,B.; Yang,C.T.: Analysis of power spinning of cones. Trans. Amer. Soc. mech. Engrs.,Ser.B, J.Engng. for Ind. 82 (1960) 231-245.

7 Kobayashi,S.; Hall,I.K.; Thomsen,E.G.: A theory of shear spinning of cones. Trans. Amer. Soc. mech. Engrs.,Ser.B., J.Engng.for Ind. 83 (1961) 485-495.

8 Kobayashi,S.; Thomsen,E.G.: Theory of spin-forging. CIRP Annalen, 10 No.2 (1964) 114-123.

9 Joorabchian,A.: Spin-forging of sheet metal cones. Ph.D. Thesis, The City University (1976).

10 Slater,R.A.C.;Joorabchian,A.: An experimental study of the spin-forging of sheet metal cones using a mandrel of constant cone angle. Proc. 17th Intern. Mach. Tool Des. and Res. Conf. University of Birmingham, Vol.III, Forming of Machine Tools, (20-24 September,1976) 531-537.

11 Slater,R.A.C.;Joorabchian,A.: Spin-forging of sheet metal cones having various cone angles and an upper-bound estimate for the tangential component force exerted at the workpiece-roller interface. Proc. 17th Intern. Mach. Tool Des. and Res. Conf. University of Birmingham, Vol.III, Forming of Machine Tools, (20-24 September,1976) 521-529.

12 Johnson,W.: Estimation of upper-bound loads for extrusion and coining operations. Proc. Instn. mech. Engrs. 173 (1951) 61.

13 Kudo,H.: An upper-bound approach to plane strain forging and extrusion-I. Intern. J. mech. Sci. 1 (1960) 57.

14 Kudo,H.: Some analytical and experimental studies of axisymmetric cold forging and extrusion-I. Intern.J.mech.Sci.2(1960) 102.

15 Green,A.P.: Calculation on the theory of sheet drawing. B.I.S.R.A. Report No. MW/B/7/52 (1952).

APPENDIX 1 Approximate upper-bound estimate for the dimensionless tangential component force during the spin-forging of a cone assuming axisymmetric deformation with straight lines of tangential velocity discontinuity on a median plane

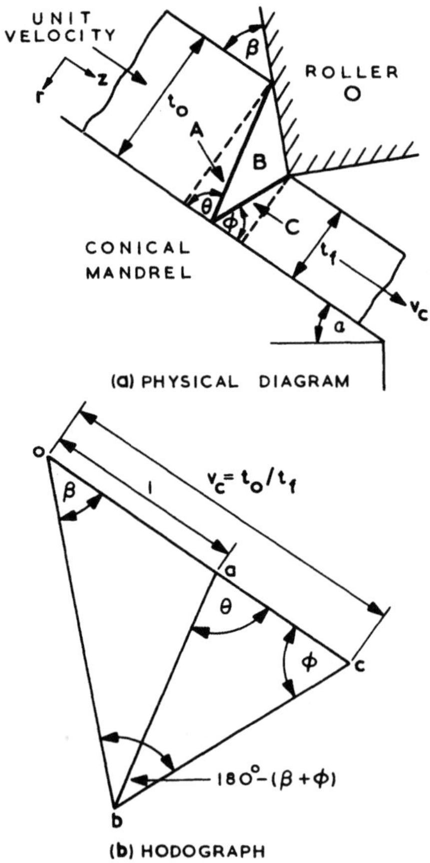

Figure A.1. Spin-forging of a sheet metal cone assuming axisymmetric deformation; (a) Physical diagram comprising a single triangular field of tangential velocity discontinuities on a median plane (b) The corresponding hodograph

The proposed velocity field comprising a single triangular pattern of tangential velocity discontinuities on a median plane is shown in the physical diagram of figure A.1(a) such that four regions 0,A,B and C may be considered. Unit velocity parallel to the roller axis is impressed on the whole system such that region 0, the roller, becomes stationary and region A then moves as a rigid body with unit velocity parallel to the roller axis. Region C is also assumed to move as a rigid body. The corresponding hodograph is given in figure A.1(b).

If the workpiece material is assumed to be incompressible the velocity of region C parallel to the roller axis for unit circumferential width is given by

$$v_c t_f = v_a t_o$$

Assuming the initial wall thickness, t_o, to be unity and since $v_a = 1$

$$v_c = 1/t_f \tag{A.1}$$

For unity width and unity initial wall thickness, t_o, the equations for the areas of the surfaces on which the tangential velocity discontinuities occur are:

$$\left.\begin{aligned} S_{AB} &= t_o/\sin\theta = 1/\sin\theta \\ S_{BC} &= t_f/\sin\phi \\ S_{OB} &= (t_o - t_f)/\sin\beta = R/\sin\beta \end{aligned}\right\} \tag{A.2}$$

The total rate of energy dissipation is given by

$$dE/dt = k\left[\int_V \dot{\gamma}\,dV + \int_S V^* dS\right] \tag{A.3}$$

$$= \dot{E}_1 + \dot{E}_2$$

The maximum shear strain rate in an element of volume dV is $\dot{\gamma}$, V^* is the magnitude of the tangential velocity discontinuities on the surfaces S, k is the yield shear stress of the deforming material, $\dot{E}_1$ is the rate of energy dissipation of a continuously deforming region and, $\dot{E}_2$ is the rate of internal energy dissipation occurring on the surfaces of velocity discontinuity S.

$$\dot{E}_1 = \int_0^L \int_{t_f}^{t_o} \dot{\bar{\varepsilon}}\, 2\pi\, r\, \bar{\sigma} dt\, dL = 2\bar{\sigma}\pi\, Lr \int_{t_f}^{t_o} \dot{\bar{\varepsilon}}\, dt \qquad (A.4)$$

where, $\bar{\sigma}$, is the average flow stress = 2k, L is the axial length of the cone and, $\dot{\bar{\varepsilon}}$, is the effective strain rate which assuming the von Mises' yield criterion to be applicable is given by

$$\dot{\bar{\varepsilon}} = (2/\sqrt{3})\left\{\tfrac{1}{2}\dot{\varepsilon}_{ij}\dot{\varepsilon}_{ij}\right\}^{\frac{1}{2}} = (2/\sqrt{3})\left\{\tfrac{1}{2}(\dot{\varepsilon}_r^2 + \dot{\varepsilon}_\theta^2 + \dot{\varepsilon}_z^2) + \tfrac{1}{4}(\dot{\gamma}_{r\theta}^2 + \dot{\gamma}_{\theta z}^2 + \dot{\gamma}_{zr}^2)\right\}^{\frac{1}{2}} \qquad (A.5)$$

where $\dot{\varepsilon}_r$, $\dot{\varepsilon}_\theta$, $\dot{\varepsilon}_z$, $\dot{\gamma}_{r\theta}$, $\dot{\gamma}_{\theta z}$, and $\dot{\gamma}_{zr}$ are the components of the strain rate tensor $\dot{\varepsilon}_{ij}$ which may be determined for region B in the velocity field of the physical diagram of figure A.1(a). However, for simplicity, homogeneous deformation will be assumed such that

$$\dot{\bar{\varepsilon}} = d\varepsilon/dt \approx \bar{\varepsilon}(\text{axial velocity of cone /length of cone})$$

$$\approx (1/L)\ \ell n(t_o/t_f) \qquad (A.6)$$

Combining equations (A.4) and (A.6) produces

$$\dot{E}_1 \approx 2\pi\bar{\sigma}\ rL(1/L) \int_{t_f}^{t_o} \ell n(t_o/t_f) dt \approx 2\pi\bar{\sigma}\ r \left[t_f\left\{1 - \ell n(1/R)\right\}\right] \qquad (A.7)$$

The rate of internal energy dissipation occurring on the surfaces of velocity discontinuity is

$$\dot{E}_2 = k\left\{(S_{AB}\cdot \overline{ab}) + (S_{BC}\cdot \overline{bc}) + \psi(S_{OB}\cdot \overline{ob})\right\} \qquad (A.8)$$

where $\overline{ab}$, $\overline{bc}$, and $\overline{ob}$ are the tangential velocity discontinuities on the surfaces S_{AB}, S_{BC} and S_{OB} respectively and, ψ, is the fraction for frictional resistance which becomes equal to unity if the workpiece-roller interface is assumed to be perfectly rough.
From the hodograph of figure A.1(b):

$$\left.\begin{aligned} \overline{bc} &= \overline{oc}\ \sin\beta/\sin(\beta+\phi) = t_o\sin\beta/t_f\sin(\beta+\phi) \\ &\qquad\qquad \text{since } v_c = t_o/t_f \\ \overline{ob} &= \overline{bc}\ \sin\phi/\sin\beta \quad = t_o\sin\phi/t_f\sin(\beta+\phi) \\ \overline{ab} &= \overline{bc}\ \sin\phi/\sin\theta \quad \approx t_o\sin\beta\sin\phi/t_f\sin^2(\beta+\phi) \end{aligned}\right\} \qquad (A.9)$$

if Green's assumption that $\theta \approx \phi + \beta$ is valid in this axisymmetric upper-bound approach.

Combining equations (A.2), (A.8) and (A.9) it may then be deduced that

$$\dot{E}_2 = 2\pi rk \left[(1/\sin\theta)\left\{\sin\phi\sin\beta/t_f\sin^2(\beta + \phi)\right\} + (t_f/\sin\phi)\left\{\sin\beta/t_f\sin(\beta + \phi)\right\} + (R/t_f\sin\beta)\left\{\sin\phi/\sin(\beta + \phi)\right\}\right]$$
$$= 2\pi rk \left[\left\{\sin\phi\sin\beta/t_f\sin^3(\beta + \phi)\right\} + \left\{\sin\beta/\sin\phi\sin(\beta + \phi)\right\} + \left\{R\sin\phi/t_f\sin\beta\sin(\beta + \phi)\right\}\right] \quad (A.10)$$

Adding equations (A.7) and (A.10), dividing by $2\pi r$ and equating to the rate of external work done by the axial component force, F_a, produces

$$F_a/\bar{\sigma}t_o f = (1/2)\left[\left\{\sin\phi\sin\beta/t_f\sin^3(\beta + \phi)\right\} + \left\{\sin\beta/\sin\phi\sin(\beta+\phi)\right\} + \left\{R\sin\phi/t_f\sin\beta\sin(\beta + \phi)\right\} + 2\, t_f\left\{1 - \ln(1/R)\right\}\right] \quad (A.11)$$

If it is now assumed that the normal stress acting on the contact surface between the workpiece and roller is uniformly distributed during a spin-forging process then it can be shown that

$$F_t = F_a\,(A_t/A_a) \quad (A.12)$$

where, A_t, and, A_a, are the projected areas of the contact surface between the workpiece and roller in the tangential and axial planes respectively. These areas may be approximated to triangular areas such that

$$\left.\begin{aligned} A_t &\approx \tfrac{1}{2}R^2\cot\alpha \\ A_a &\approx \tfrac{1}{2}fR\cot\alpha \end{aligned}\right\} \quad (A.13)$$

where, α, is the mandrel cone semi-angle, R is the fractional reduction in wall thickness and, f , is the axial feed rate which is unity. Expressing, F_a, in terms of, F_t, given by combining equations (A.12) and (A.13) then $F_t \approx R$ and substituting $t_f = 1 - R$ it can be shown that

$$F_t/\bar{\sigma}t_o f = (R/2)\left[\left\{\sin\phi\sin\beta/(1 - R)\sin^3(\beta + \phi)\right\} + \left\{\sin\beta/\sin\phi\sin(\beta+\phi)\right\} + \left\{R\sin\phi/(1-R)\sin\beta\sin(\beta + \phi)\right\} + 2\left\{(1-R)(1+\ln R)\right\}\right] \quad (A.14)$$

The least upper-bound estimate for the dimensionless tangential component force given by equation (A.14) is obtained by determining the optimum value of the angle ϕ corresponding to various values of fractional reduction in wall thickness R. In the present study these values were obtained with the aid of a computer. For the particular roller used in the spin-forging operations, the angle of inclination of the land, $\beta = \pi/4$.

Residual Stresses in Deep Drawn Cups and Sunk Tubes

K. SAITŌ and Y. SHIMAHASHI

University of Osaka Prefecture, Japan

Summary

It is explained that the predominant cause of the formation of circumferential residual stresses in deep drawn cups and sunk tubes is the axial unbending of the wall element of the product at the exit of the die. The theoretical distributions of residual stresses through the wall thickness are calculated and compared with the experimental results of mild steel products.

1. Introduction

The residual stresses in deep drawn cups and sunk tubes, known well in relation to season cracking phenomena of brass products, have been studied for many years [1-11]. The first crack which appears in such a cylindrical product generally has been observed to be in the longitudinal direction. Therefore, the cause of the crack must be circumferential stress in the wall. In this paper, the mechanism of the creation of the circumferential residual stress will be explained on the basis of the theory of plasticity. Also, the result of the approximate solution of the theoretical stress distribution will be compared with the corresponding experimental data. As a result, typical patterns of these residual stresses are determined.

2. Basic consideration on the creation of residual circumferential stress

Papers on the creation of the longitudinal residual

stress in the wall of the cylindrical drawn cup have already been published [8-11], and it has been known that the cause is the axial unbending of the wall at the exit of the die (Fig.1). But we cannot find any paper studying the creation of the circumferential residual stress. Here we would like to state the basic concept of our consideration about the creation of that residual stress. At the exit of the die profile radius, the element of material of walls with the same curvature as that of the die profile radius is straightened in a short range, where in general the magnitude of the circumferential strain increment is neglegibly small as compared with that of the longitudinal strain increment, because the displacement of the element in radial direction should be neglegibly small in the short range of the exit of the die. An assumption that the axial unbending is under the plane strain condition, where no dimensional change occurs in circumferential direction, makes it easy for us to calculate the approximate value of the stresses after passing through the exit. We know that, in plane strain bending of a wide and flat plate, longitudinal bending stresses occur accompanied by transverse normal stresses. By the application of this knowledge to the axial unbending at the exit of the die in deep drawing or in tube-

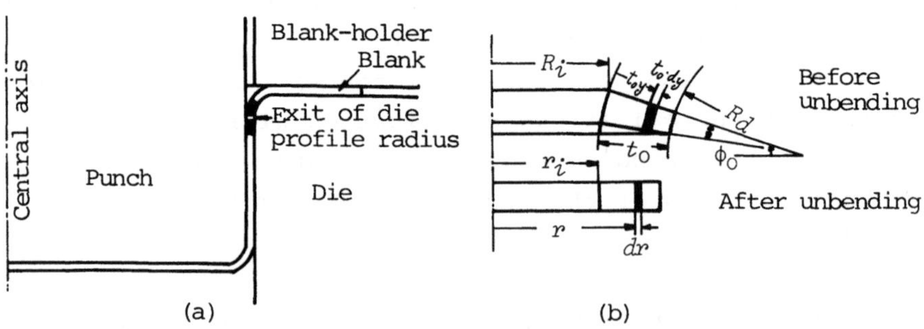

Fig.1. Axial unbending at the exit of the die profile radius; (a) the straightening of wall element in deep drawing, (b) material element and notation for stress analysis.

sinking, we can easily understand that the circumferential stresses are induced in cylindrical products concurrently with the axial bending stresses and are confined in the walls of the products. When the external constraints are removed after unbending, residual stresses develop in the product. Therefore, we can say that the predominant cause of the formation of the circumferential stress must be the axial unbending, and it follows immediately that the longitudinal stress pattern and the circumferential one are similar. The mechanism of the creation of residual stresses may also be applied to sunk (drawn without internal support) tubes, especially to thin-walled tubes.

3. Theoretical stress analyses

3.1 Calculation of unbending stress state

The application of the deformation theory of plasticity to the calculation makes it easy for us to obtain the practical stress state after unbending or straightening at the exit of a die. Now, we assume the material to be incompressible and to be elastic and perfectly plastic. Then, the principal strains $\varepsilon_l, \varepsilon_t$ and ε_r in unbending of a small element of material which was inside at the distance $t_0 \cdot y$ apart from the inner surface ($y=0$) of the wall are represented as follows:

$$\varepsilon_l = \varepsilon_i - l_n\{1 - t_0 y \;/\; (Rd + t_0)\} \tag{1}$$

$$\varepsilon_t = l_n\{(r_i + t_0 \int_0^y e^{\varepsilon r} dy)/(R_i + t_0 y \cos\phi_0)\} \tag{2}$$

$$\varepsilon_r = -(\varepsilon_l + \varepsilon_t) \tag{3}$$

Here, the subscripts l, t and r denote the longitudinal, the circumferential and the radial directions of a circular cylinder, respectively, and the term ε_i represents the longitudinal strain at the inner surface ($y = 0$). If the strain ε_i in unbending is given, a strain distribution through the wall thickness is determined by means of eqs.(1)-(3). Neglecting the elastic components of strains during unbending, we can obtain the deviatoric stress state calculated from the plastic strain components adopting the Hencky stress-strain equations. From a distribution of deviatoric principal stresses through the wall thickness, that of principal stresses can be determined by satisfying both the boundary conditions ($\sigma_r = 0$ at $r = r_i$ and $r = r_0$) and the equilibrium condition of radial direction $d(r\sigma_r)/dr = \sigma_t$. Therefore, by means of numerical analysis we can find the unbending stress distribution letting the total of the longitudinal stress distribution correspond to the external force (that is, a punch load or a sinking load). When the external constraint is removed, the unbending stresses are redistributed to the residual stresses.

3.2 Calculation of residual stress state

If the deformation of spring-back by removal of external force were perfectly elastic, we could superpose the stress change which depends on the elastic recovery on the unbending stresses under external force. However, in this case, plastic deformation takes place near the inside surface of the hollow cylinder. Since superposition is not possible, we must trace the variation of the stress state by numerical analyses, corresponding to subsequent external forces relieved. In the case of a sunk tube, a removed external force is only an axial load. So, the procedure for calculating residual stresses is equivalent to that of solving the elastic-plastic problem of a prestressed hollow circular cylinder subjected to a decreasing axial load by means of the flow theory of plasticity. As a result of tracing the variation of stresses and strains, we can finally obtain the theoretical distribution of residual stresses in a sunk tube corresponding to the stress state of zero axial load. The same way of calculating residual stresses may also be applied to the calculation of residual stresses at the middle height of a cup wall. But for the stress analysis of the portions near the rim and near the bottom of a drawn cup, we should consider the effect of additional deformation discussed later.

3.3 Theoretical distributions of residual stress

3.3.1 Case of a sunk tube

Figure 2 shows the theoretical distributions of unbending and residual stresses obtained by the calculation method described above [12]. The calculated condition corresponds to the data of experimental tube sinking described below in section 4·1. In the numerical analysis, we assumed $\phi_0=0$ and substituted the wall thickness of the original tube for t_0 (the wall thickness at the exit of the die before unbending). It will be found that the difference between the longitudinal and the circumferential stress patterns in the compressive stress region in Fig.2 is comparable to the experimental results.

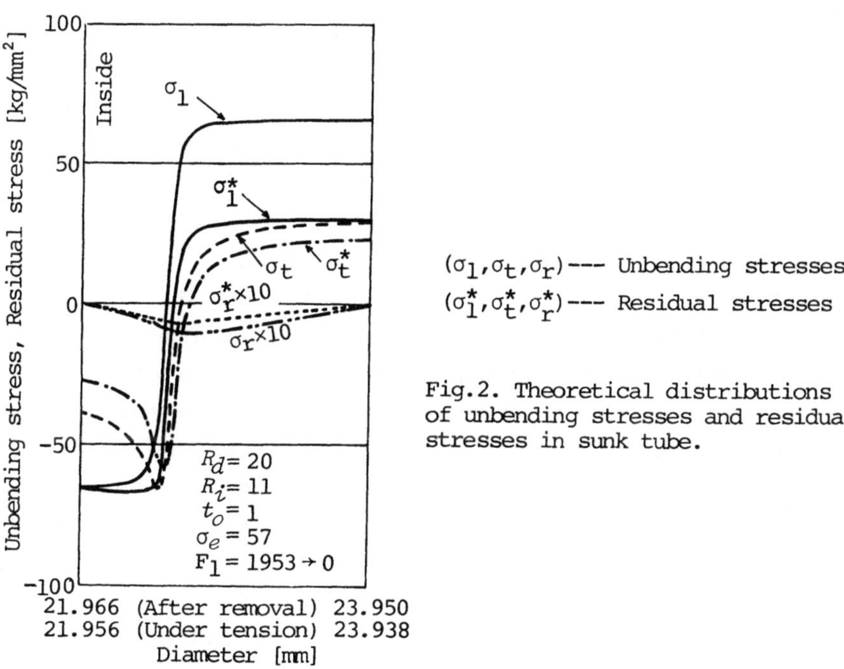

Fig.2. Theoretical distributions of unbending stresses and residual stresses in sunk tube.

3.3.2 Case of a drawn cup

The stress patterns in the cup walls at the time when all of the material of a circular blank has already been drawn into a die hole in the drawing process as shown in Fig.3a, are similar to those of residual stresses shown in Fig.2, since the axial force (i.e., punch load) has vanished. The stress patterns calculated for our experiment in section 4.2 are shown in Fig.4a [13]. We have estimated the magnitude of the equivalent strain of each considered portion from the magnitudes of the circumferential strain and the bending strain in deep drawing, and then obtained the mean value of yield strength. In order to get the value of the axial tensile force under which the considered portion of the wall was unbent at the exit of the die, we adopted the experimental punch load-travel diagram. Then we must evaluate the additional stresses resulting from taking the cup out of the die, but the exact evaluation is not easy. However, we can explain the stress patterns of a cup taken out of the die on the basis of the result (Fig.4a) qualitatively as follows: The main external forces still acting on a drawn cup in the die, as shown in Fig.3a, are the external pressure P_1 at the rim of the cup (i.e., die pressure) and the internal pressure P_2 (i.e., punch pressure) at the cup wall near the bottom. Considering that the deformation of a cylindrical cup wall due to the removal of the external forces P_1 and P_2 is analogous to that due to the action of bending moments uniformly distributed along the circumferences as shown in Fig.3b, we can easily estimate the additional stresses created in the measuring portions of residual stress. For the drawn cup in our experiment described later, the evaluation of the additional deformation by the theory of elastic shell [14] resulted in Fig.3c, where the deformation appeared at the upper and the lower measuring portions and not at the middle portion of the cup wall. Here we can explain the

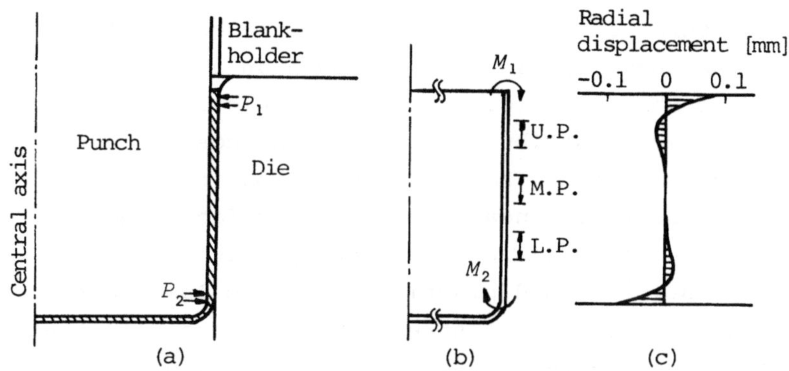

Fig.3. Estimation of additional stresses on taking a cup out of the die; (a) external forces P_1 and P_2, (b) uniform bending moments M_1 and M_2 to evaluate the additional deformation on removal of P_1 and P_2, (c) radial displacement of cylindrical cup wall.

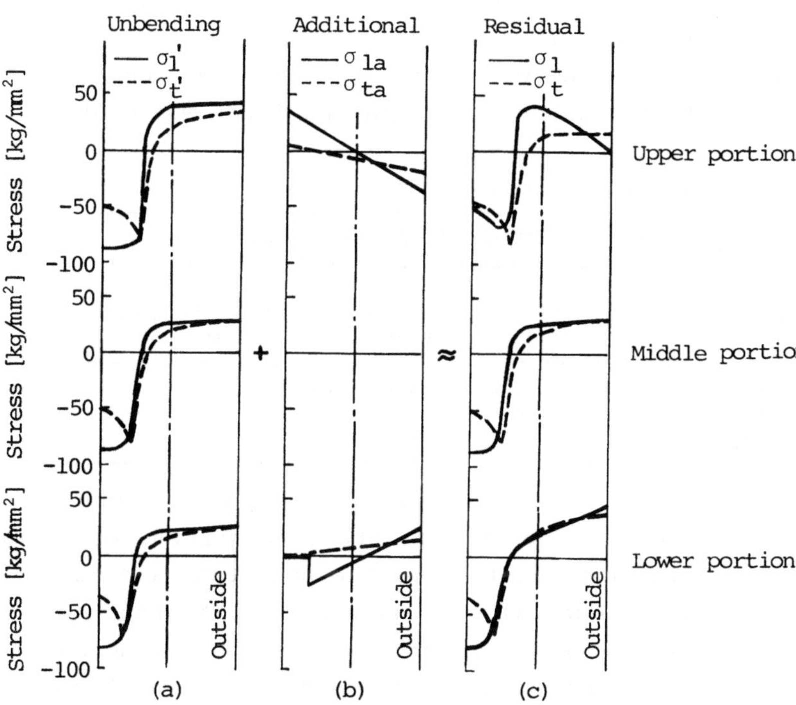

Fig.4. Considerration of residual stress distribution in cup wall.

typical patterns of residual stress in the cup wall schematically as shown in Fig.4. When such an additional deformation as shown in Fig.3c occured in a cylindrical cup wall with stress destributions in Fig.4a whose stresses have already satisfied the yield condition in the compressive stress region near the inside surface, the cup wall deforms elastically at the upper potion of the cup wall and elastically and plastically at the lower portion. Then the additional stress patterns are illustrated in Fig.4b. And therefore, the typical patterns of the distribution of residual stresses at the considered portions of cup wall result in the stress patterns shown in Fig.4c.

4. Comparison of theoretical and experimental results

4.1 Residual stress in sunk tubes

Sinking of steel tubes was performed under the condition shown in table 1, and then the drawing force was measured.

Table 1. Condition of tube sinking

Original tube	0.15 % Carbon Steel Tube (as-drawn) Outside diameter 30mm, Wall thickness 1mm
Die	Die hole dia. 24 mm, Half angle of die profile 20° Die profile radius at the exit 20 mm
Lubricant	Machine oil mixed with M_oS_2
Drawing speed	1m/min.

In order to get an information about the yield stress of the material in the calculation of unbending stress, we carried out the uniaxial tensile test of the sunk tube and obtained a yield stress value of 57 kg/mm^2. The theoretical distribution of residual stress obtained by the numerical analysis corresponding to this experimental condition is shown in Fig.2. Measuring residual stresses by Sachs' method, we obtained the experimental results by means of

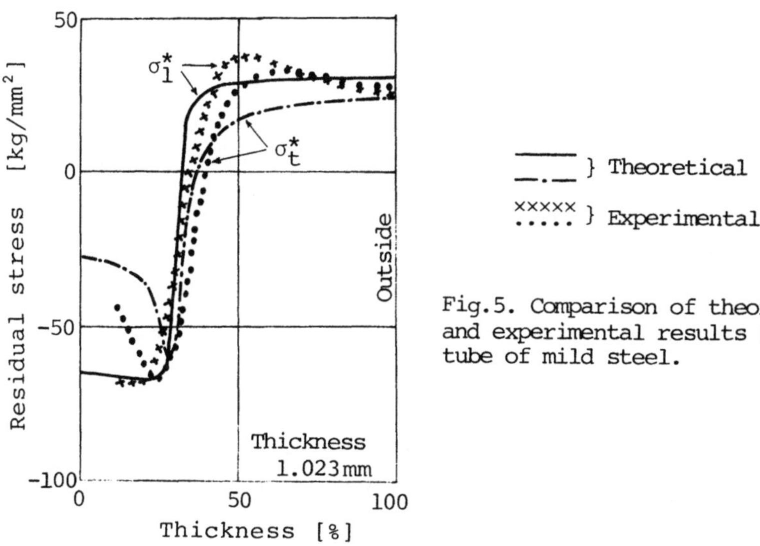

Fig.5. Comparison of theoretical and experimental results in sunk tube of mild steel.

external and internal etching, respectively. A comparison of the theoretical and the experimental results obtained by the external etching method, in which the effect of plastic deformation taking place during residual stress measurement on the evaluation is negligible, is shown in Fig.5. It can be seen that the theoretical stress distribution pattern agrees considerably with the experimental results.

4.2 Residual stresses in drawn cups

Cylindrical cups of mild steel (0.23 % carbon content) were obtained by deep drawing under the conditions shown in table 2. The drawn cup with attached electric strain gauges for residual stress measurement is shown in Fig.6.

Table 2. Condition of deep drawing

Die throat diameter	66.4mm	Die profile radius	3mm
Punch diameter	64.0mm	Punch profile radius	3mm
Blank diameter	120mm	Blank thickness	1mm
Blank holding force	5tons	Maximum punch load	9tons
Lubricant: Liquidised disulfide molybdenum (M_oS_2)			

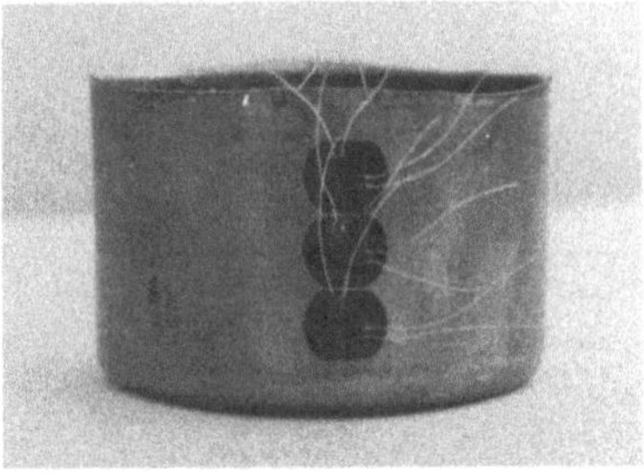

Fig.6. Drawn cup of mild steel with electric strain gauges at different heights of the cup wall.

In order to catch the patterns of residual stresses in the cup wall, Sachs' boring method was adopted to the short cylindrical specimen cut from the cup, and measurements by strain gauges were taken at the three portions of different heights of the cup wall. Figs.7 and 8 show the experimental results, the pattern of which agreed generally with that of the theoretical distribution of residual stresses already shown in Fig.4. Fig.9 shows a comparison of stress patterns at the middle portion of cup wall, magnitudes of which were nearly constant throughout the experimental results, although the magnitudes of residual stresses in the upper portion near the rim of the cup varied greatly with the specimen. It was verified that the predominant cause of the residual stresses was the axial unbending at the exit of the die profile radius. The typical patterns of residual stresses in cup walls were almost clarified.

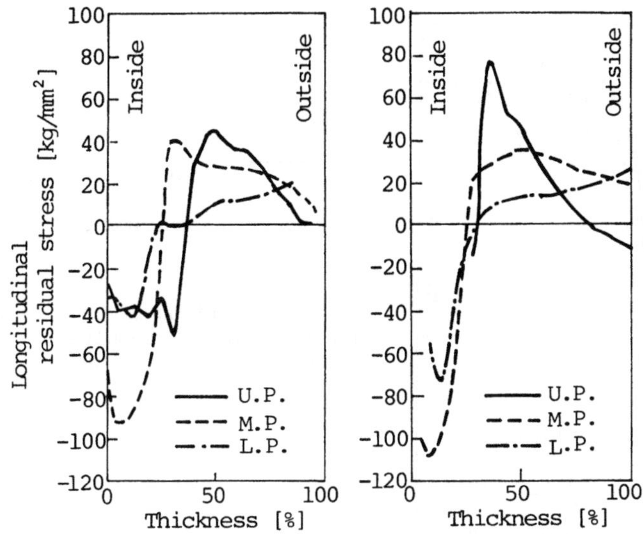

(a) Internal etching (b) External etching

Fig.7. Comparison of longitudinal residual stress distribution through the wall thickness (——: upper portion in height, - - - -: middle portion, —·—: lower portion).

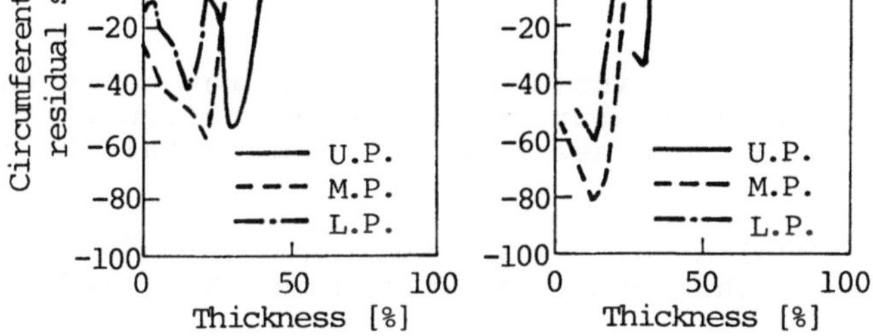

(a) Internal etching (b) External etching

Fig.8. Comparison of circumferential residual stress distribution through the wall thickness.

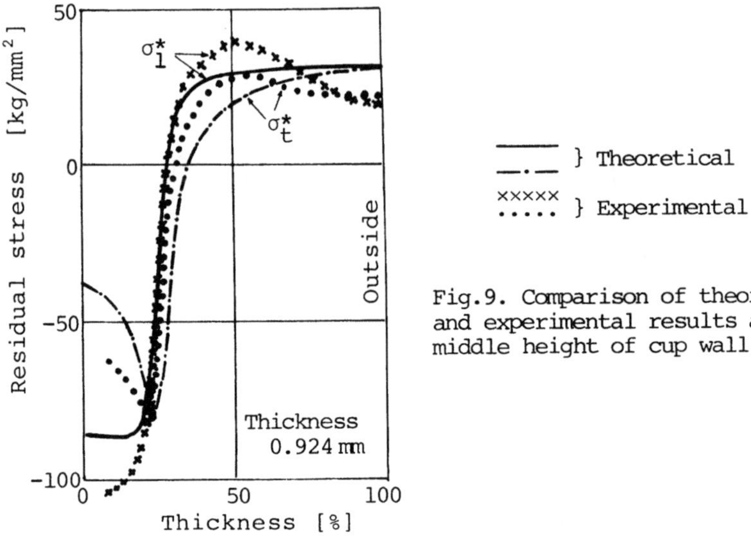

Fig.9. Comparison of theoretical and experimental results at the middle height of cup wall.

References

1 Peiter, A.: Eigenspannungen I. Art. Michael Triltsch Verlag Düsseldorf 1966.

2 Sachs, G.; Espey, G.: Residual stress in sunk cartridge-brass tubing. Inst. Metals Div. A.I.M.& M.E. (1941) 74-89.

3 Saitō, K.; Umeki, S.: Plastic deformation of thick-walled tube subjected to internal and external pressures and longitudial load. J. JSTP 3-19 (1962) 521-526.

4 Meadows, B.J.: The influence of cold drawing on the magnitude and distribution of residual circumferential stresses in 70/30 brass tube. J. Inst. Metals 93 (1964-65) 353-357.

5 Dahl, W.; Mühlenweg, H.: Eigenspannungen und Verfestigung beim Rohrziehen. Stahl u. Eisen 84 (1964) 1250-1260.

6 Bühler, H.: Kreher, P.J.: Eigenspannungen in kalt gezogenen Rohren. Arch. Eisenhüttenwes. 39 (1968) 353-359.

7 Misra, S.k.; Polakowski, N.H.: In-Process control of residual stress in drawn tubing. Trans. ASME Ser. D 91-4 (1969) 810-815.

8 Willis, J.: Deep drawing. Butterworths 1954.

9 Siebel, E.; Mühlhäuser, W.: Eigenspannungen beim Tiefziehen. Mitt. Forsch.-Ges. Blechverarb. 21 (1954) 241-244.

10 Baba, A.: Residual stresses in deep drawing (1st report). J. Faculty of Engineering, University of Chiba 6-10 (1955) 45-52.

11 Baba, A.; Kawada, K.: Residual stresses in deep drawing (2nd report). ibid. 9-15 (1958) 57-65.

12 Shimahashi, Y.; Saitō, K.: Residual stresses in sunk-tubing and drawn cups. Trans. JSME 40-329 (1974) 78-88.

13 Shimahashi, Y.; Saitō, K.: Distributions of residual stresses through the wall thickness in deep drawn cups of mild steel. J. JSTP 18-202 (1977) 909-915.

14 Timoshenko, S.P.; Woinowsky-Krieger, S.: Theory of Plates and shells. McGraw-Hill 1959.

On the Complete Numerical Solution of the Axisymmetrical Deep-Drawing Problem

B. KAFTANOĞLU

Mechanical Engineering Department
Middle East Technical University, Ankara, TURKEY.

Summary

A theoretical method of solution to the problem of axisymmetrical deep-drawing is developed. The theory incorporates the effects of plastic anisotropy in the thickness direction, non-linear strain-hardening and friction between sheet-metal and the tools. Strain history is also taken into account by the use of incremental laws of plasticity. The problem is treated as a non-steady state, multi-region and a multiple moving boundary one, with two-point boundary conditions. The previously obtained numerical solution for the regions over the punch head and between the die and punch is extended to cover the regions over the die. The treatment of the boundary conditions is complete and a "recursive matching" technique is used to satisfy them. Together with the method of determining the onset of plastic instability according to the Strain Propogation criterion, the strains and stresses up to the instability point can be computed. This enables one to calculate whether deep-drawing will be successful or not. Hence limiting blank size can be computed. Partial results obtained so far have been encouraging.

Introduction

Deep-drawing is an industrial process in which material properties in the plastic range, frictional conditions and geometry of tools interact in a complicated way and determine the success or failure of the product to be drawn. The process is generally limited by the instability taking place over the punch head leading to fracture and the failure of the product. It is therefore of interest to be able to predict the stage and location of instability, the corresponding blank diameter, the distribution of strains, and the drawing load.

As seen in Fig.1, the plastic deformation takes place in several

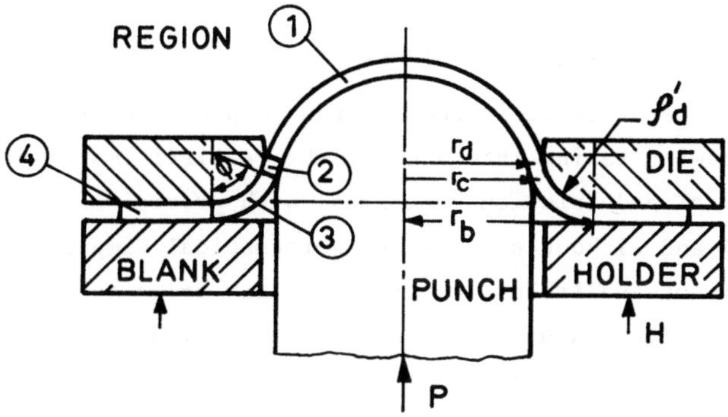

Fig.1: Main Features of Deep-Drawing

regions: a) over the punch head (1) b) between the punch and die (2) c) over the die profile (3) and d) over the die (4). The plastic deformation in each region is connected to that in the neighbouring one by means of the boundary conditions at the fixed, and moving boundaries like the contact point of the sheet metal with die and the punch.

The early researches in the field of deep-drawing were generally concentrated in the experimental and theoretical investigations of one or two of the regions. Reviews by Alexander [1] and Fukui et al [2] give comprehensive accounts of the early researches. Later, attempts were made to get numerical solutions for the whole of deep-drawing. Woo [3] extended his work for stretch-forming to cover deep-drawing. Present author's comments about his work has been published previously. [4,5,6] It is also felt that the location of the moving boundary where the sheet-metal leaves the die surface could not have been calculated theoretically without making assumptions. Instability analysis and results on limiting blank sizes were not included. Odell and Clausen [7] investigated the effect of bending stress in deep-drawing using a special numerical technique. They showed the effect of bending stresses on the punch load, but did not report any results on stress and strain distributions and instability. Kobayashi and

Kim [8] **developed** a finite element solution for deep-drawing for rigid-plastic materials. Their analysis did not incorporate plastic instability and hence limiting blank diameters cannot be computed.

From the survey of literature it is seen that a complete theoretical analysis of deep-drawing taking into account the realistic material parameter, geometrical and friction variables, with provisions to predict instability and without significant mathematical simplifications, is still not available. The present paper tries to fullfil this gap and offers a numerical procedure eliminating the previous assumptions and simplifications.

Since the size of the computer program, memory required are large and the development is difficult, in this paper the complete numerical procedure is given together with partial results.

Basic equations

The necessary equations for the numerical solution are nearly the same for each of the four regions except for small variations. In the present analysis the flow rules due to Hill [9] are used. Using these equations the plastic anisotropy in the thickness direction is taken into account. The extensive derivations of the equations are given in previous publications [6,10] on stretch forming and hydraulic bulging.

Equilibrium of forces:

From the radial and vertical equilibrium of forces shown on an element of shell wall in Fig.2 the following equations can be derived:

for curved parts of the shell,

$$\frac{d}{dr}(\sigma_1 t)=\sigma_1 t\left[\frac{x-1}{r}+\mu\left(\frac{1}{\rho_1 \cos\theta}+\frac{x\tan\theta}{r}\right)\right] \qquad (1)$$

$$p=\frac{t}{\rho_1' r'}(\sigma_1 r+\sigma_2 \rho_1 \sin\theta) \qquad (2)$$

for flat parts of the shell

$$\frac{d}{dr}(\sigma_1 t) = \frac{\sigma_1 t}{r}(x-1) \tag{3}$$

where $x = \frac{\sigma_2}{\sigma_1}$

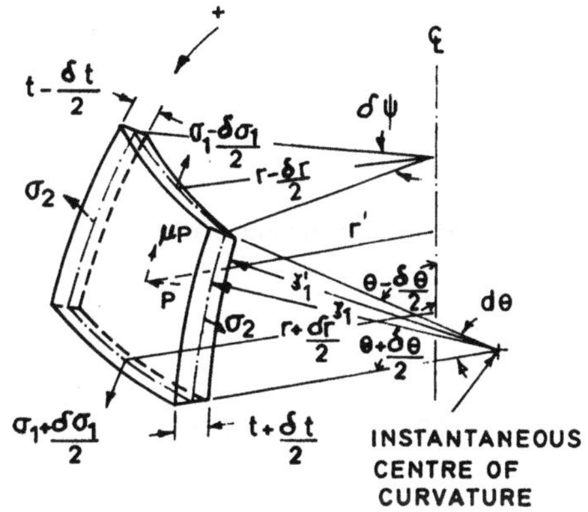

Fig.2: Stresses on element of shell wall

Strain compatibility:

For large plastic strains, the compatibility between the instantaneous strains on the shell wall is:

$$\frac{d\varepsilon_2}{dr} = \frac{1}{r}\left(1 - \frac{e^{2\varepsilon_2+\varepsilon_3}}{\cos\theta}\right) \tag{4}$$

Yield criterion:

$$\bar{\sigma} = \left(\frac{3[x^2+1+2k(k-1-x)+R(1-2x+x^2)]}{2(2+R)}\right)^{1/2}\sigma_1 \tag{5}$$

where $k = \frac{\sigma_3}{\sigma_1} \approx -\frac{p}{2\sigma_1}$ and $R = \frac{\varepsilon_w}{\varepsilon_t}$ = constant in uniaxial tension

Stress-strain relationships:

$$d\varepsilon_1^* = d\lambda'[R(1-x)+(1-k)]$$

$$d\varepsilon_2^* = d\lambda'[(x-k)+R(x-1)]$$

$$d\varepsilon_3^* = d\lambda'[2k-1-x]$$

between the above equations eliminating the scalar parameter $d\lambda'$:

$$d\varepsilon_3^* = \frac{2k-1-x}{(R+1)x-R-k}\, d\varepsilon_2^* \tag{6}$$

where $k = -\dfrac{t}{2\rho_1' r'}(r + x\rho_1 \sin\theta)$

where (*) indicates the strain increments following an element

Equivalent strain-increment:

$$d\bar{\varepsilon}^* = \frac{[\frac{2}{3}(2+R)]^{1/2}}{1+2R}\,[(1+3R+2R^2)d\varepsilon_3^{*2}+2(2R+1)(d\varepsilon_2^{*2}+d\varepsilon_2^* d\varepsilon_3^*)]^{1/2} \tag{7}$$

Strain hardening law:

$$\bar{\sigma} = [\frac{3(R+1)}{2(R+2)}]^{1/2}\, A(B+[\frac{3(R+1)}{2(R+2)}]^{1/2}\,\bar{\varepsilon})^n \tag{8}$$

where A, B, and n are constants for a given material.

Compatibility of increments in time and space:

in the increasing r direction -

$$V_{(m-1),n}+\delta V_n = V_{m,(n-1)}+\delta V_m^* \tag{9}$$

in the decreasing r direction -

$$V_{m,n}-\delta V_n = V_{(m-1),(n-1)}+\delta V_{(m-1)}^* \tag{10}$$

where V represents variables like stress, strain etc.

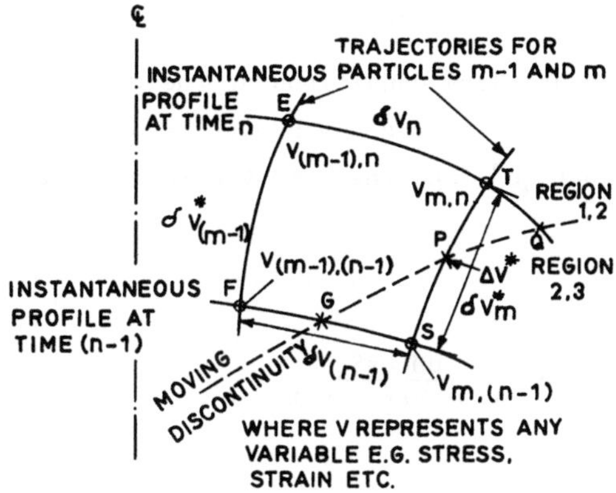

Fig. 3: Increments of variables in time and space

Geometrical relationships:

The punch and die may have different geometrical shapes. The punches may be of hemispherical, ellipsoidal or flat headed in shape. The most common geometry for the die profile is a toroidal shape defined by the following equation:

$$r' = r_b - \rho'_d \sin\theta \tag{11}$$

In the above equations when the required expressions for $r=f(\theta)$ are substituted, the necessary equations can be obtained for the particular applications.

Boundary conditions

In deep-drawing there are two types of boundary conditions:
a) Initial conditions as functions of radius before the occurence of any plastic deformation, i.e. initial values of $\bar{\varepsilon}$, θ as functions of r, which have to be fed into the solution as data.
b) Boundary conditions which have to be calculated for each stage at fixed and moving boundaries. Fixed boundaries-pole, start of die curvature. Moving boundaries-contacts of sheet metal with

punch and die curvature, rim of blank.

Detailed analysis of the boundary conditions at the pole and at the contact point of sheet-metal with the punch have been presented in detail in a previous paper [6], and they are the same for deep-drawing.

At the rim-

$$(\varepsilon_2)_{m,n} = \ell n \frac{r_{m,n}}{r_{m,o}} \quad \text{and} \quad \sigma_1 = \left(\frac{\mu H \cos\theta}{\pi r t}\right)_{m,n} \tag{12}$$

Then the values for x, ε_3, $\bar{\varepsilon}$ can be calculated from the basic equations, together with their derivatives.

At the start of the die curvature-

It can be proved that ε_2, ε_3, ε_1, $\bar{\varepsilon}$, $\bar{\sigma}$ and $\bar{\sigma}_1$ are continious. The values of σ_2, k, and p can be discontinious if the geometry is discontinious. The values for the continious and discontinious variables and their derivatives can be calculated similar to procedure given in [6].

At point of departure of sheet metal from die-

This point seperates regions 2 and 3. Again the σ_2, k and p are discontinious and the others are continious. The procedure in calculating the boundary values of the variables are similar to that given in [6] for the boundary between regions 1 and 2. However, it is not posible to calculate the value of $r_c(\phi)$ at which the sheet-metal leaves the die. It is calculated as a result of boundary value matching procedure at the end of the numerical analysis.

Numerical solution

The numerical solution for the deep-drawing problem consists of the solutions of each of the four regions and matching of the boundary conditions between them. The numerical solutions for regions 1 and 2 have been given previously [6]. Here the numerical procedures for regions 4 and 3, together with the boundary value matching algorithms for the complete solution of the deep-drawing problem will be given.

Region 4-

1. The boundary conditions at the rim are calculated according to the above procedure. The value of $(\varepsilon_2)_{m,n}$ or $\delta\varepsilon_2^*$ is used as a time variable and can be chosen arbitrarily. It will only affect the accuracy of the results. In the deep-drawing problem the value of ε_2 at the rim and at the pole, and the value of θ at the die profile should be compatible and unique. This is accomplished later during the boundary-value matching procedure.

Initial values for $\bar{\varepsilon}$, θ, ε_3 are set for the blank.

2. The value of suffix m is decreased by one, and the starting values for δx_n and $(\delta\varepsilon_2)$ are set according to those values in the previous stages or can be assumed.

Beginning of outer loop

3. From eqn. (10):

$$x_{m,n} = x_{(m+1),n} + \delta x_n \tag{13}$$

Mean stress ratio following an element:

$$x^* = 0.5(x_{m,n} + 0.5\ x_{m,(n-1)}) \tag{14}$$

Beginning of the inner loop:

4. From eqn. (10):

$$(\varepsilon_2)_{m,n} = (\varepsilon_2)_{(m+1),n} + (\delta\varepsilon_2)_n \tag{15}$$

Steplength between the two nodes:

$$\ell = r_{m,o}\ \exp(\ (\varepsilon_2)_{m,n}) - r_{(m+1),o}\ \exp(\ (\varepsilon_2)_{(m+1),n}) \tag{16}$$

5. From eqn. (10):

$$(\delta\varepsilon_2)^*_m = (\varepsilon_2)_{(m+1),n} + (\delta\varepsilon_2)_n - (\varepsilon_2)_{m,(n-1)} \tag{17}$$

6. Using $(\delta\varepsilon_2)^*_m$ from eqn(17), and x* from eqn. (14), $(d\varepsilon_3)^*_m$ can be calculated from eqn. (6).

7. Eqn. (4) can be integrated using the Runga-Kutta technique with $\theta=f(r)$ expressed by means of Lagrangian polynomials for curved dies. $\varepsilon_3=g(r)$ also expressed by Lagrangian polynomials. Steplength ℓ is used from eqn.(16). and $(\varepsilon_2)_{m,n}$ is therefore calculated.

$$(\delta\varepsilon_2)_n = (\varepsilon_2)_{m,n} - (\varepsilon_2)_{(m+1),n} \tag{18}$$

$$\text{Error, } \Delta=(\delta\varepsilon_2)_n[\text{step } 7]-[\delta\varepsilon_2]_n[\text{step } 1] \tag{19}$$

If $|\Delta|<0.005$ % , the strain compatibility equation is considered to be solved, if not by linear interpolation a new $(\delta\varepsilon_2)_n$ is calculated to give $\Delta=0$ and calculation returns to step 1 until the accuracy limit is satisfied.

End of inner loop

8. Using $(\delta\varepsilon_2)^*_m$ and $(\delta\varepsilon_3)^*_m$ obtained in steps (5), (6) $(d\bar{\varepsilon}^*)_m$ is evaluated from eqn.(7) and therefore

$$(\bar{\varepsilon})_{m,n}=\sum_{o}^{n}(\delta\bar{\varepsilon}^*)_m \tag{20}$$

9. With $(\bar{\varepsilon})_{m,n}$ from eqn.(20), $\bar{\sigma}_{m,n}$ can be calculated from eqn.(8).

10. Using $x_{m,n}$ from eqn.(13) in eqn.(5) $(\sigma_1)_{m,n}$ is computed.

11. It is now possible to compute $(\sigma_1 t)_{m,n}$ from steps (10) and (6).

12. The equation of equilibrium (3) or (1) can now be integrated using a Runga-Kutta technique. In cases of curved dies ρ_1, θ, and x are expressed as functions of r by means of Lagrangian polynomials. As a result of numerical integration $(\sigma_1 t)_{m,n}$ is obtained.

13. If the assumed or the later modified value of δx_n is correct then the values of $(\sigma_1 t)_{m,n}$ from steps (11) and (12) would be

equal. If they are different, then the value of δx_n is modified and calculation returns to step (3). The modification to δx_n is carried out such that the solution which gives the minimum plastic work is selected.

End of outer loop

14. When the two loops are satified, then all the variables would have correct values. The calculation then can proceed to the next point by setting m=m-1 and restarting from step (2). The calculation will end where $r=r_b$, which will be the starting point for region 3.

Region-3

At the start of the die curvature the boundary conditions can be calculated as explained above. For these calculations, the values of the variables obtained from region 4 will be used.

The numerical procedure will be the same as in region 4, with the exception of the use of eqn.(11) in steps (7) and (12). The solution will continue until the contact point of the die with sheet metal. This point c and its radius r_c will be calculated as a result of boundary value matching as given in the next section.

Boundary-value matching in deep-drawing:

At any stage of deep-drawing, the following boundary conditions are not known:

a) ε_2 at the pole in region (1). During the solution this can be regarded as a measure of time, and therefore it can be the only independent variable.

b) Location of the moving boundary (r_d) between regions (1) and (2).

c) Location of the moving boundary (r_c-ϕ) between regions (2) and (3).

d) ε_2 at the rim in region (4).

To find the three unknowns in the last three items the following procedure can be used.

1. ε_2 at the pole can be assumed initially by using:

$$(\varepsilon_2)_p = - 0.024\mu + 0.03$$

Beginning of the outer loop

2. The solution in region (1) terminates at an initially assumed or later modified value of r_d.
3. The solution of region (2) terminates at a point such that the slope of the bulge is tangent to that of the die at the same point.

Beginning of the inner loop

4. The solution of region (4) is started by an initially assumed or a later modified value of ε_2 at the rim-
5. Solution of region (4) progresses to region (3) and terminates at a point where the solution of region (2) terminates as mentioned in step (3). However, it should be noted that the suffix m for the end points of the two solutions may not be the same, although the current radii for both may be the same.
6. The values of $Q=2\pi r\sigma_1+\sin\theta$ are calculated from the results of steps (3) and (5) and compared. They must be equal to satisfy the conditions of vertical equilibrium of forces at the boundary. If they are different more than an allowed inaccuracy, then ε_2 at the rim in region (4) is changed in a way as to reduce this difference, and calculation returns to step (4).

End of inner loop

7. Two values of the suffix m (or $r_{m,o}$) obtained from the solutions in steps (3) and (5) are compared. They must be the same if the assumed location of the boundary between regions (1) and (2) is correct.
8. When $(m)_2 \neq (m)_3$, then the location of the moving boundary between regions one and two is varied such that $(m)_2 = (m)_3$ is obtained. Hence with a modified location of the moving boundary between regions (1) and (2) calculations return to step (2).

End of outer loop

9. When all the boundary conditions are satisfied, the solution can proceed to the next stage by incrementing $(\varepsilon_2)_p$ at step (1). When the SP instability is reached possibly in region (1) [6] the calculation terminates indicating the onset of fracture, and thus the failure of the deep-drawn component. If blank diameter is reduced in small steps, the diameter for which SP

instability will not be found will correspond to the limiting blank diameter.

Discussion of preliminary results

Deep-drawing tests are carried out on soft copper with 1.2 mm thickness and having the following material parameters:

A = 471 MPa , B = 0.01 , n = 0.38 , R = 0.8

The tool geometry is as follows:
Punch diameter = 33 mm, die throat diameter = 36.4 mm, die profile radius = 10.6 mm.
Experimentally found limiting drawing ratio = 2.359.
An oversize blank by one increment with 7.87 cm in diameter is drawn corresponding to a drawing ratio of 2.384 using a graphite grease. The strain distributions are measured and the curves are shown in Fig.4.

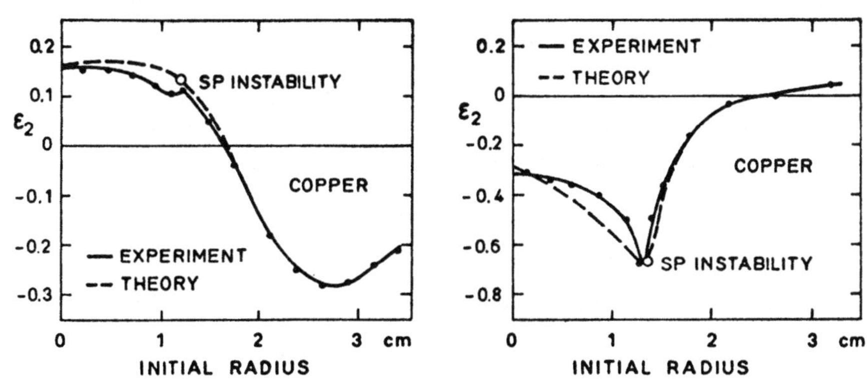

Fig.4: Circumferential and thickness strain in deep-drawing.

The experiment is stopped just before the occurance of fracture.

For the same conditions the numerical calculations are partially carried out. Since the failure occurs in region one, the solutions for regions (1) and (2) are obtained and the boundary conditions between regions (2) and (3) are simulated. From the theoretical results in Fig.4 it can be seen that the strain distributions are predicted fairly closely. The prediction of the location and the magnitude of the plastic instability using the Strain Propagation criterion have been successful. Due to limitations on the computer facilities full solutions including the boundary-value matching have not yet been obtained.

Conclusions

The theory developed for the complete numerical solution of the deep-drawing problem incorporates the real material parameters, uses the incremental plasticity theory and allows for friction. A procedure is provided for the boundary-value matching so that compatibility between different regions can be maintained. Strain Propagation criterion is used for the prediction of instability. The preliminary results obtained show that the predictions of strain distributions and instability are satisfactory.

References

1 Alexander, J.M.: An appraisal of the theory of deep-drawing. Metallurgical Reviews, Vol. 5, No. 19 (1960) 349-411.

2 Fukui, S., Yoshida, K., and Kobayashi, S.: Sheet metal forming research in Japan. Journal of Engineering for Industry, Trans. ASME, Series B, Vol. 88, No. 1 (1966) 101-110.

3 Woo, D.M.: On the complete solution of the deep-drawing problem. Int. J. Mech. Sci. 12 (1970) 1065-1084.

4 Kaftanoğlu, B.: Communications. J. Mech. Sci., 7 (1965) 123.

5 Kaftanoğlu, B.: Letter to the editor. Int. J. Mech. Sci. 7 (1965) 447.

6 Kaftanoğlu, B., Alexander, J.M.: On quasistatic axisymmetrical stretch forming. Int. J. Mech. Sci. 12 (1970) 1065-1084.

7 Odell, E.I., Clausen, W.E.: Numerical solution of a deep-drawing problem. Paper presented at ASME Winter Annual Meeting, Dec. 5-10 (1976) No. 76-WA/Prod-3.

8 Kobayashi S., Kim, J.H.: Deformation analysis of axisymmetric sheet metal forming processes by the rigid-plastic finite element method. Paper presented at the Mechanics of Sheet Metal Forming Symposium sponsored by General Motors, Oct. 17-18, 1977, Detroit, U.S.A.

9 Hill, R.: The mathematical theory of Plasticity. Oxford (Clarendon Press) (1950).

10 Kaftanoğlu, B.: Plastic instability of thin shells deformed by rigid punches and by hydraulic pressure. Trans. ASME J. Eng. Met. and Tech. Series H, 95 (1973) 36-40.

Application of the Stereophotogrammetric Methods to the Analysis of the Kinematics and Dynamics of Shells

T.BEDNARSKI

Institute of Metal Forming Technology

Warsaw Technical University, Poland

SUMMARY

The paper presents the methods of analysing the kinematics and dynamics in the cold drawing processes by means of stereophotogrammetry. A short description of the stereophotogrammetric methods is given. These methods are presented as being used for measuring displacements of material points while bulging cilcular membranes under static and dynamic loads. As a result of numerical analysis of respective pairs of stereoscopic photographs the space co-ordinates of marked points on the shell have been obtained as a function of pressure or time. This rendered it possible to determine changes in shell shape, displacement fields and displacement rates fields in the course of the deformation process. The knowledge of kinematics allowed in turn to solve the equilibrium equation for the deformed shell, to calculate the stress field, to determine the stress-strain characteristics of material as well as to reproduce the dynamics of the process.

INTRODUCTION

It is known that an analysis of the kinematics is possible if we have determined a motion function of material points of a body in the vector form $f/\vec{r}_i,t/$ or in the parametrical form $c_{ij} = c_{ij}/t/$, where $\vec{r}_i$ is a radius-vector of the material point which is designated by index i = =1,2,3,...,I, t - a time, j is a number of cartesian co-cordinate /Fig.1/. In each case we must introduce a reference co-ordinate system making use of fixed points denoted subsequently by the index w = 1,2,3...,W.

Those fixed points make it possible to reproduce the reference co-ordinate system in the course of numerical analysis /Fig.1/. While investigating the deformation processes of membranes bulging the fixed points are disposed on two mutually parallel planes on the drawing ring.

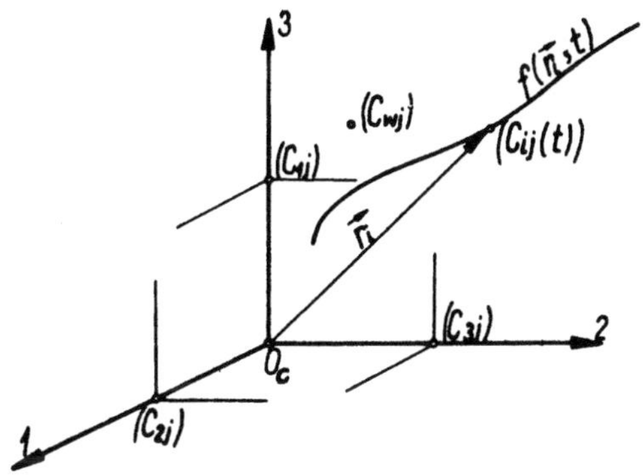

Fig.1. Materialization of a reference co-ordinates system by fixed points c_{wj}.

The motion function can only be directly obtained from measurements utilizing stereophotogrammetric methods. All remaining kinematic and dynamic magnitudes must be calculated on the basis of membrane shells theory [1] and theory of plasticity.

THE STEREOPHOTOGRAMMETRIC METHODS

The stereophotogrammetric methods are based on stereoscopic photographs, that means on the photographs taken from two different points s^m, m = 1,2, in space. From geometrical point of view we obtain two central projections of the process on the plates or films. The points s^m, which determine centres of projections, must be chosen in such a way that the given elements of the object are visible on both pictures /Fig.2/. For the analysis of the stereoscopic photographs their

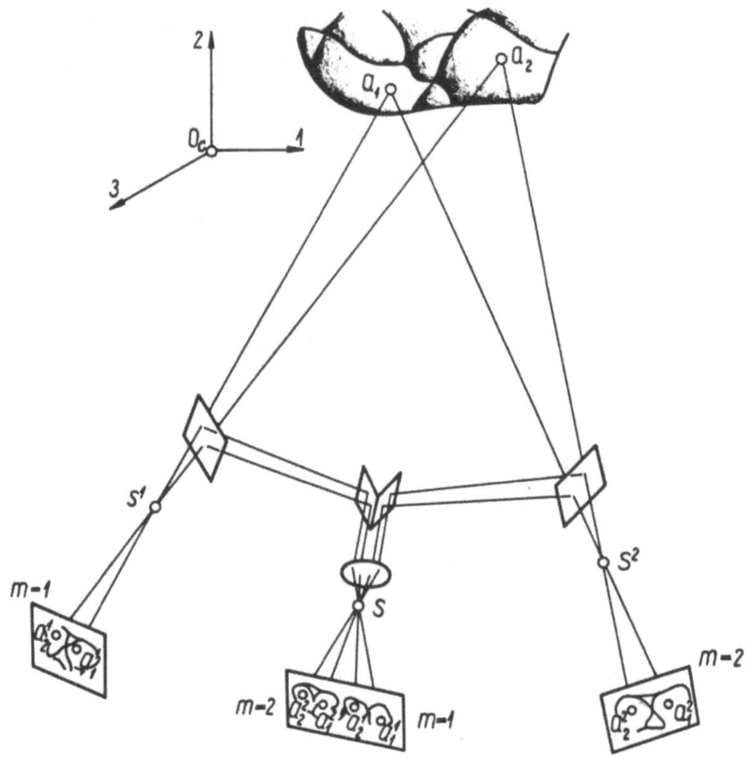

Fig.2. Scheme of the stereoscopic system with two cameras or one camera with stereoscopic attachment.

elements of inner and outer orientation must be known A set of quantities making possible the localization of the centres of projections s^m, m = 1,2, in relation to plates or frames to be reproduced are called the elements of inner orientation of the photographs. It is so as if we had new co-ordinate systems $\{z_j^m\}$ determined with origins in the centres of projections s^m and the axes z_3^m perpendicular to the adequate plates of photographs /Fig.3/.
The elements of the outer orientation of the stereoscopic system consist of translation vector u_{sj}^m of the co-or-

dinate systems $\{z_j^m\}$, and angles α_j^m, m = 1,2, j = 1,2, ,3, between the directions of the axes z_j^m and the axes u_j of the assumed reference co-ordinate system $\{u_j\}$ /Fig.3/.

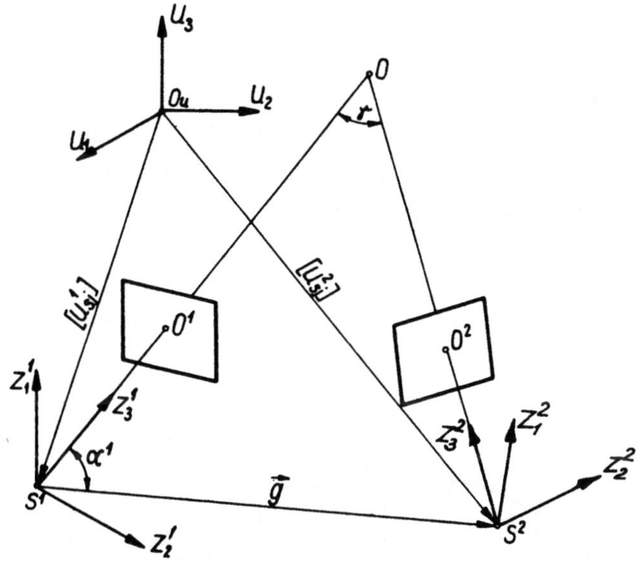

Fig.3. Geometric interpretation of the inner and outer orientation elements for the stereoscopic system

Two methods of the stereoscopic photographs analysis are used. One of them is the classical method commonly used in photogrammetry [2,3,4,5], and the other, based on the projective transformations, is a new one [6,7,8,9].

To make use of the classical method of stereophotogrammetry it is necessary to know the elements of the inner and outer orientation of the stereoscopic system in the course of the process investigation.

The projective method does not require knowledge of the orientation elements what causes it to be very useful. In this method the orientation elements are calculated in the course of numerical analysis of the fixed points co-ordinates.

The adaptation of the stereophotogrammetric methods to the measurements of the material points displacement is described in detail in papers [8-14].

BULGING PROCESSES

The stereophotogrammetric methods are used in the analysis of the kinematics in the plastic bulging processes of the circular membranes /Fig.4./. The membranes are clamped at their circumference and subjected to loading with pressure p/ ρ ,t/. The pressure is exerted in two different ways - quasi-statically and dynamically. In both processes on the surface of shells grids were placed by means of the chemigraphic method.

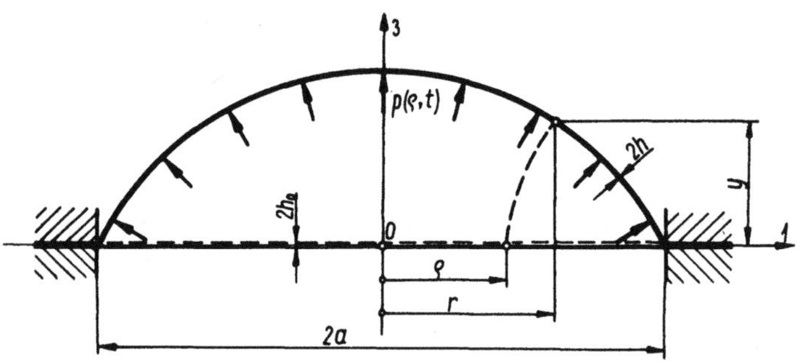

Fig.4. Scheme of the membrane bulging.

STATICAL MEMBRANE BULGING

The process of plastic bulging of the circular membrane under quasi-statical loading conditions was analyzed in all its aspects. The disc-shaped specimen had a diameter of 160 mm and was made of 0.915 mm thic aluminium sheet. The grid consists of four radial lines crossed with transverse lines about 5 mm from each other.
The lines of the grid were about 0.2 mm thick. The general views of the specimen before and after deformation

Fig.5. The specimen before and after deformation.

are shown in Fig.5. The plastic bulging process was carried on in a way schematically illustrated in Fig. 6. For photographing the progress of deformation process two cameras of Pentacon Six TL type were used. As a result of this duplicity, photographs taken in pairs

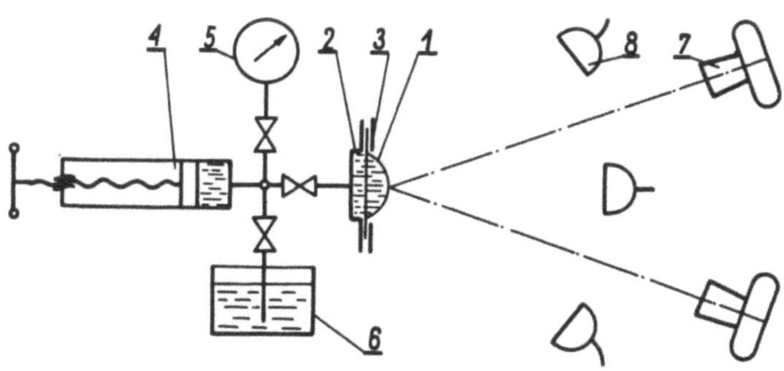

Fig.6. Scheme of measurements, 1.specimen, 2.oil chamber, 3.drawing ring, 4.hand pump, 5.pressure gauge, 6.tank, 7.photo cameras, 8.reflectors.

after each successive pressure increment $\Delta p = 1$ kG/cm^2 were obtained. In this way 15 pairs of photographs were obtained, including those before starting the deformation and after fracture of the specimen. The pair of photographs before fracture of the specimen is shown in Fig.7.

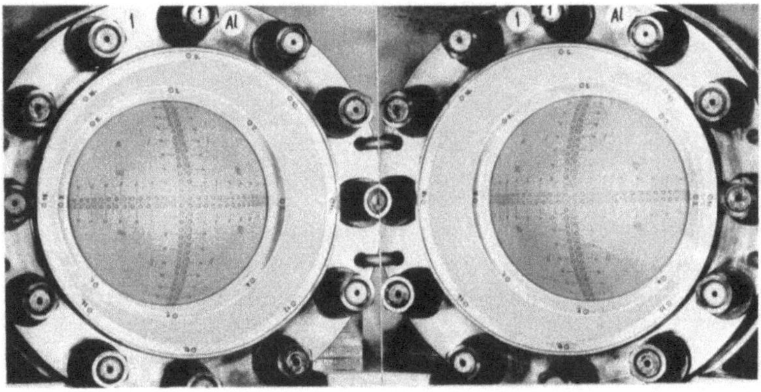

Fig.7. Part of stereogram of photographing process.

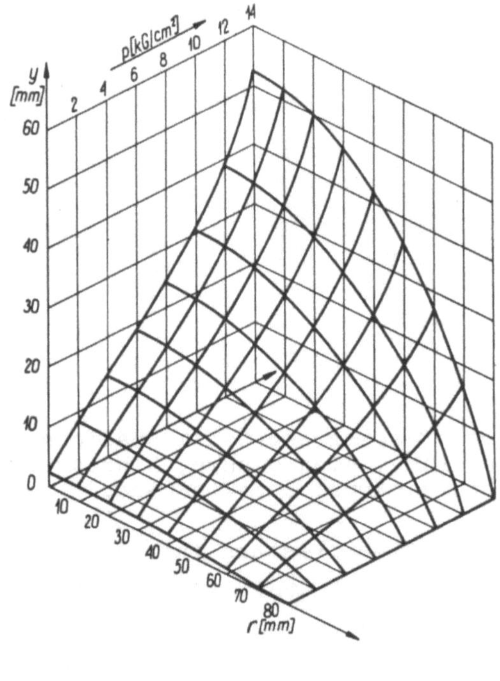

Fig.8. Change of the shell shape y/r,p/ in the course of deformation.

Pairs of stereoscopic photographs obtained in this way were measured using toolmaker's microscope and results of measurements were subsequently analyzed using the projective method. As a result of calculations the motion functions in the parametrical form r/ϱ,p/ and y/ϱ, p/ were obtained. The change of the shell shape y/r,p/ was determined by interpolation of the motion function. The results obtained in that way are illustrated in Fig.8.

The numerical transformation of the results obtained in the form of motion functions rendered it possible to determine the strain field of the shell during plastic deformation. Principal components of strain are calcu-

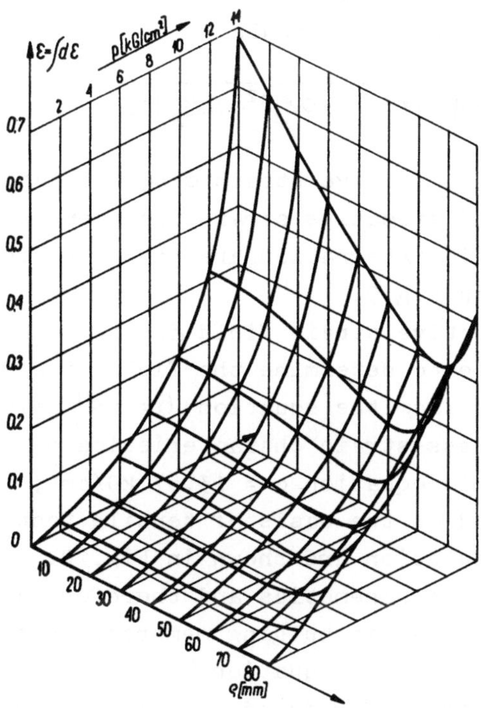

Fig.9. Distribution of the effective strain field $\mathcal{E}$/ϱ,p/ in the course of shell deformation process.

lated as meridional component

$$\varepsilon_1 = \ln\sqrt{(r')^2+(y')^2}, \qquad \text{or as} \tag{1}$$

circumferential component

$$\varepsilon_2 = \begin{cases} \varepsilon_1, & \rho = 0, \\ \ln \frac{r}{\rho}, & \rho > 0, \end{cases} \tag{2}$$

whereas the thickness strain component is obtained from the incompressibility condition

$$\varepsilon_3 = -\varepsilon_1 - \varepsilon_2. \tag{3}$$

Increments of strain components $d\varepsilon_i/\rho,p/$, i = 1,2,3, are also calculated for determining increment of effective strain using the formula

$$d\varepsilon = \frac{2}{\sqrt{3}}\sqrt{(d\varepsilon_1)^2+(d\varepsilon_2)^2+(d\varepsilon_3)^2}. \tag{4}$$

Then the effective strain $\varepsilon/\rho,p/$ is calculated by integration of effective strain increment /4/ and the results of this integration are presented in the graphical form in Fig.9.

When the effective strain field $\varepsilon/\rho,p/$ is known then it is possible to determine the field of effective stress $\sigma/\rho,p/$ from the equation $\sigma = c\varepsilon^n$ which approximates the stress-strain characteristics of shell material. In this equation c = 13.67 kG/cm^2, n = 0.24 are determined on the basis of tensile test. The results of calculation are shown in Fig.10.

The knowledge of kinematics and therefore also of geometry of the plastic bulging membrane enables us to solve the equilibrium equation [1,9]. The solution of this equations is reduced to solving the ordinary differential equation with regard to the meridional stress component

$$\frac{d\sigma_1}{d\rho} - \left(\frac{r'r'' + y'y''}{(r')^2 + (y')^2} - \frac{1}{\rho} - \frac{y''}{y'}\right)\sigma_1 + \frac{r'r''}{2h_o y'}\left((r')^2 + (y')^2\right)p = 0 \qquad /5/$$

at the boundary condition for $\rho = 0$, $\sigma_1 = -\frac{/r\,/^4}{4h_o y}\,p$, and to solving the algebraic equation with regard to the circumferential component of stress

$$\sigma_2 = \begin{cases} \sigma_1, & \rho = 0, \\ -\frac{r y''}{r' y'}\sigma_1 - \frac{r^2}{2h_o \rho y'}\left((r')^2 + (y')^2\right)p, & \rho > 0. \end{cases} \qquad /6/$$

As a result of solving the equation /5/ and /6/ we obtain the fields of stress components $\sigma_1/\rho, p/$ and $\sigma_2/\rho, p/$, where p designates loading in the form of pressure, which in the equations /5/ and /6/ is con-

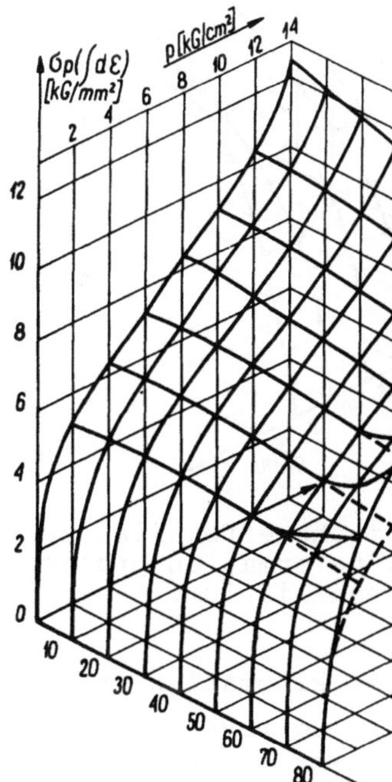

Fig.10. Distribution of the effective stress field σ_p.

sidered as parameter. The effective stress field $\sigma/\rho,p/$ is calculated from the Huber-Mises criterion and is presented in Fig.11. The stress component fields were also determined by means the associated flow rule.

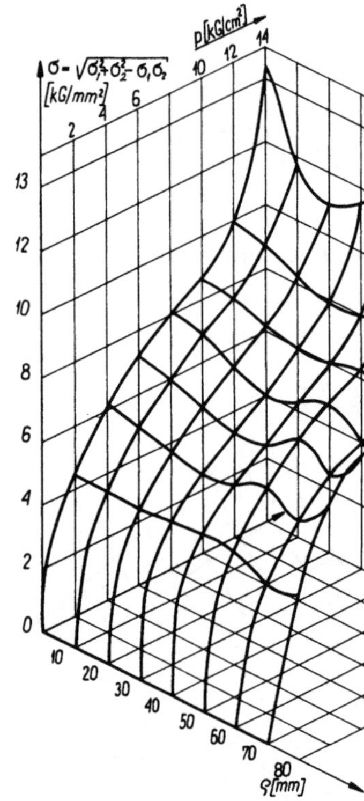

Fig.11. Distribution of the effective stress $\sigma/\rho,p/$ in the course of shell deformation process calculated from the equations of equilibrium.

The knowledge of process kinematics and physical relationship $\sigma/\varepsilon/$ enables us to reverse the problem and to determine the field loading in the form of hydrostatic pressure $p/\rho,t/$ which causes the deformation. By inverting the equation /6/ the following relationship is obtained

$$p = \begin{cases} -4h_o \sigma \dfrac{y''}{(r')^4}, & \rho = 0, \\ -2h_o \dfrac{ry'\sigma_1 + r'y'\sigma_2}{r^2 r' ((r')^2 + (y')^2} \rho, & \rho > 0, \end{cases} \qquad /7/$$

where σ_1 and σ_2 are calculated by means of the associated flow rule. The results of calculations are presented in graphical form in Fig.12. Since they were obtained from the knowledge of kinematics then they determine the upper bound loading.

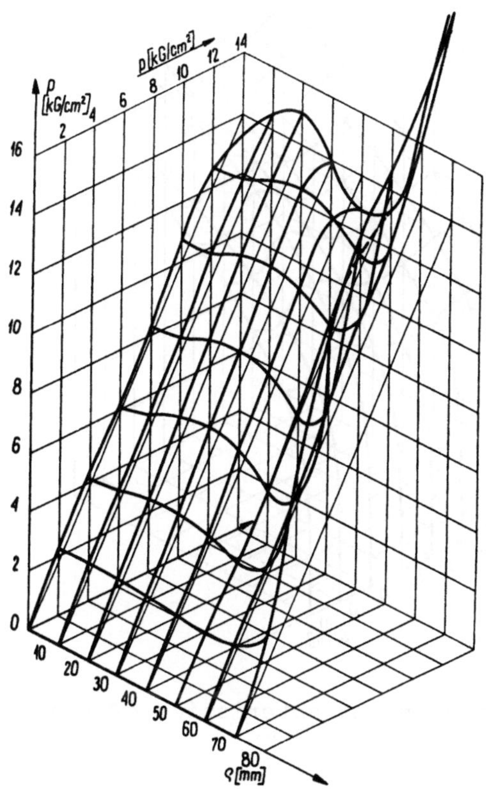

Fig.12. Destribution of the loading pressure p/ρ,p/ in the course of shell deformation process.

The knowledge of kinematics and loading in the course of the plastic bulging of membrane from measurements renders it possible to consider this process as a tensile test under plane-stress conditions in order to determine the stress-strain characteristics of the shell material. The results obtained were approximated using the $\sigma_p = c\varepsilon^n$ formula and are presented in the form of diagrams in Fig.13 for each material point ρ of the shell.

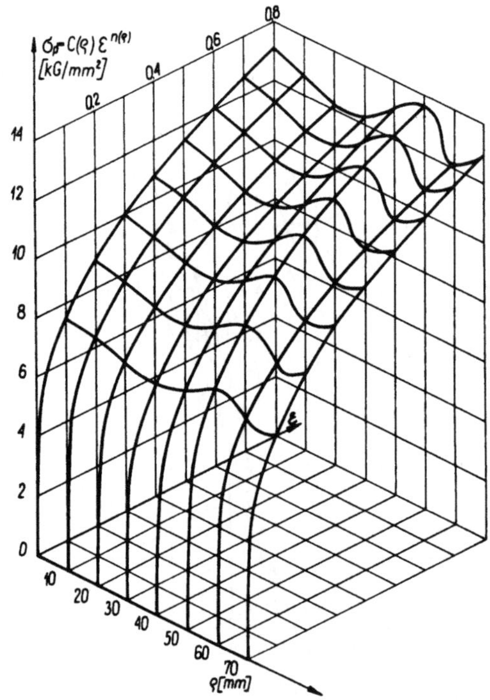

Fig.13. Stress-strain characteristics of the shell material.

DYNAMIC MEMBRANE BULGING

The process of bulging the circular membrane was investigated also under dynamic loading conditions obtained by detonation of explosive in water. The disc-shaped specimen had a diameter of 200 mm and was made of 0.91 mm thick aluminium sheet. The grid consists of four radii crossed with transverse lines spaced about 10 mm from each other. The grid lines were about 0.4 mm thick. The general views of specimen before and after deformation are shown in Fig.14. The process of plastic bulging was conducted in accordance with the schematic illustration thereof being shown in Fig. 15.

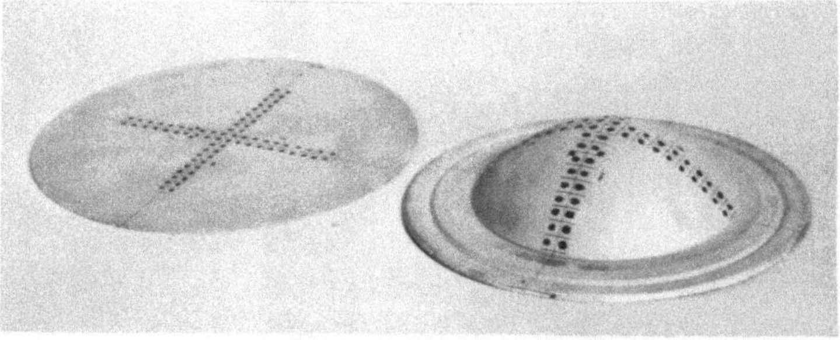

Fig.14. The specimen before and after deformation.

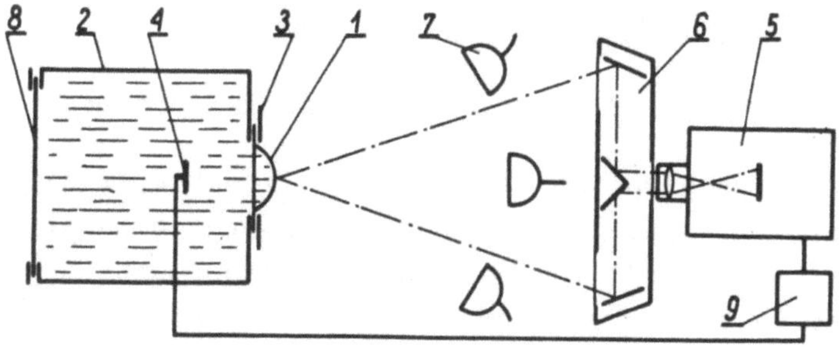

Fig.15. Scheme of measurements, 1.specimen, 2.tank, 3.drawing ring, 4.explosive, 5.high-speed film camera, 6.stereoscopic attachment, 7.reflectors, 8.rubber membrane, 9.timer.

The high-speed film camera /Pentazet 35/ equipment with a special stereoscopic attachment was used in the investigation of the dynamic process. As a result the series of picture pairs taken at the rate of 7666 frames per second was obtained. In this way the entire process was filmed on 34 frames. A portion of the filmed process is shown in Fig.16.

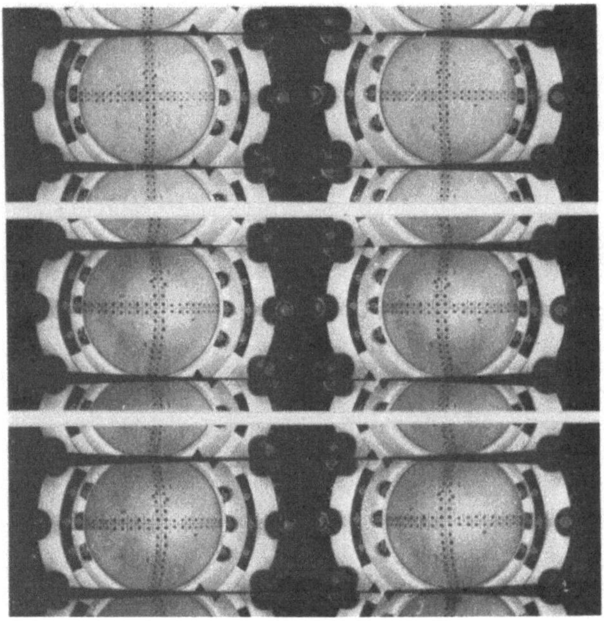

Fig.16. The part of stereogram of filming process.

The measurement on the film previously cut apart into two strips was made by means of the stereocomparator /Zeiss Stecometer/. The results of measurements were recorded in tabulated form and on the eight-channel tape. These measurements consisted in measuring the co-ordinates of collimating points, fixed points and gird points, the latter ones being material points of the shell. The results of measurements were analyzed by means of the classical stereophotogrammetric method. As a result of numerical calculations the motion function was obtained in the parametrical

form r/ρ,t/ and y/ρ,t/.
The function y/r,t/ of changes in the shell shape during the deformation process was obtained as a result of interpolating the motion functions r/ρ,t/ and y/ρ,t/ and is presented in Fig.17. The displacement rate fields $\dot{r}$/ρ,t/ and $\dot{y}$/ρ,t/ determined by differentiating the motion functions r/ρ,t/ and y/ρ,t/ towards the time were also calculated. Furthermore the acceleration fields of shell material points $\ddot{r}$/ρ,t/ and $\ddot{y}$/ρ,t/ were also calculated as secondary time derivatives of the motion functions. The results of measurements obtained in the form of motion functions were then numerically transformed and as

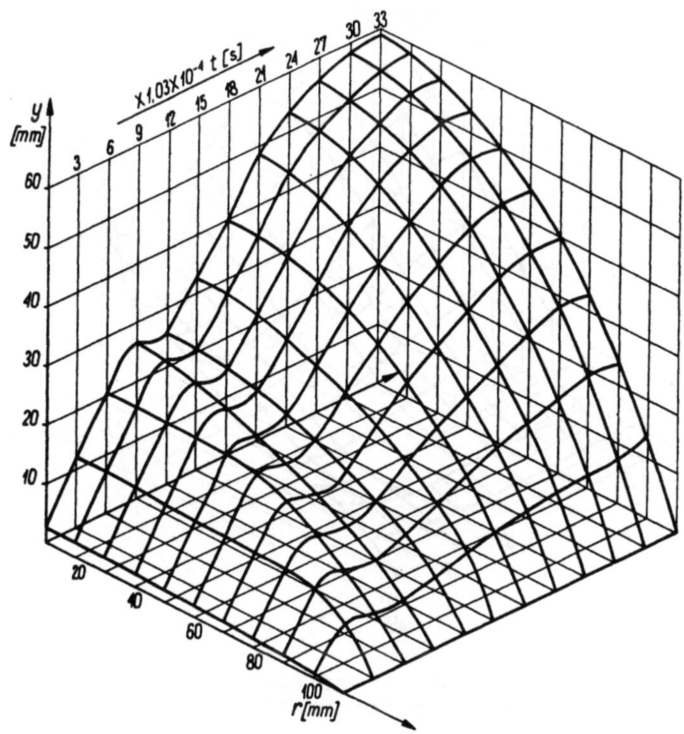

Fig.17. Change of the shell shape y/r,t/ in the course of deformation.

a result the strain field in the course of deformation process was determined. The fields of principal strain components $\varepsilon_i/\rho,t/$ from the formulae /1/, /2/, /3/ as well as the field increments of strain components $d\varepsilon/\rho,t/$ and the field increments of effective strain $d\varepsilon_i/\rho,t/$ from the formula /4/ were also calculated. The field of effective strain $\varepsilon/\rho,t/$ calculated by integrating the increments of effective strain $d\varepsilon/\rho,t/$ in the course of deformation process is shown in Fig.18.

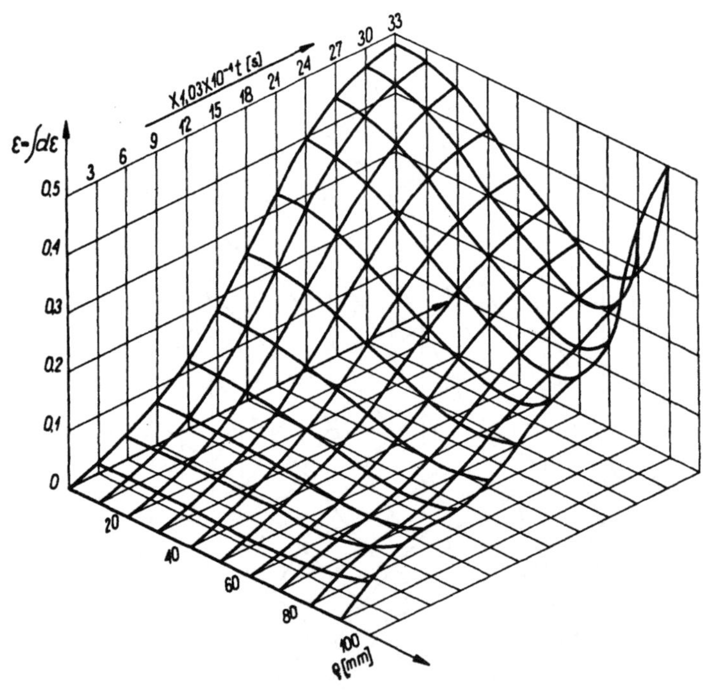

Fig.18. Distribution of the effective strain $\varepsilon/\rho,t/$ in the course of dynamic deformation process.

The fields of deformation rate components $\dot{\varepsilon}_i/\rho,t/$ were calculated for the meridional component

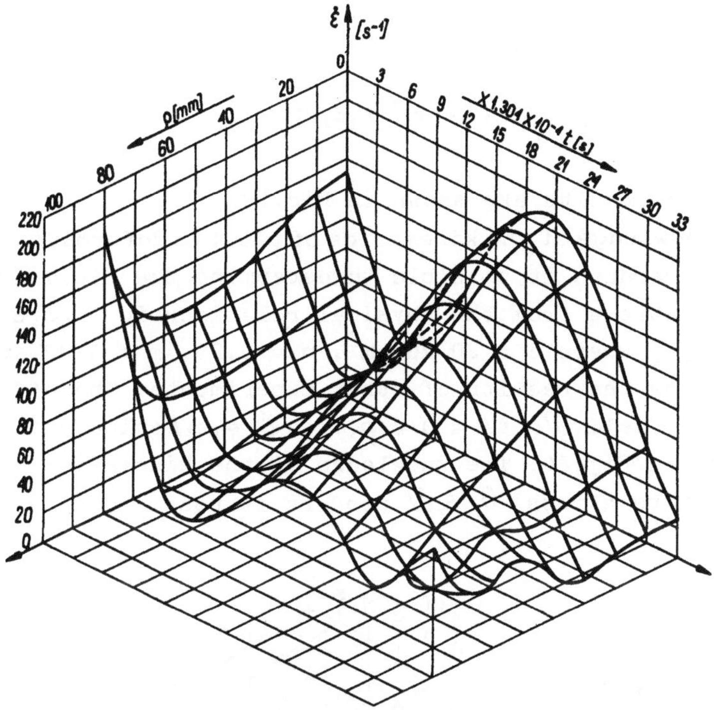

Fig.19. Distribution of the effective strain rate field $\dot{\varepsilon}/\rho,t/$ in the course of dynamic deformation process.

$$\dot{\varepsilon}_1 = \frac{r'\dot{r}' + y'\dot{y}'}{(r')^2 + (y')^2}, \qquad /8/$$

circumferential component

$$\dot{\varepsilon}_2 = \begin{cases} \dot{\varepsilon}_1, & \rho = 0, \\ \frac{\dot{r}}{r}, & \rho > 0, \end{cases} \qquad /9/$$

and thickness component obtained from the incompressibility condition

$$\dot{\varepsilon}_3 = -\dot{\varepsilon}_1 - \dot{\varepsilon}_2. \qquad /10/$$

The field of effective strain rate $\dot{\varepsilon}/\rho, t/$ was calculated from the formula

$$\dot{\varepsilon} = \frac{2}{\sqrt{3}}\sqrt{(\dot{\varepsilon}_1)^2 + (\dot{\varepsilon}_2)^2 + (\dot{\varepsilon}_3)^2} \qquad /11/$$

and the results are shown in Fig.19.

The knowledge of the kinematics of plastic deformation of the shell renders it possible to determine the field of effective stress from the physical relationship $\sigma/\varepsilon, \dot{\varepsilon}/ = c\varepsilon^n \dot{\varepsilon}^m$, where c = 20 kG/cm^2, n = 0.25, m = 0.02 as rendered by the results of other experiments. The effective stress field $\sigma/\rho, t/ = \sigma/\varepsilon/\rho, t/, \dot{\varepsilon}/\rho, t//$ calculation is shown in Fig.20. The knowledge of the kinematics and effective stress $\sigma/\varepsilon, \dot{\varepsilon}/$ renders it possible to calcu-

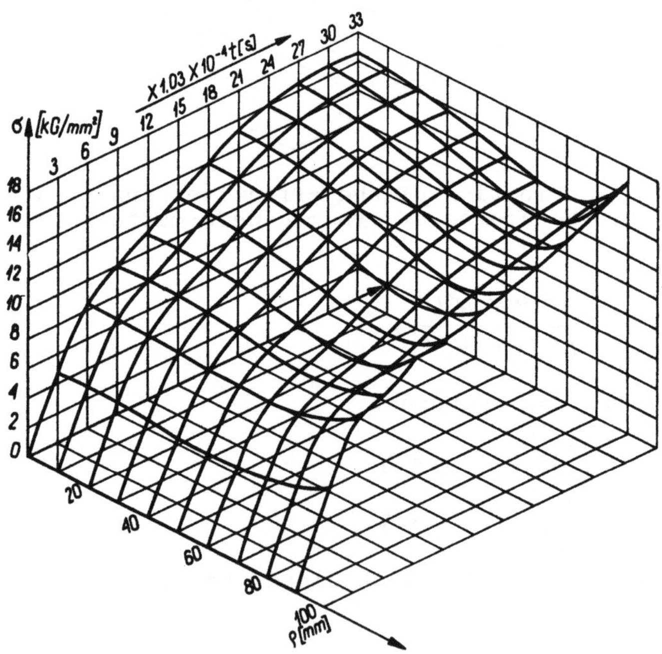

Fig.20. Distribution of the effective stress field $\sigma_p/\rho, t/ = c\varepsilon^n \dot{\varepsilon}^m$ in the course of dynamic deformation process.

late the fields of stress components $\sigma_1/\rho,t/$ and $\sigma_2/\rho,t/$ from the associated flow rule.

The load field $p/\rho,t/$ in the dynamic process of membrane bulging was calculated from the relationship obtained by inverting the motion equations [1,9], similarly as in the case of the static process

$$p=\begin{cases}-4h_0\sigma\dfrac{y''}{(r')^4}+\delta\dfrac{\ddot{y}}{(r')^2}, & \rho=0,\\ -\dfrac{2h_0}{(r')^2+(y')^2}\left(\dfrac{y''}{rr'}\sigma_1+\dfrac{y'}{r^2}\sigma_2+\dfrac{\delta}{r}\left(-y'\ddot{r}+r'\ddot{y}\right)\right), & \rho>0,\end{cases} \qquad /12/$$

where the kinematic magnitudes r, r', r,y', y', y are known from measurements. The stress components σ_1 and σ_2 are calculated using the associated flow rule,

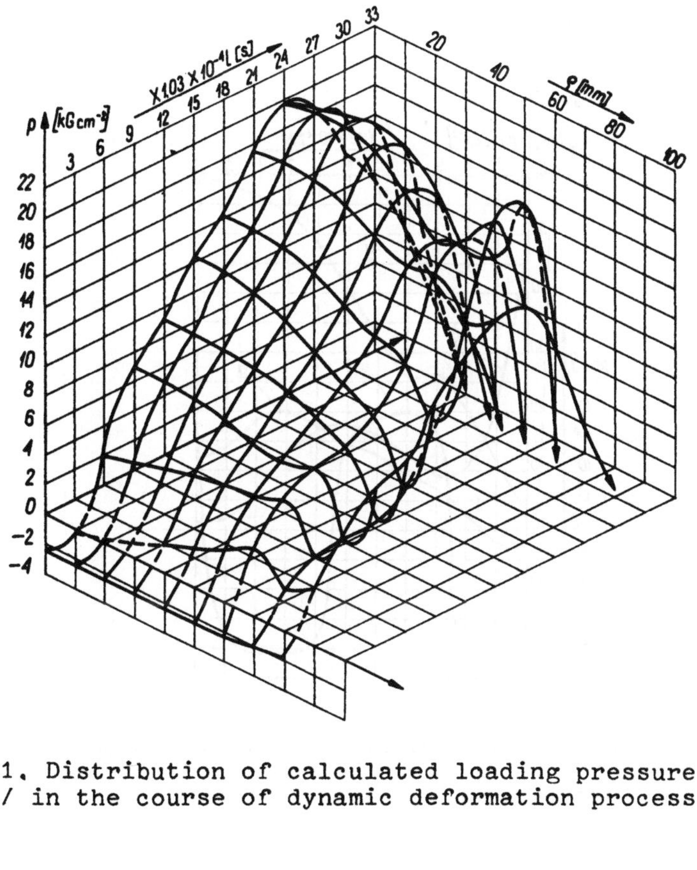

Fig.21. Distribution of calculated loading pressure $p/\rho,t/$ in the course of dynamic deformation process.

$\sigma = /\varepsilon, \dot{\varepsilon}/$ is the effective stress determined from the above physical relationship, and δ the density of shell material. The results of calculations are presented graphically in Fig.21. It should be noted that in the measurements the very beginning of the process is not included. By comparing the results of calculations for the displacement component $y/\rho,t/$ the deformation rate $\dot{y}/\rho,t/$ and the acceleration $\ddot{y}/\rho,t/$ with pressure $p/\rho,t/$ for the point $\rho = 0$ /pole/ we are able to reproduce the process run on the basis of these functions /Fig.22/. It appears clearly from this figure that the range of measurements being made does not cover the very beginning of the membrane loading with a pressure surge but its starting point is placed a little later in the time. At first

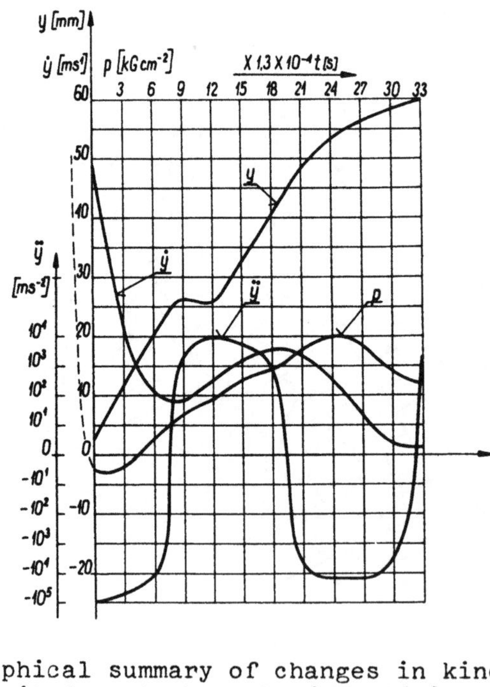

Fig.22. Graphical summary of changes in kinematic and dynamic magnitudes at the pole $/\rho = 0/$ in the course of dynamic deformation process.

the membrane is deformed as it receives an initial impulsive velocity. The central area of the shell remains flat within that period. After the momentum decay and pressure increase the displacement rate increases again and the membrane undergoes a deformation over its area until pressure drops sufficiently.

ACCURACY OF MEASUREMENT

Accuracies of displacement measurements are different for both methods and depend on many factors. They were determined by checking the accuracy of measuring the fixed points disposed on the drawing ring with the original accuracy ± 0.01 mm. Therefore the co-ordinates of fixed points c_{wj} are known. These co-ordinates were measured using stereophotogrammetric methods and were designated with c_{wj}. Then the measuring accuracy is obtained by calculating the difference between these co-ordinates

$$\Delta c_{wj} = C_{wj} - c_{wj}, \quad j = 1,2,3, \quad w = 1,2,3,\ldots,W \qquad /13/$$

Actual measuring accuracies calculated in this way are ± 0.01 mm for the static process analyzed by means of the projective method but ± 0.1 mm for the dynamic process analyzed in classical way.

CONCLUSIONS

Application of stereophotogrammetric methods for investigating plastic forming processes as presented in this paper show that these methods render it possible to get information on all aspects of processes investigated under various conditions. One of them was typically static in its nature whereas another represented the dynamic type of plastic forming process. The duration of the first was four hours whereas the second lasted 4.3×10^{-3} seconds only.

It has been demonstrated that the stereophotogrammetric methods enable us to determine directly the kinematics and indirectly also the dynamics of plastic deformation of a shell as well as to determine the static stress-strain characteristics in the plane-stress state. These methods were used for examining framework structures, rock crushing processes, the cumulation phenomenon and other processes occuring in the three-dimensional space and in the time.

The projective method renders it possible to carry on the analysis of measuring results without any necessity to know in advance any orientation elements. This facilitates substantially the use of this method. The possibility of making measurements on photographs using the toolmaker's microscope renders the measurements more accurate and the method itself easier availabe for circles of mechanicians.

Displacement measurements made by means of stereophotogrammetric methods are of contact-less type and therefore can be carried on without interfering with the process itself while assuring at the same time, complete safety for both the equipment and the operating personnel which property is particularly essential while investigating various dynamic processes.

REFERENCES

1 Green, E.; Adkins, J.E.: Large Deformation and Non-Linear Continuum Mechanics, Oxford 1960.

2 Piasecki, M.B.: Fotogrametria naziemna i lotnicza. Warszawa 1968.

3 Wilczkiewicz, E.: Zasady zdjęć fotogrametrycznych /Fotogrametria naziemna i lotnicza/, Lwów 1930.

4 Dubovik, A.S.: Fotografičeskaja registracija bystroprotekajuščich processov. Moskva 1964.

5 Lobanov, A.N.: Fototopografija. Moskva 1965.

6 Thompson, E.H.: Space Resection without Interior Orientation. Photogrammetric Record 7/37/ 1971 39-45.

7 Bednarski, T.; Majde, A.: The Spatial Restitution of Terrestrial Photographs on the Basis of Projective Transformation. Societe Francaise de Photogrammetrie Bulletin 42 /1971/ 55-62.

8 Bednarski, T.: Wyznaczanie kinematyki procesów deformacji metodą stereofotogrametryczno-rzutową. Mechanika Teoretyczna i Stosowana 3 13 /1975/ 433-456.

9 Bednarski, T.: Zastosowanie stereofotogrametrycznych metod do analizy kinematyki i dynamiki powłok. IFTR Raports, Warsaw 1977.

10 Bednarski, T.: Plastyczne wypuklanie cienkiej płyty kołowej impulsem ciśnienia. Rozprawy Inżynierskie 1 17 /1969/ 69-100.

11 Bednarski, T.: The Dynamic Deformation of a Circular Membrane. Int. J. Mech. Sci. 12 11 /1969/ 949-959.

12 Bednarski, T.: Pomiar odkształceń plastycznych membrany kołowej obciążonej impulsem ciśnienia. Mechanika Teoretyczna i Stosowana 3 8 /1970/ 295-318.

13 Bednarski, T.; Majde, A.: Adaptation of the Film Camera to the Stereophotogrammetric measurements. Societe Francaise de Photogrammetrie Bulletin 42 /1971/ 35-54.

14 Bednarski, T.: Zastosowanie stereophotogrametrycznych metod do analizy procesów kształtowania powłok. Obróbka Plastyczna 1 /1978/.

Deformations finies en coordonnées locales et application à un problème d'emboutissage

P. LE NEVEZ
LABORATOIRE DE MECANIQUE ET TECHNOLOGIE
E.N.S.E.T.
61, Avenue du Président Wilson 94230 CACHAN (FRANCE).

RESUME

L'étude que nous présentons ici est divisée en deux parties. Dans la première partie, nous reprenons la théorie des grandes déformations au sens de E.H. LEE et J. MANDEL pour laquelle nous formulons les déformations en coordonnées locales.

Dans ce but, nous introduisons, d'une part, un ensemble matériel, qui permet de définir le terme "convectif", et, de donner une définition simple de la dérivée convective, utile pour les taux de déformations. D'autre part, la notion de repère permet d'écrire l'application tangente entre deux configurations comme un composé d'applications. Il est facile alors de montrer que les déformations sont représentées par une même matrice, dans l'état initial et dans l'état actuel.

Dans la seconde partie, nous appliquons cette théorie à un problème d'emboutissage axisymétrique, pour lequel la matrice des déformations est construite, sur un repère local. Puis nous définissons, pour ce problème particulier, le champ de déformations, en tenant compte du frottement tôle-poinçon, et d'un encastrement non parfait de la tôle.

Enfin, nous comparons les résultats théoriques, à ceux d'une expérience réalisée sur des tôles en acier inoxydable.

INTRODUCTION

Dans le domaine de la mise en forme et particulièrement en emboutissage, une des préoccupations principales de l'ingénieur, est de prévoir si le produit travaillé supportera ou non les déplacements imposé, tout en conservant les caractéristiques nécessaires à son emploi.

Les courbes limites d'emboutissage ont actuellement un certain succès à cet égard ; mais elles demandent la connaissance des déformations en tous points. Ceci conduit, lors de la mise au point d'un produit, l'industriel à des essais expérimentaux pour les déterminer.

Nous pensons qu'il est possible, moyennant un minimum de calculs, d'obtenir des renseignements théoriques, permettant à l'ingénieur d'effectuer un choix dès le stade du bureau d'études , et, de diminuer ainsi en partie, des essais longs et coûteux.

Dans ce but, nous nous sommes attachés à définir la méthode la plus simple pour déterminer les déformations, et ce, en coordonnées locales, de façon à faciliter la comparaison avec l'expérience.

Pour illustrer cette étude, nous présentons l'exemple simple d'une tôle encastrée, déformée sous l'action d'un poinçon hémisphérique.

1. PARTIE 1

1.1 Définitions et notations

Nous utilisons par la suite, les définitions suivantes :

- particule p : élément macroscopique d'un milieu matériel m, assimilable à un point.
- espace matériel $\mathcal{B}$: variété affine différentiable, dont les éléments sont les particules du milieu matériel m.
- espace de configuration $E^{\bullet}$: espace affine dans lequel sont effectuées les observations. Les éléments sont appelés positions.
- configurations a(t) : variété différentiable plongée dans $E^{\bullet}$. Les éléments sont les positions, à la date t, du milieu m.
- évolution ψ(t) : application de l'espace $\mathcal{B}$ dans la configuration a(t).
- repère : isomorphisme de $\mathbb{R}^n$ dans E (espace vectoriel).
- produit scalaire : si x et y $\in$ E, alors nous noterons :

$$<x \mid y> = g_E(x)(y) = \lambda \in \mathbb{R}$$

- application adjointe (notée :+—) : si y=f(z) avec z∈F, alors :

$$g_E(x)(f(z)) = g_F(\overset{+}{f}(x))(z)$$

- matrice de Gram : si $\mathcal{S}_t$ est un repère, alors la matrice de Gram s'exprime par : $G_t = \overset{+}{\mathcal{S}}_t \circ \mathcal{S}_t$
- l'application tangente entre les espaces tangents T_p et T_{M_t} sera notée : $\Psi(t)(p)$

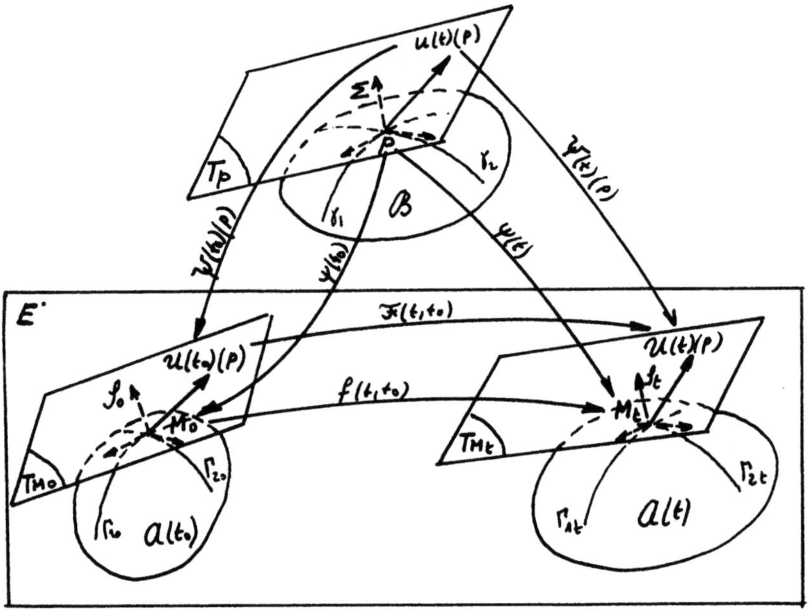

Figure n° 1

1.2. Vecteurs, repères et convection

Soit un vecteur $u(t)(p) \in T_p$, son image dans T_{M_t} est définie à l'aide de l'application tangente par :

$$\mathcal{U}(t)(p) = [\Psi(t)(p)](u(t)(p)) \qquad (1)$$

De même, si $\Sigma(t)(p)$ est un repère de T_p en p, alors, son image dans T_{M_t} au point M_t, est définie par le repère :

$$\mathcal{S}(t)(p) = \Psi(t)(p) \circ \Sigma(t)(p) \qquad (2)$$

Il est possible d'exprimer tout vecteur de T_p à l'aide du repère

$\Sigma(t)(p)$:

$$u(t)(p) = [\Sigma(t)(p)](U(t)(p)) \ , \qquad (3)$$

où $U(t)(p) \in \mathbb{R}^n$ représente les composantes du vecteur $u(t)(p)$.
De (1) nous obtenons alors :

$$\mathcal{U}(t)(p) = [\mathcal{S}(t)(p)](U(t)(p)) \ , \qquad (4)$$

c'est-à-dire, si $\mathcal{U}(t)(p)$ est l'image de $u(t)(p)$ et si, $\mathcal{S}(t)(p)$ est l'image de $\Sigma(t)(p)$, alors les deux vecteurs ainsi définis ont même composantes dans $\mathbb{R}^n$.

DEFINITION :

Nous appellerons vecteur convectif, tout vecteur de T_{M_t} , dont le vecteur image dans T_p est indépendant du temps. Cette définition s'étend ipso-facto aux repères, endomorphismes, tenseurs, etc.....

1.3. Cinématique :

Nous appellerons vitesse intrinsèque de la particule p , le vecteur de E :

$$v^E(t)(p) = \lim_{h \to 0} \left[\frac{M(t+h) - M(t)}{h} \right] = \dot{\psi}(t)(p) \qquad (5)$$

définissant le champ : $v^E(t) = \dot{\psi}(t) : \mathcal{B} \longrightarrow E$

L'expression $p = [\psi(t_0)]^{-1}(M_0)$ permet de modifier l'écriture précédente de la façon suivante :

$$v^E(t)(p) = [\dot{\psi}(t) \cdot \psi(t_0)^{-1}](M_0) = V_L^E(t)(M_0) \qquad (6)$$

définissant le champ Lagrangien : $V_L^E(t) : a(t_0) \longrightarrow E$

De même, à partir de $p = [\psi(t)]^{-1}(M_t)$, nous obtenons le champ Eulerien : $V_E^E(t) = \dot{\psi}(t) \cdot \psi(t)^{-1} : a(t) \longrightarrow E$

1.4. Gradient de vitesse

A partir du champ intrinsèque des vitesses :

$$v^E(t)(p + dp) = v^E(t)(p) + D[v^E(t)](p)(dp) + O(dp) \qquad (7)$$

est défini le gradient intrinsèque noté :

$$\ell(t)(p) = D[v^E(t)](p) \qquad (8)$$

$$\begin{array}{ccc} T_p & \xleftarrow{\ \ell(t)(p)\ } & T_p \\ \Psi(t)(p) \downarrow & & \downarrow \Psi(t)(p) \\ T_{M_t} & \xleftarrow{\ \mathcal{L}(t)(p)\ } & T_{M_t} \end{array}$$

De même le gradient Eulérien est

donné par :

$$\mathcal{L}(t)(p) = D[V_t^{\epsilon}(t)](M_t) \qquad (9)$$

Le diagramme commutatif ci-contre donne la liaison entre les deux champs. $\mathcal{L}(t)(p)$ possède les propriétés suivantes :

$$\mathcal{L}(t)(p) = \dot{\psi}(t)(p) \cdot [\psi(t)(p)]^{-1} = -\psi(t)(p) \cdot [\dot{\psi}(t)(p)]^{-1} \qquad (10)$$

$$[\mathcal{L}(t)(p)]^{-1} = -\mathcal{L}(t)(p) \quad \text{et} \quad [\mathcal{L}(t)(p)]^2 = -I_{T_{M_t}} \qquad (11)$$

$$\mathcal{L}(t)(p) = \dot{\mathcal{S}}_t \circ \mathcal{S}_t^{-1} = -\mathcal{S}_t \cdot \dot{\mathcal{S}}_t^{-1} \qquad (12)$$

Cette dernière expression, n'est valable que si $\mathcal{S}_t$ est un repère convectif. On peut alors, en tenant compte de cette remarque, exprimer la partie symétrique de $\mathcal{L}(t)(p)$, notée :

$$\mathcal{D}(t)(p) = 1/2\,[\mathcal{L}(t)(p) + \overline{\mathcal{L}(t)(p)}^{+}] \qquad (13)$$

sur un repère convectif :

$$\mathcal{D}(t)(p) = \mathcal{S}_t \,.\, D \,.\, \mathcal{S}_t^{-1} = \overset{+}{\mathcal{S}}_t^{-1} \,.\, D_* \,.\, \mathcal{S}_t^{-1} \qquad (14)$$

et déterminer D et D_* à l'aide de la matrice de Gram de l'espace :

$$D = 1/2\; G_t^{-1} \cdot \dot{G}_t \qquad \text{et} \qquad D_* = G_t \cdot D = 1/2\; \dot{G}_t \qquad (15)$$

1.5. Dérivée convective

Définissons maintenant la dérivation temporelle dans un repère local en mouvement.

1.5.1. Champ des vecteurs

Soit un champ de vecteurs $u(\cdot)(p) : t \longmapsto u(t)(p) \in T_p$; en projetant ce vecteur dans T_{M_t}, nous obtenons un vecteur tel que (1). Posons alors par définition la dérivée convective de ce vecteur :

$$D_c(\psi)(\mathcal{U}(t)(p)) = [\psi(t)(p)](\dot{u}(t)(p)) \in T_{M_t} \qquad (16)$$

c'est-à-dire, la dérivée convective est le transport dans T_{M_t}, par $\psi(t)(p)$, de la dérivée temporelle ordinaire dans T_p (remarquons que la dérivée convective d'un vecteur convectif est nulle, c'est une conséquence de la définition).

En utilisant (1)et(10) :

$$D_c(\psi)(\mathcal{U}(t)(p)) = \dot{\mathcal{U}}(t)(p) - [\mathcal{L}(t)(p)](\mathcal{U}(t)(p)) \qquad (17)$$

Le résultat précédent s'exprime sur un repère convectif $\mathcal{S}_t$ (en tenant compte de (4)):

$$\mathcal{D}_c(\psi)(\mathcal{U}(t)(p)) = \mathcal{S}_t(\dot{\mathcal{U}}(t)(p)) \qquad (18)$$

Ceci montre que, sur un repère local convectif, la dérivée convective s'exprime par la dérivée temporelle des composantes du vecteur ainsi défini.

1.6 Déformation

Il est possible de définir un champ d'application (voir Figure N°1)

$$f(t,t_0) = \psi(t) \cdot \psi(t_0)^{-1} \qquad (19)$$

qui, à la position à l'instant t_0 de la particule p , fait correspondre la position à l'instant t de cette même particule.

$$M_t = \psi(t)(p) = f(t,t_0)(M_0) \qquad (20)$$

Ce champ est appelé champ de déplacement.

De même entre les espaces tangents T_{M_0} et T_{M_t} , il existe une application :

$$\mathcal{F}(t,t_0) = \Psi(t)(p) \circ [\Psi(t)(p)]^{-1} \qquad (21)$$

telle que :

$$\mathcal{U}(t)(p) = [\Psi(t)(p)](u(t)(p)) = \mathcal{F}(t,t_0)(\mathcal{U}_0(t)(p)) \qquad (22)$$

avec les relations :

$$\mathcal{F}(t,t_0) = \mathcal{D}[f(t,t_0)](M_0) \quad \text{et} \quad \mathcal{L}(t)(p) = \dot{\mathcal{F}}(t,t_0) \circ [\mathcal{F}(t,t_0)]^{-1} \qquad (23)$$

$$\mathcal{F}(t,t_0) = \mathcal{S}_t \circ \mathcal{S}_{t_0}^{-1} \qquad (24)$$

La relation (24) est importante pour définir la matrice des déformations.

1.6.1. Définition des déformations et relations

La définition des déformations est faite à partir de la différence de produit scalaire, entre les vecteurs de T_{M_t} et ceux de T_{M_0} , quand une référence temporelle t_0 est choisie :

$$\lambda^2 = g_{T_{M_t}}(\mathcal{U}_t)(\mathcal{V}_t) - g_{T_{M_0}}(\mathcal{U}_0)(\mathcal{V}_0) \qquad (25)$$

- Dans T_p nous pouvons définir un endomorphisme en écrivant :

$$g_{T_{M_t}}(\mathcal{U}_t)(\mathcal{V}_t) = g_{T_p}(c(t)(u))(v) \qquad (26)$$

avec $c(t) = {}^{t}\overline{\Psi(t)(p)} \cdot \Psi(t)(p)$. De même, en posant :

$c(t_0) = {}^{t}\Psi(t_0)(P) \cdot \Psi(t_0)(P)$, (25) devient :

$$\lambda^2 = g_{T_P}\left(\left[C(t) - C(t_0)\right](u)\right)(v) \tag{27}$$

L'expression suivante sera appelée : endomorphisme intrinsèque des déformations :

$$c(t,t_0) = 1/2\left[\,c(t) - c(t_0)\right] \tag{28}$$

- Dans T_{M_0} la relation (25) devient :

$$\lambda^2 = g_{T_{M_0}}\left(\left[\mathscr{C}(t,t_0) - I_{T_{M_0}}\right](\mathcal{U}_0)\right)(\mathcal{V}_0) \tag{29}$$

avec $\mathscr{C}(t,t_0) = {}^{t}\mathcal{F}(t,t_0) \cdot \mathcal{F}(t,t_0)$. Alors par définition l'endomorphisme Lagrangien des déformations est :

$$\mathscr{E}(t,t_0) = 1/2\left[\,\mathscr{C}(t,t_0) - I_{T_{M_0}}\right] \tag{30}$$

- Si $\mathscr{B}(t,t_0) = \mathcal{F}(t,t_0) \cdot {}^{t}\mathcal{F}(t,t_0)$, alors l'expression (25) est définie par :

$$\lambda^2 = g_{T_{M_t}}\left(\left[\,I_{T_{M_t}} - \mathscr{B}(t,t_0)^{-1}\right](\mathcal{U}_t)\right)(\mathcal{V}_t) \tag{31}$$

et dans T_{M_t} , nous avons comme endomorphisme Eulérien des déformations la quantité :

$$\mathscr{A}(t,t_0) = 1/2\left[\,I_{T_{M_t}} - \mathscr{B}(t,t_0)^{-1}\right] \tag{31 bis}$$

Le diagramme commutatif ci-contre donne les principales relations entre les endomorphismes ainsi définis.

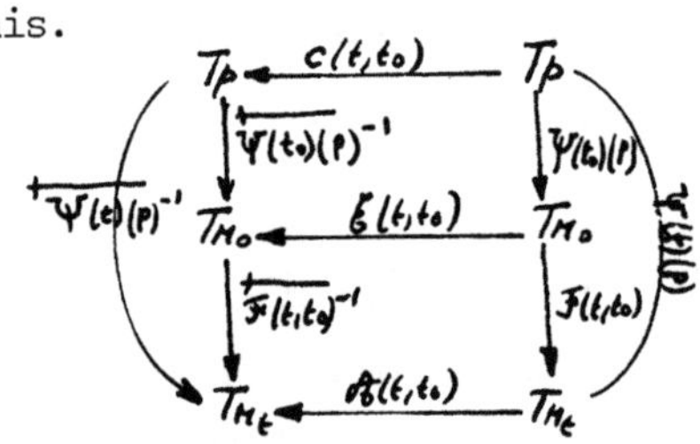

Ces différents endomorphismes sont représentés sur les repères convectifs par :

$$c(t,t_0) = {}^{t}\Sigma^{-1} \cdot C \cdot \Sigma^{-1} \tag{32}$$

$$\mathscr{E}(t,t_0) = {}^{t}\mathcal{S}_{t_0}^{-1} \cdot E \cdot \mathcal{S}_{t_0}^{-1} \tag{33}$$

$$\mathscr{A}(t,t_0) = {}^{t}\mathcal{S}_t^{-1} \cdot A \cdot \mathcal{S}_t^{-1} \tag{34}$$

soit, en utilisant les relations (2) et (24) on montre :

$$C = E = A = \tfrac{1}{2}\left[G_t - G_{t_0}\right] \tag{35}$$

Ainsi sur ces repères locaux, les matrices des déformations ont une expression unique ; calculée directement à partir des matrices de Gram des espaces initial et actuel. Il est alors inutile d'utiliser la méthode habituelle suivante.

1.6.2. <u>Etude des déformations à partir du vecteur déplacement</u>

Dans l'espace affine de la géométrie classique, nous avons :

$$M_t = M_0 + u(t)(M_0) \tag{36}$$

Le champ de déformation s'exprime alors par :

$$\mathcal{F}(t,t_0) = I_E + D[u(t)](M_0) \tag{37}$$

et, la déformation Lagrangienne, par exemple, a pour expression

$$\mathcal{E}(t,t_0) = \tfrac{1}{2}\left[D[u(t)](M_0) + {}^{t}\overline{D[u(t)](M_0)} + {}^{t}\overline{D[u(t)](M_0)} \cdot D[u(t)](M_0) \right] \tag{38}$$

soit en posant :

$$u(t)(M_0) = \mathcal{S}_{t_0}(V) = {}^{t}\mathcal{S}_{t_0}^{-1}(V) \tag{39}$$

nous obtenons la dérivée covariante :

$$D[u(t)](M_0) = \mathcal{S}_{t_0} \cdot \nabla V \cdot \mathcal{S}_{t_0}^{-1} \tag{40}$$

de composantes :

$$\nabla V = \Gamma * V + \frac{\partial V}{\partial \varphi} \qquad \text{avec} \qquad \Gamma = \mathcal{S}_{t_0}^{-1} * \frac{\partial \mathcal{S}_{t_0}}{\partial \varphi} \tag{41}$$

Si nous tenons compte de $\nabla V = G_{t_0} \cdot \nabla V$, l'expression (35) devient :

$$C = E = A = \tfrac{1}{2}\left[\nabla V + {}^{t}\overline{\nabla V} + {}^{t}\overline{\nabla V} \cdot \nabla V \right] \tag{42}$$

Ceci fournit, une second approche pour déterminer la matrice des déformations ; mais cette méthode est plus compliquée que la précédente.

II. PARTIE 2

2.1. Données du problème

Nous allons envisager maintenant, l'opération d'emboutissage d'une tôle, à l'aide d'un poinçon hémisphérique. Les paramètes sont précisés sur la figure N° 2. Nous déterminerons la matrice des déformations en choisissant un repère adapté à la géométrie.

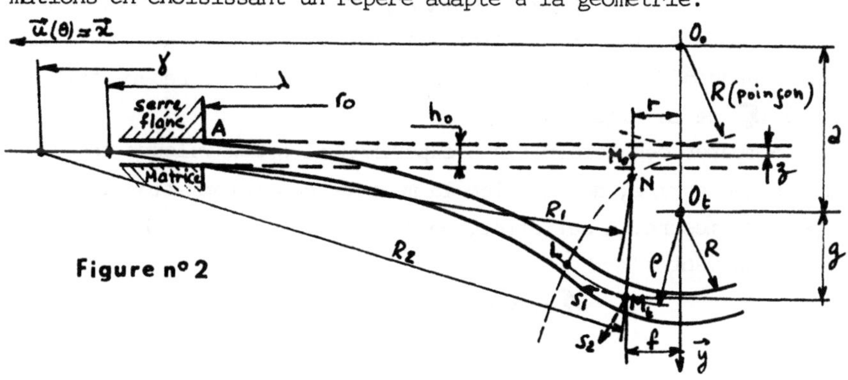

Figure n° 2

soit dans $a(t_o)$: $M_0 = O_0 + r\vec{u}(\theta) + (R+z)\vec{z}$, (43)

et dans $a(t)$: $M_t = O_t + f\vec{u}(\theta) + g\vec{z}$ (44)

HYPOTHESE : Le problème est considéré comme axisymétrique isotrope, donc $f = f(r,z,a)$ et $g = g(r,z,a)$ sont indépendants de θ .

2.2. Matrice des déformations

Le repère de T_{M_0} est défini par une carte $\mathcal{C}(t_0) : \mathbb{R}^3 \longrightarrow a(t_0)$. C'est-à-dire $\mathcal{C}(t_0) : \varphi_0 \longmapsto M_0$ avec $\varphi_0 = (r,\theta,z)$ (avec $\mathcal{C}$ au moins C^1 sur $\mathbb{R}^3 - \{0\}$). De (43) il vient :

$$\mathcal{S}_{t_0} = D[\mathcal{C}(t_0)](\varphi_0) = [\underbrace{\vec{u}(\theta)}, \underbrace{r\vec{u}(\theta+\tfrac{\pi}{2})}, \vec{z}] \quad (45)$$

et, le repère de T_{M_t} est défini de la même manière par :

$$\mathcal{S}_t = D[\mathcal{C}(t)](\varphi_0) = [\underbrace{a\vec{u}(\theta)+c\vec{z}}, \underbrace{f\vec{u}(\theta+\tfrac{\pi}{2})}, \underbrace{b\vec{u}(\theta)+d\vec{z}}] \quad (46)$$

avec : $a = \partial f/\partial r$, $b = \partial f/\partial z$, $c = \partial g/\partial r$, $d = \partial g/\partial z$.

Les matrices de Gram dans T_{M_0} et T_{M_t} sont alors :

$$G_{t_0} = {}^t\mathcal{S}_{t_0} . \mathcal{S}_{t_0} = \begin{bmatrix} 1 & 0 & 0 \\ 0 & r^2 & 0 \\ 0 & 0 & 1 \end{bmatrix} ; \quad G_t = {}^t\mathcal{S}_t . \mathcal{S}_t = \begin{bmatrix} u^2 & 0 & w \\ 0 & f^2 & 0 \\ w & 0 & v^2 \end{bmatrix} \quad (47)$$

avec $a^2 + c^2 = u^2$, $b^2 + d^2 = v^2$, $ab - cd = w$. Ceci donne la matrice des déformations suivantes :

$$c = E = A = \begin{bmatrix} \tfrac{1}{2}(u^2-1) & 0 & \tfrac{1}{2}w \\ 0 & \tfrac{1}{2}(f^2-r^2) & 0 \\ \tfrac{1}{2}w & 0 & \tfrac{1}{2}(v^2-1) \end{bmatrix} \quad (48)$$

Les déformations expérimentales, ont été déterminées à partir de la méthode des cercles, et calculées en utilisant la quantité :

$$\tfrac{1}{2}[1 - (l_0/l)^2] \quad (49)$$

ou l_o , est le diamètre initial des cercles, et l, l'axe de l'ellipse après déformation. Le repère de mesure est le repère orthonommé S_t de la figure N° 2.

Pour comparer ces déformations expérimentales aux valeurs théoriques, il est nécessaire, de calculer la matrice des déformations, sur ce repère de mesure. A cet effet, nous avons utiliser le changement de repère suivant :

$$\mathcal{S}_t = S_t . H_t \quad (50)$$

et

$$H_t = \begin{bmatrix} u & 0 & w/u \\ 0 & f & 0 \\ 0 & 0 & t/u \end{bmatrix} \quad \text{avec} \quad \begin{aligned} t &= ad-bc \\ s &= f^2/r^2 \end{aligned} \tag{51}$$

Ainsi, $\mathcal{A} = \overleftarrow{\mathcal{S}_t}^{-1}.A.\mathcal{S}_t^{-1} = \overleftarrow{S_t}^{-1}.\overleftarrow{H_t}^{-1}.A.H_t^{-1}.S_t^{-1} = \overleftarrow{S_t}^{-1}.\mathbb{A}.S_t^{-1}$ (52)

D'où, la matrice des déformations cherchée :

$$\mathbb{A} = \begin{bmatrix} \frac{1}{2}\left(1-1/u^2\right) & 0 & w/2tu^2 \\ 0 & \frac{1}{2}\left(1-1/s^2\right) & 0 \\ w/2tu^2 & 0 & \frac{1}{2}\left(1-\frac{u^4-w^2}{t^2u^2}\right) \end{bmatrix} \tag{53}$$

2.3. <u>Détermination de f(r,z,a) et g(r,z,a)</u>

Pour terminer, il reste à définir les fonctions f=x et g=y-R+a. Cette étude est partagée en trois :

2.3.1. <u>Etude de la partie en contact avec le poinçon (sans frottement.</u>

La trajectoire d'un point M_t a été choisie en prenant un réseau de cercles à point limite :

$$(x-\lambda^*)^2+(y-z)^2 = R_1^2 \quad \text{avec} \quad \lambda^* = \frac{r^2+r_0^2}{2r} \quad \text{et} \quad R_1 = \lambda^*-r \tag{54}$$

Pour tenir compte d'un encastrement non parfait, nous avons modifié le réseau précédent de la façon suivante :

$$(x-\lambda)^2+(y-z)^2 = R_1^2 \quad \text{avec} \quad \lambda = e\lambda^* \quad \text{et} \quad R_1 = \lambda - r \tag{55}$$

Le paramètre e est ajusté en fonction des expériences. La déformée de la tôle est définie localement par un cercle d'équation :

$$(y+R-a)^2 + x^2 = \rho^2 \quad \text{avec} \quad \rho = R + h(r,z,a) \tag{56}$$

En utilisant l'hypothèse d'incompressibilité plastique, nous avons défini l'épaisseur :

$$h(r,z,a) = -\frac{rz}{R}\frac{\partial y_0}{\partial r} \tag{57}$$

y_0 étant l'intersection de (54) et (56) pour ρ = R.

2.3.2. <u>Etude de la partie en contact avec le poinçon (avec frottement.</u>

Pour l'étude avec frottement, nous avons dans un premier temps, choisi la courbe limite de contact de la tôle avec le poinçon :

$$y = k\left[\frac{\rho}{2} \pm \sqrt{\frac{\rho^4}{4(\rho^2 - x^2)} - x^2}\right] \qquad (58)$$

le paramètre k étant ajusté, en tenant compte des résultats expérimentaux. Ensuite, nous avons modifié la trajectoire d'un point (voir figure N° 2) :

$$(x-\gamma)^2 + (y-z)^2 = R_2^2 \quad \text{avec} \quad R_2 = R_1\, e^{f\left(\frac{r_2 - r}{r_2}\right)} \qquad (59)$$

avec $\gamma = x_N + \sqrt{R_2^2 - R_1^2 + (\lambda - x_N)^2}$. f caractérisant le frottement c'est-à-dire, si $f=0 \Rightarrow R_2=R_1$ et $f \to \infty \Rightarrow R_2 \to \infty$).

2.3.3 Etude de la partie libre de la tôle

Pour cette partie de la tôle, comprise entre le poinçon et l'encastrement, nous avons conservé le réseau de cercles à point limite précédent, et, modifié seulement la déformée, en prenant pour équation celle d'une parabole :

$$Ay^2 + Bxy + Cx^2 + Dy + Ex + 1 = 0 \quad \text{avec} \quad B^2 - 4AC = 0 \qquad (60)$$

en imposant à cette parabole, de passer par le point limite L de coordonnées (x_L , y_L) et par le point d'encastrement A de coordonnées (r_0 , 0) (voir figure N° 2). La tangente en L étant commune à la parabole et au cercle poinçon. La tangente à l'encastrement est choisie comme suit :

$$tg\,A = a^n / r_0 \qquad (61)$$

(n étant définit par les essais). Il est alors possible de définir les coefficients de l'équation (60).
L'épaisseur de la tôle en un point courant a été établie de la même manière que précédemment, en utilisant le cercle osculateur à la parabole en ce point.

2.4 CONCLUSION

Un programme de calcul a permis d'établir les courbes de déformations en fonctin de r , pour différents enfoncements poinçons (le rayon de poinçon est de 75 mm).
La figure N° 3 montre le champ de déplacement et les trajectoires,

Figure n° 3

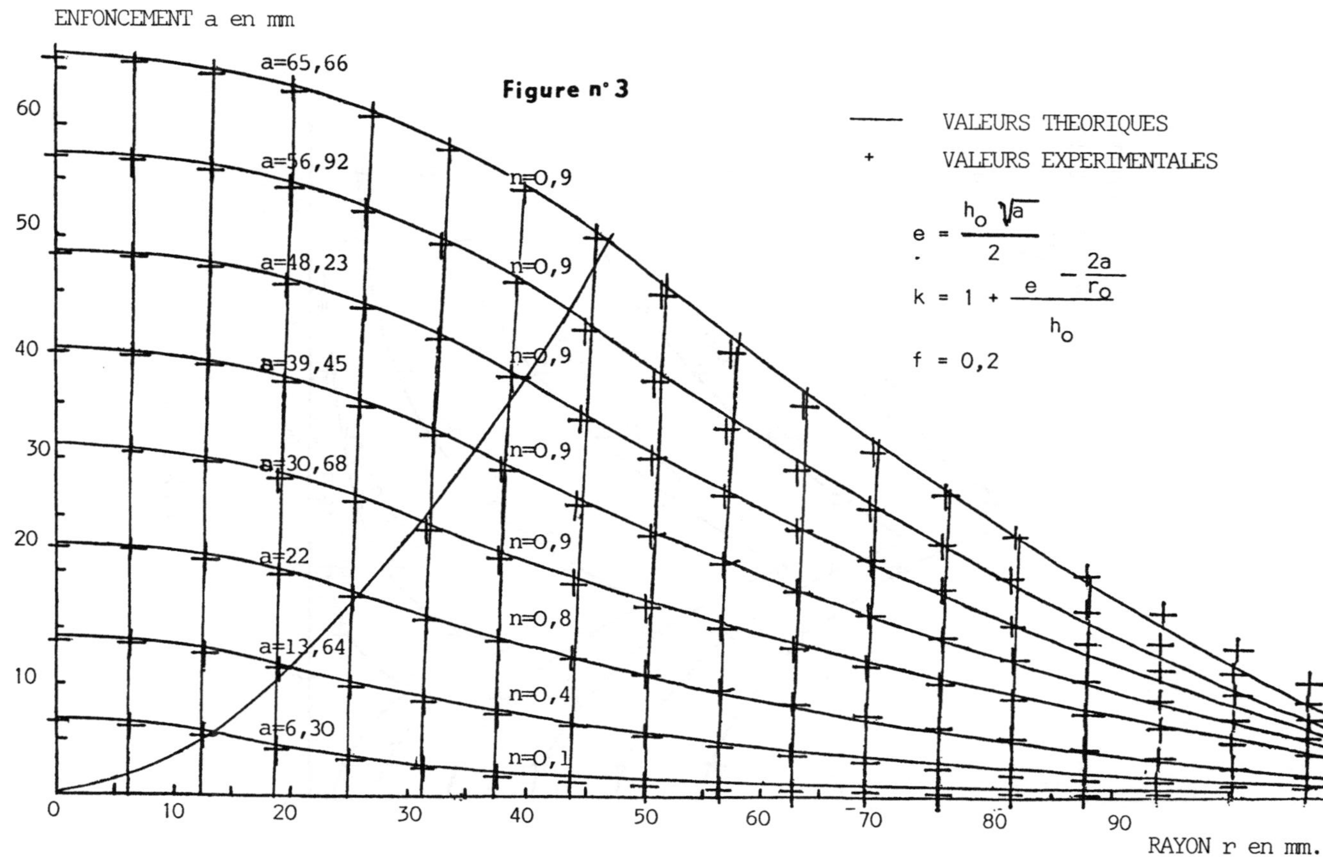

$$e = \frac{h_o \sqrt{a}}{2}$$

$$k = 1 + \frac{e^{-\frac{2a}{r_o}}}{h_o}$$

$f = 0,2$

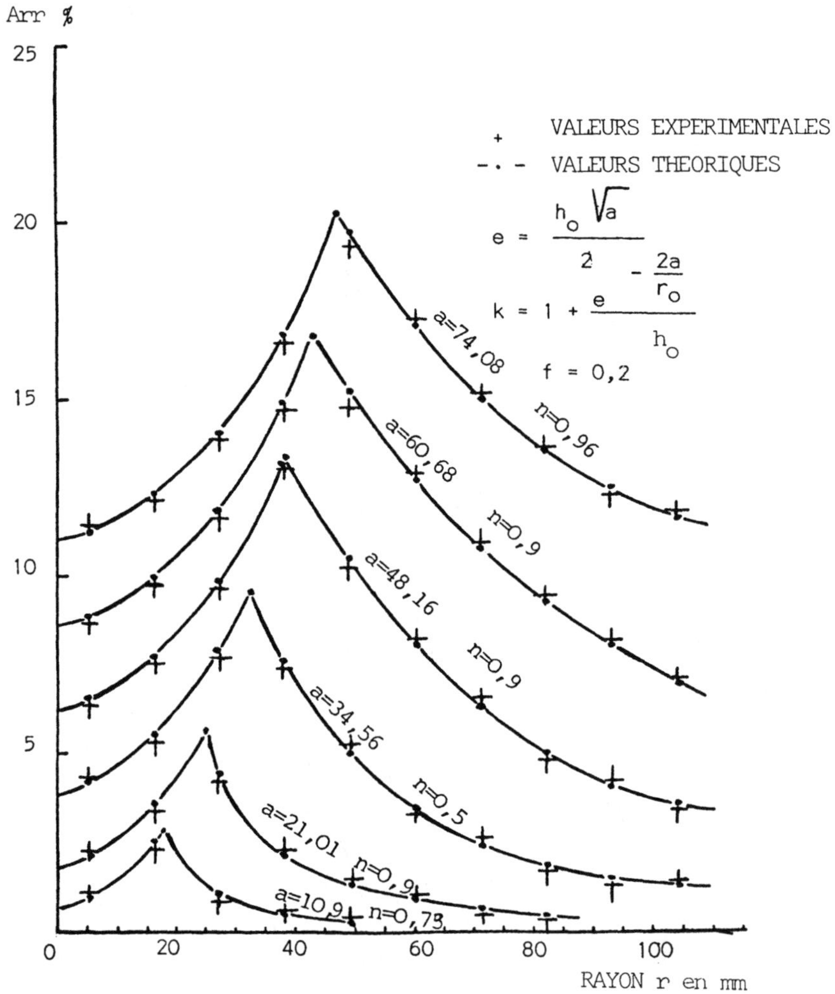

Arr %
25
20
15
10
5
0
20
40
60
80
100
RAYON r en mm
+ VALEURS EXPERIMENTALES
-.- VALEURS THEORIQUES
e = h_o √a / 2 - 2a / r_o
k = 1 + e / h_o
f = 0,2
a=74,08
n=0,96
a=60,68
n=0,9
a=48,16
n=0,9
a=34,56
n=0,5
a=21,01
n=0,9
a=10,9
n=0,73

Figure n°4

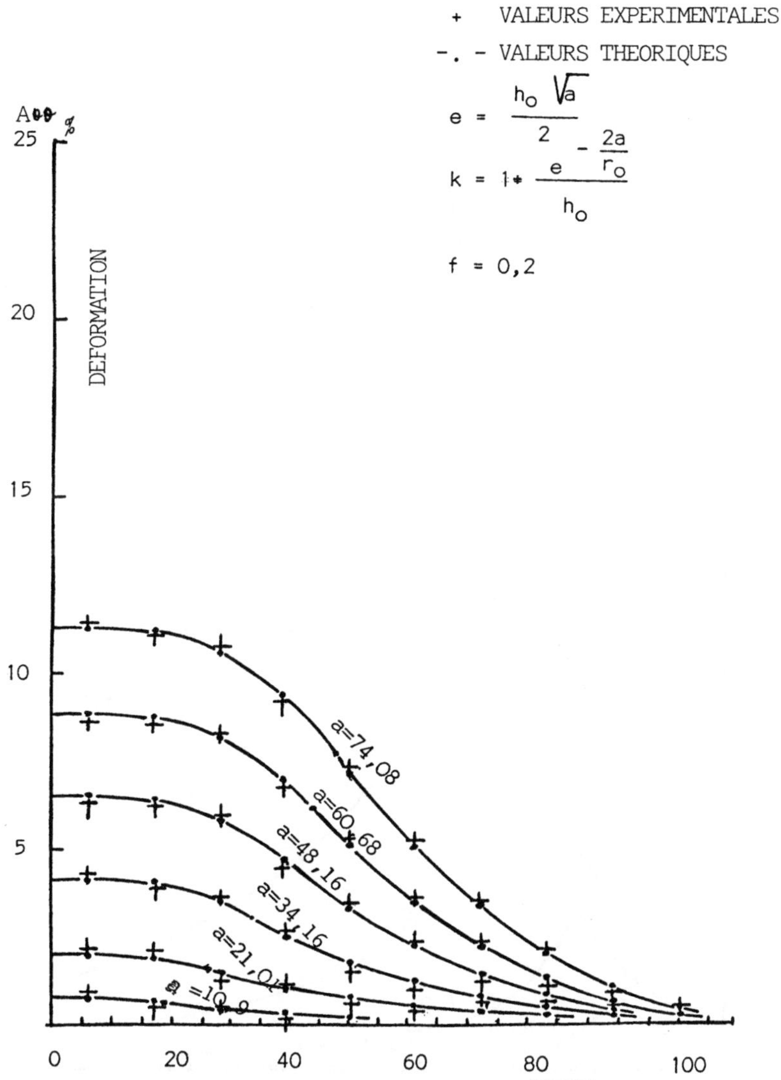
+ VALEURS EXPERIMENTALES
-. - VALEURS THEORIQUES
$e = \frac{h_o \sqrt{a}}{2}$
$k = 1 + \frac{e - \frac{2a}{r_o}}{h_o}$
f = 0,2
A %
DEFORMATION
25
20
15
10
5
0
20
40
60
80
100
RAYON en mm
a=74,08
a=60,68
a=48,16
a=34,16
a=21,04
a =10,9

Figure n° 5

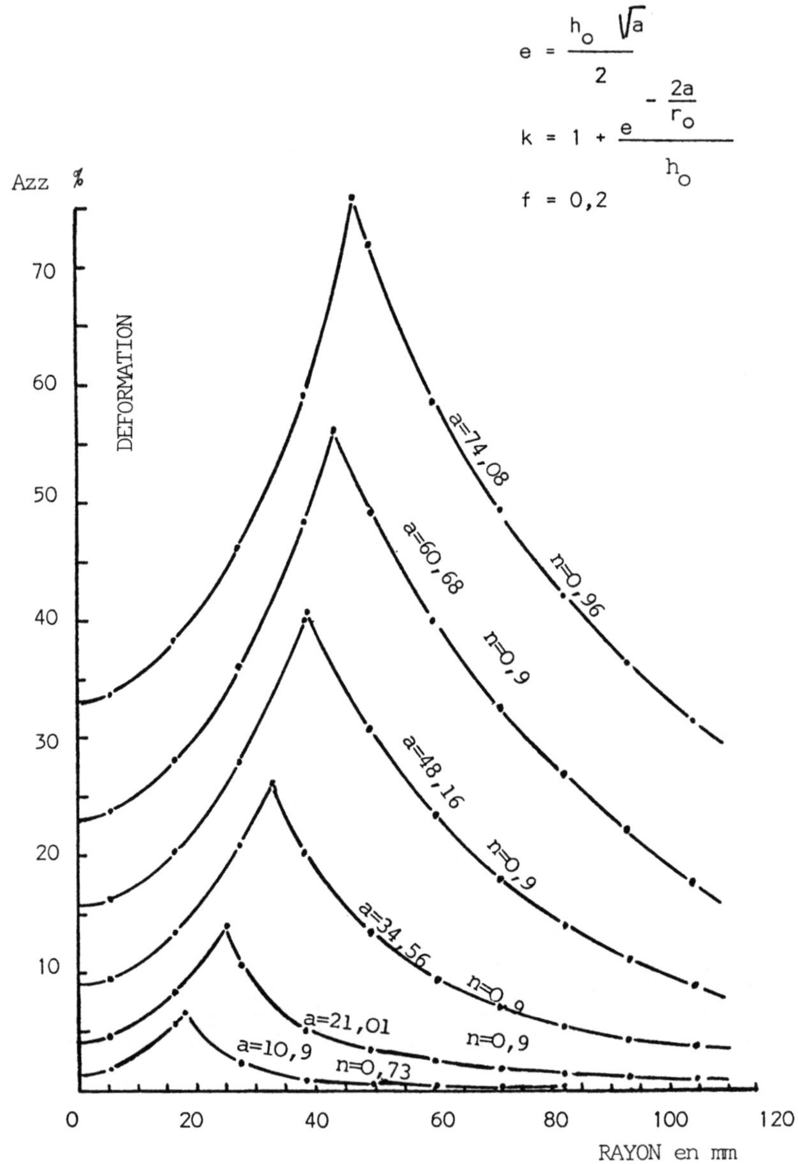
e = h_o √a / 2
k = 1 + e^(-2a/r_o) / h_o
f = 0,2
Azz %
DEFORMATION
70
60
50
40
30
20
10
0
20
40
60
80
100
120
RAYON en mm
a=74,08
n=0,96
a=60,68
n=0,9
a=48,16
n=0,9
a=34,56
n=0,9
a=21,01
n=0,9
a=10,9
n=0,73

Figure n°6

pour une tôle d'épaisseur h_0 = 0,5 mm, et de rayon r_0 = 125 mm en ayant choisi les coefficients d'ajustement suivants :

$$e = \frac{h_0 \sqrt{a}}{2} \quad , \quad k = 1 + \frac{e^{-\frac{2a}{r_0}}}{h_0} \quad \text{et} \quad f = 0,2 \qquad (62)$$

Les figures N°s 4, 5 et 6 comparent les déformations théoriques et expérimentales. Les résultats coïncident à quelques pour cent près.

BIBLIOGRAPHIE

1 Freund L.B. : Constitutive equation for elastic-plastic materials at finite strain. Int. Journ. Sol. Struct. 6 (1970) 1193-1209.

2 Germain P. : Cours de Mécanique des Milieux Continus. Tome 1, Théorie Générale. Masson et Cie. 1973.

3 Kratochvil J. Finite strain theory of crystalline elastic-inelastic materials. J. App. Physics 42 (1971) 1104-1108.

4 Lee H.E. : Elastic-plastic deformation at finite strains. J. App. Mech. 36 (1969) 1-6.

5 Mandel J. : Plasticité classique et visco-plasticité. Cours au CISM Udine 1971;
Cours de Plasticité 1974. (E.N.S.M. PARIS).

6 Naghdi P.M. : Trapp J.A. : On finite elastic-plastic deformation of metals. J. Appl. Mech. 41; (1974) 254-260.

7 Sedov L. : A course in continuum Mechanics, Vol. 1, Wolters-Noorthoff Publishing 1971.

8 Sidoroff F. : The geometrical concept of intermediate configuration and elastic-plastic finite strain. Arch. of Meet. 25.2. (1973).

9 Truesdell C. : Introduction à la Mécanique rationnelle des Milieux Continus. Masson et Cie. 1974.

Optimum Deformation Rate in Large Inelastic Deformations

O. BRUHNS and TH. LEHMANN

Institute of Mechanics,
Ruhr-University Bochum, F.R. Germany

Introduction

Metal forming processes are considered as thermomechanical processes governed by material-independent general equations, constitutive equations of the material, and boundary conditions of the process representing the independent process variables. Within the frame of classical continuum mechanics and thermodynamics a phenomenological theory of inelastic deformations of solid bodies is presented which can take into account solid phase transformations, recrystallisation etc. by means of internal variables.

Kinematics

Let us consider two sets of coordinate systems:

(i) an arbitrary (in general curvilinear) space-fixed system x^α with base vectors $\underset{\sim}{g}_\alpha(x^\rho)$ and a metric tensor $g_{\alpha\beta}(x^\rho)$ and

(ii) a body-fixed (convected) system ξ^i with base vectors $\underset{\sim}{g}_i(\xi^r,t)$ and a metric tensor $g_{ik}(\xi^r,t)$.

The material points of the body then may be identified either by the body-fixed coordinates ξ^i or by their position in the space at time $\overset{o}{t}$ denoted by $\overset{o}{x}{}^{\alpha}(\xi^i,\overset{o}{t})$. A motion of these material points is described by the

$$x^{\alpha} = x^{\alpha}(\xi^i,t) = x^{\alpha}(\overset{o}{x}{}^{\rho},t) \quad . \qquad (2.1)$$

It may be mentioned here that in general it has some advantages to choose both sets of coordinate systems such that they coincide instantaneously. The deformations of the body under consideration can now be described in different ways. One possibility is to follow here the most common way in deriving the deformations from the displacements of the material points in the space. We therefore introduce the deformation gradient

$$\underset{\sim}{F} = \frac{\partial x^{\alpha}}{\partial \overset{o}{x}{}^{\beta}} \underset{\sim}{g}_{\alpha} \overset{o}{\underset{\sim}{g}}{}^{\beta} \qquad (2.2)$$

where $d\overset{o}{x}{}^{\beta} \overset{o}{\underset{\sim}{g}}_{\beta}$ is related to the initial position of the material point, whereas $dx^{\alpha} \underset{\sim}{g}_{\alpha}$ corresponds to the current state.

In order to gain appropriate constitutive equations for an elastic-plastic (in general inelastic) material it is convenient to decompose the total deformations into elastic (recoverable) and inelastic parts, respectively. This decomposition which e.g. has been described by E.H. Lee [1] is based on the reasonable assumption that the total deformations can be considered as a sequence of two partial deformations, namely an inelastic deformation from the initial configuration into an intermediate state and a recoverable deformation from this intermediate state into the current configuration. This commonly is expressed by

$$\underset{\sim}{F} = \underset{\sim}{F}_e \cdot \underset{\sim}{F}_i = \underbrace{\underset{\sim}{g}_\alpha \frac{\partial x^\alpha}{\partial \overset{*}{x}^\rho} \overset{*}{\underset{\sim}{g}}^\rho}_{\underset{\sim}{F}_e} \cdot \underbrace{\overset{*}{\underset{\sim}{g}}_\lambda \frac{\partial \overset{*}{x}^\lambda}{\partial \overset{o}{x}^\beta} \overset{o}{\underset{\sim}{g}}^\beta}_{\underset{\sim}{F}_i} \tag{2.3}$$

where the asterisk quantities refer to the intermediate state. Thus with the definition of the strain rate tensor

$$\underset{\sim}{d} = \text{sym}\,\{\dot{\underset{\sim}{F}} \cdot \underset{\sim}{F}^{-1}\} \tag{2.4}$$

immediately we get the following decomposition of $\underset{\sim}{d}$

$$\underset{\sim}{d} = \text{sym}\,\{\dot{\underset{\sim}{F}}_e \cdot \underset{\sim}{F}_e^{-1}\} + \text{sym}\,\{\underset{\sim}{F}_e \cdot \dot{\underset{\sim}{F}}_i \cdot \underset{\sim}{F}^{-1}\}$$

$$= \{\underset{(e)}{d}{}^\alpha_\beta + \underset{(i)}{d}{}^\alpha_\beta\}\underset{\sim}{g}_\alpha\, \underset{\sim}{g}^\beta \tag{2.5}$$

A second possibility describing the deformations is connected with our body-fixed (convected) system ξ^i. By introducing the metric transformation matrix [2]

$$q^i_k = \overset{o}{g}{}^{ir}\, g_{rk} \tag{2.6}$$

which according to

$$g_{ik} = q^r_i\, \overset{o}{g}_{rk} = \overset{o}{g}_{ir}\, q^r_k \tag{2.7}$$

transforms the metric of the initial state into that of the deformed state g_{ik}, again the total deformations can be decomposed into their respective parts.

$$q^i_k = \overset{o}{g}{}^{ir}\, g_{rk} = \underbrace{\overset{o}{g}{}^{ir}\, \overset{*}{g}_{rj}}_{\underset{(i)}{q}{}^i_j}\, \underbrace{\overset{*}{g}{}^{js}\, g_{sk}}_{\underset{(e)}{q}{}^j_k} = \underset{(i)}{q}{}^i_j\, \underset{(e)}{q}{}^j_k \tag{2.8}$$

And again by appropriate definition of the strain rate tensor in convected coordinates

$$
\begin{aligned}
\underset{\sim}{d} &= d^i_k \, \underset{\sim}{g}_i \, \underset{\sim}{g}^k = \frac{1}{2} \, g^{ir} \, \dot{g}_{rk} \, \underset{\sim}{g}_i \, \underset{\sim}{g}^k \\
&= \frac{1}{2} \, (q^{-1})^i_r \, (\dot{q})^r_{.k} \, \underset{\sim}{g}_i \, \underset{\sim}{g}^k
\end{aligned}
\tag{2.9}
$$

this quantity can be decomposed into an elastic (recoverable) and an inelastic part

$$
\underset{\sim}{d} = \{ \underset{(e)}{d}{}^i_k + \underset{(i)}{d}{}^i_k \} \, \underset{\sim}{g}_i \, \underset{\sim}{g}^k \tag{2.10}
$$

Moreover in many cases [3] this latter decomposition is simply correlated to the logarithmic (or Hencky) strain measure which in this notation becomes

$$
\underset{\sim}{\varepsilon} = \frac{1}{2} \, (\ln q)^i_k \, \underset{\sim}{g}_i \, \underset{\sim}{g}^k \tag{2.11}
$$

and which has been found a reasonable strain measure since it simply allows to decompose the strain into volume changes and distortions

$$
\begin{aligned}
\varepsilon^i_k &= \frac{1}{3} \, \varepsilon^m_m \, \delta^i_k + \gamma^i_k \\
\varepsilon^m_m &= \varepsilon = \ln \frac{\overset{o}{\rho}}{\rho}
\end{aligned}
\tag{2.12}
$$

where $\overset{o}{\rho}$ and ρ are density of the material at initial and current state, respectively. This correlation now is due to

$$
d^i_k = \varepsilon^i_{k|o} \tag{2.13}
$$

and hence

$$
\underset{(e)}{d}{}^i_k = \underset{(e)}{\varepsilon}{}^i_{k}|_o , \quad \underset{(i)}{d}{}^i_k = \underset{(i)}{\varepsilon}{}^i_{k}|_o \tag{2.14}
$$

where $(\ )|_o$ means corotational (or Jaumann) derivative with respect to time defined as [4]

$$a^{i}_{.k|o} = (\dot{a})^{i}_{.k} + d^{i}_{r}\, a^{r}_{.k} - d^{r}_{k}\, a^{i}_{.r} \qquad (2.15)$$

in a body-fixed coordinate system ξ^i.

Both aforementioned decompositions in general do not lead to identical results [5]. The difference, however, is on principle very small and is moreover of neglegible extent e.g. for most metals when the recoverable elastic strains can be assumed small.

In what follows we will now prefer the description of the body in convected coordinates. This will have some advantages in the definition of constitutive equations. In particular it may be more expedient to relate the quantities to the undeformed state.

<u>Stress and mechanical work</u>

The stress is measured by the true (or Cauchy) stress tensor

$$\underset{\sim}{\sigma} = \sigma^{i}_{k}\, \underset{\sim}{g}_{i}\, \underset{\sim}{g}^{k} \qquad (3.1)$$

In thermodynamics it has some advantages introducing a weighted true stress defined by

$$\underset{\sim}{s} = \frac{\overset{o}{\rho}}{\rho}\, \underset{\sim}{\sigma} \qquad (3.2)$$

The rate of work per unit mass then becomes

$$\dot{w} = \frac{1}{\rho}\, \sigma^{i}_{k}\, d^{k}_{i} = \frac{1}{\overset{o}{\rho}}\, s^{i}_{k}\, d^{k}_{i} \qquad (3.3)$$

which according to (2.13) also can be expressed by

$$\dot{w} = \frac{1}{\overset{o}{\rho}} s^i_k \, \varepsilon^k_{i|o} \tag{3.4}$$

Whereas now the decomposition of the total strain into an elastic and inelastic part is only restricted to very particular cases, the rate of the mechanical work always can be split into a recoverable part and into an inelastic part

$$\dot{w} = \frac{1}{\overset{o}{\rho}} s^i_k \underset{(e)}{d}{}^k_i + \frac{1}{\overset{o}{\rho}} s^i_k \underset{(i)}{d}{}^k_i = \underset{(e)}{\dot{w}} + \underset{(i)}{\dot{w}} \tag{3.5}$$

Another possibility of decomposition is to split $\dot{w}$ into the rate of work due to change of volume and due to distortions, respectively,

$$\begin{aligned} \dot{w} &= \frac{1}{\overset{o}{\rho}} \frac{1}{3} s^r_r \, d^m_m + \frac{1}{\overset{o}{\rho}} t^i_k \left(d^k_i - \frac{1}{3} d^m_m \, \delta^k_i\right) \\ &= \underset{(\varepsilon)}{\dot{w}} + \underset{(\gamma)}{\dot{w}} \end{aligned} \tag{3.6}$$

which again with some restrictions can be written as

$$\dot{w} = \frac{1}{\overset{o}{\rho}} s_m \, \dot{\varepsilon} + \frac{1}{\overset{o}{\rho}} t^i_k \, \gamma^k_{i|o} \tag{3.7}$$

where with standard definitions t^i_k and s_m are deviatoric stresses and mean principal stress

$$\begin{aligned} s_m &= \frac{1}{3} s^r_r \\ t^i_k &= s^i_k - s_m \, \delta^i_k \end{aligned} \tag{3.8}$$

Thermodynamics of inelastic deformations

Inelastic deformations including solid phase transformations, recrystallisation etc. represent thermomechanical irreversible processes in an extended sense. Our thermomechanical considerations of such processes will be based on the assumption that each element of the body has the properties of a local thermodynamical system whose state is uniquely defined by a set of (external and internal) state variables [3, 6, 7]. This assumption allows introducing state functions like free energy, entropy etc. and state equations for the material elements even for non-equilibrium states - provided the respective process is not too far from equilibrium.

These thermodynamical state equations, however, only provide the general frame for an appropriate description of the considered thermomechanical process. Moreover a proper process description is needed which fits into this frame. The aim of such process descriptions is delivering the evolution laws of the dependent process variables as functionals of the internal state and the history of the independent process variables. Under usual conditions of e.g. smoothness these functionals can be expressed by a system of first order differential equations (possibly completed by some auxiliary conditions).

The thermodynamic statements are based on the laws of thermodynamics. The first step then consists in the definition of a suitable set of state variables which uniquely determines the thermodynamic state of a material element and which finally will take into account that our considerations will be restricted to metals.

We therefore may assume:

(i) the elastic behaviour is isotropic and independent of the inelastic deformations,

(ii) the inelastic deformations on the other hand are incompressible and

(iii) the solid phase transformations do not influence the elastic distortions.

This latter assumption implies that the volume changes which are connected with the phase transformations will be governed by a state equation. A reasonable set of state variables then consists of:

$$\{\varepsilon,\ \underset{(e)}{\gamma}{}^{i}_{k},\ T;\ P_{(\nu)},\ \kappa,\ \beta^{i}_{k},\ \ldots\} \tag{4.1}$$

with

ε dilatation

$\underset{(e)}{\gamma}{}^{i}_{k}$ elastic (recoverable) distortions

T temperature

$P_{(\nu)}$ mass fractions of solid phases (ν = 1,...n-1) and

κ, β^{i}_{k}... some further internal variables not yet specified.

The first law of thermodynamics states

$$\dot{u} = \dot{w} + r - \frac{1}{\rho}\, q^{i}{}_{|i} = \dot{w} + \dot{q} \tag{4.2}$$

where u denotes the specific internal energy and r and q^i are with usual definitions heat supply and heat flux vector, respectively.

According to our assumptions the rate of mechanical work can be split into

$$\dot{w} = \underbrace{\dot{w}_{\varepsilon} + \dot{w}_{\gamma e}}_{\dot{w}_e} + \underbrace{\dot{w}_D + \dot{w}_H}_{\dot{w}_i} \tag{4.3}$$

with

w_{ε} volumetric work

$w_{\gamma e}$ elastic distortional work

w_D dissipated work

w_H hardening work.

The evolution of specific entropy s is given by

$$T\dot{s} = \underbrace{r - \frac{1}{\rho} q^i{}_{|i}}_{\dot{q}} + \dot{w}_D + \dot{\eta}_D \tag{4.4}$$

which also may be written in the form

$$T\dot{s} = r - \frac{T}{\rho} \left(\frac{q^i}{T}\right)_{|i} + T \underset{(i)}{\dot{s}} \tag{4.5}$$

where η_D denotes the dissipation of energy due to changes of internal structure and $\underset{(i)}{\dot{s}}$ designates the irreversible entropy production.

The second law of thermodynamics then states

$$T \underset{(i)}{\dot{s}} \geqq 0 \tag{4.6}$$

The next step now consists in defining the specific free energy $\psi = u - Ts$ as an appropriate function of state variables. This will be done such that in accordance with our assumptions we find

$$\Psi = \Psi_1 \{ \underset{(e)}{\gamma}{}^i_k , T\} + \Psi_2 \{\varepsilon, T, p_{(\nu)}\} + \Psi_3 \{T, p_{(\nu)}, \kappa, \beta^i_k\} \tag{4.7}$$

Introducing the above relations into (4.6) we gain

$$\begin{aligned} T \underset{(i)}{\dot{s}} &= \dot{w} - \dot{\Psi} - \dot{T}s - \frac{1}{\rho T} q^i T_{,i} \\ &= \underline{\{\frac{1}{\overset{o}{\rho}} s_m - \frac{\partial \Psi_2}{\partial \varepsilon}\}} \dot{\varepsilon} + \underline{\{\frac{1}{\overset{o}{\rho}} t^i_k - \frac{\partial \Psi_1}{\partial \underset{(e)}{\gamma}{}^k_i}} \; \underset{(e)}{\gamma}{}^k_{i|o} \\ &- \underline{\{s + \frac{\partial \Psi}{\partial T}\}} \dot{T} + \dot{w}_D + \dot{w}_H - \frac{1}{\rho T} q^i T_{,i} \\ &- \sum_{\nu=1}^{n-1} \frac{\partial \Psi}{\partial p_{(\nu)}} \dot{p}_{(\nu)} - \frac{\partial \Psi_3}{\partial \kappa} \dot{\kappa} - \frac{\partial \Psi_3}{\partial \beta^i_k} \beta^i_{k|o} \geqq 0 \end{aligned} \tag{4.8}$$

Since herein ε, γ^i_k and T have been assumed independent state variables the underlined terms in brackets must vanish. This yields the state equations

$$s_m = \overset{o}{\rho} \frac{\partial \Psi_2}{\partial \varepsilon} \qquad t^i_k = \overset{o}{\rho} \frac{\partial \Psi_1}{\partial \underset{(e)}{\gamma}{}^k_i} \qquad s = - \frac{\partial \Psi}{\partial T} \tag{4.9}$$

Furthermore there remains

$$\dot{w}_H = \sum_{\nu=1}^{n-1} \frac{\partial \Psi}{\partial p_{(\nu)}} \dot{p}_{(\nu)} + \frac{\partial \Psi_3}{\partial \kappa} \dot{\kappa} + \frac{\partial \Psi_3}{\partial \beta^i_k} \beta^i_{k|o} + \dot{\eta}_D \tag{4.10}$$

with the restriction that

$$\dot{w}_D + \dot{\eta}_D - \frac{1}{\rho T} q^i T_{|i} \geqq 0 \qquad (4.11)$$

The terms $\dot{p}_{(\nu)}$, $\dot{\kappa}$ and $\beta^i_{k|o}$ as well as $\dot{w}_D$ and $\dot{\eta}_D$ in (4.1o) and (4.11) now have to be specified by evolution laws. These laws together with the law for the heat flux then form the connection between process description and the thermodynamical statements whereas the above relations provide the general frame for these considerations.

As mentioned before this frame has to be filled with detailed statements for the evolution laws of the (external and internal) dependent process variables in agreement with experimental results and in order to define a complete system of constitutive equations.

Elastic-viscoplastic processes

As an example we now define the constitutive equations for an elastic-viscoplastic body. We therefore assume that the state of the body and initial values of the state variables are known. Furthermore it is presumed that the state function which determines the free energy Ψ is given.

Under these assumptions we may choose the stresses s^i_k and temperature T as independent (external) process variables, which e.g. can be measured by experiments. Dependent process variables then are e.g. the total strains ε^i_k, the applied heat q and the internal variables $p_{(\nu)}$, κ and β^i_k.

Concerning viscoplastic materials it is usually assumed

that the acting stresses s^i_k can be decomposed into a plastic and a viscous part according to

$$s^i_k = \bar{s}^i_k + \bar{\bar{s}}^i_k = \bar{s}^i_k + (s^i_k - \bar{s}^i_k) \qquad (5.1)$$

The plastic stresses herein are assumed to obey the plastic yield condition

$$\bar{F}\ (\bar{s}^i_k,\ T,\ p_{(\nu)},\ \kappa,\ \beta^i_k) = 0 \qquad (5.2)$$

and thus can be eliminated. For the present we disregard solid phase transformations and any kind of kinematic hardening, which latter will be caused by the internal variables β^i_k, assuming that Ψ is given in the special form

$$\Psi = \Psi_1\ (\underset{(e)}{\gamma}{}^i_k) + \Psi_2\ (\varepsilon,\ T) + \kappa + g\ (T) \qquad (5.3)$$

Further we assume that the dissipation of energy is defined by

$$\dot{\eta}_D = 0 \text{ and}$$

$$\dot{w}_D = \underbrace{\gamma \frac{1}{\overset{o}{\rho}} \bar{s}^i_k \underset{(i)}{d}{}^k_i}_{\dot{w}_p} + \underbrace{\frac{1}{\overset{o}{\rho}} \bar{\bar{s}}^i_k \underset{(i)}{d}{}^k_i}_{\dot{w}_v} \qquad (5.4)$$

what means that the dissipated inelastic work consists of two parts, the totally dissipated viscous work w_v and the partially dissipated plastic work w_p and where γ is a factor which varies slowly from about 0.9 to unity with increasing plastic flow [8, 9].

Hence we find

$$\dot{w}_H = (1-\gamma)\,\dot{w}_p = \dot{\kappa} \tag{5.5}$$

the plastic work to be equivalent to the internal variable κ if γ is taken as a constant or as single functions of κ and w_p, respectively.

The corresponding process description now may be based on the assumption that:

(i) inelastic deformations will occur if and only if

$$F = \frac{f(s^i_k)}{k(\kappa,T)} - 1 > 0 \tag{5.6}$$

where k is the static yield stress which on the other hand satisfies the respective static yield criterion

$$\bar{F} = \frac{f(\bar{s}^i_k)}{k(\kappa,T)} - 1 = 0 \tag{5.7}$$

for the plastic stresses.

(ii) these inelastic deformations then satisfy the normality rule

$$\underset{(i)}{d}{}^i_k = \dot{\varphi}\,\phi\,(F)\,\frac{\partial F}{\partial s^k_i} \tag{5.8}$$

where herein $\dot{\varphi}$ is in general a function of state variables, which here may be taken as a material constant, and ϕ is a suitable function of the given argument. This shall be valid even for vanishing viscosity, i.e. for the plastic stresses we have

$$\underset{(i)}{d}{}^{i}_{k} = \dot{\Lambda} \frac{\partial \bar{F}}{\partial \bar{s}^{k}_{i}} \tag{5.9}$$

These assumptions now lead to the following system of constitutive relations for the process description:

$$d^{i}_{k} = d^{i}_{k} \{s^{i}_{k}, s^{i}_{k|o}, T, \dot{T}, \kappa\}$$

$$= \underset{(e)}{d}{}^{i}_{k} + \dot{\varphi}\, \Phi\, (F) \frac{\partial F}{\partial s^{k}_{i}}$$

$$\dot{\kappa} = \frac{1}{\overset{o}{\rho}} \frac{1-\gamma}{1+F} t^{i}_{k} \underset{(i)}{d}{}^{k}_{i}$$

$$\dot{q} = - T \frac{\partial^{2}\Psi}{\partial T^{2}} \dot{T} - T \frac{\partial^{2}\Psi}{\partial \varepsilon \partial T} \dot{\varepsilon} - \frac{1}{\overset{o}{\rho}} \frac{\gamma+F}{1+F} t^{i}_{k} \underset{(i)}{d}{}^{k}_{i}$$

$$= r - \frac{1}{\rho} q^{i}{}_{|i} \tag{5.10}$$

provided (5.6) holds.

When general inelastic deformations are considered these constitutive equations have to be completed by general field equations and some statements about boundary conditions. From these the equation of motion becomes

$$s^{i}_{k|i} - s^{i}_{k}\, \varepsilon_{,i} + f_{k} = \overset{o}{\rho} a_{k} \tag{5.11}$$

in convected coordinates where f_k and a_k are specific body forces and acceleration vector, respectively. Due to the fact that our constitutive equations in general contain stresses as well as stress rates it is necessary evaluating an equation of motion for the rates

$$(s^{i}_{k|o})_{|i} - s^{i}_{k|o}\,\varepsilon_{,i} - d^{i}_{m|i}\, s^{m}_{k}$$

$$+ \underbrace{[d^{m}_{k|i} - d^{m}_{i|k}]}_{\omega^{m}_{.k|i}}\, s^{i}_{m} + f_{k|o} = \overset{o}{\rho} a_{k|o} \qquad (5.12)$$

where herein $\omega^{i}_{.k}$ denotes the skew symmetric part of the velocity gradient. The heat flux vector q_i is assumed to obey e.g. Fourier's law

$$q_i = -\lambda\, T_{,i} \qquad (5.13)$$

where the coefficient of conductivity λ still may depend on the whole set of state variables.

This system of equations governing elastic-viscoplastic deformations seems to be rather complicated. It furthermore depends on the particular conditions of the problem as to which method of solution is most suitable for a given problem. We therefore restrict our further interest to a few remarks on a coupling-effect which arises when elastic-viscoplastic deformations influenced by heat conduction are considered.

Generally, we can suppose that a rising strain rate leads to an increasing temperature due to the lose of time available for the removal of heat. This leads to a decreasing yield stress. On the other hand an increasing strain rate causes an increase of the yield stress due to viscous effects. We therefore may suppose that there exists an optimal rate minimizing the inelastic work for a particular inelastic deformation.

Example

As a simple example a thin-walled cylindrical tube with initial magnitudes length L_o, radius R_o and thickness H_o is subjected to pure torsion. Due to the dissipation of inelastic work some amount of heat will be produced inside the specimen. The temperature of the surrounding of the tube is held constant. The heat flux then is considered as a (quasi-) stationary process. Furthermore, for simplicity, it is assumed that this flux is a linear function of the temperature difference $\vartheta = T - T_o$ where T and T_o are temperatures of the tube and the surrounding, respectively. Thus we get

$$m\dot{q} = - 4\pi \bar{\alpha} RL \vartheta$$

where herein $\dot{q}$ describes the heat transfer through inner and outer cylindrical surfaces of the tube to the air and R and L are the respective geometrical quantities of the current state. The coefficient $\bar{\alpha}$ approximately can be derived [10, 11]

$$\bar{\alpha}_1 \approx 1.32 \left(\frac{\vartheta}{2R}\right)^{1/4}$$

for free convection and

$$\bar{\alpha}_2 \approx \frac{6}{R} (Rv_s)^{0.786} \left[1 + \left(\frac{2R}{L}\right)^{2/3}\right]$$

when the inner surface of the tube is cooled by an air stream of velocity v_s.

Now disregarding the elastic deformations and assuming a totally dissipated plastic work ($\gamma = 1$) the constitutive relations for a viscoplastic v. Mises material become

$$d^i_k = \dot{\varphi}\, \phi\, (F) \frac{t^i_k}{\sqrt{t^r_s\, t^s_r}}$$

$$\dot{q} = c\dot{T} - \dot{w}_i$$

$$\overset{\circ}{\rho}\dot{w}_p = \frac{1}{1+F}\, \overset{\circ}{\rho}\dot{w}_i = \dot{\varphi}\, \phi\, (F)\; k\, (T,\, w_p)$$

provided

$$F = \frac{\sqrt{t^i_k\, t^k_i}}{k(T, w_p)} - 1 > 0.$$

From experimental data for mild steel [12] we derive the static yield stress

$$k\, (T,\, w_p) = \sqrt{\frac{2}{3}}\, \{y_o\, (T) + \frac{c_1\, (T)\, \overset{\circ}{\rho} w_p}{c_2\, (T) + \overset{\circ}{\rho} w_p}\}$$

with

$$y_o\, (T) = 474.1 - 389 \cdot 10^{-3}\, T$$

$$c_1\, (T) = 724.2 - 360 \cdot 10^{-3}\, T \qquad [\frac{N}{mm^2}]$$

$$c_2\, (T) = 73.5 - 80.4 \cdot 10^{-3}\, T$$

Taking now $\dot{\Psi} = 1$ [1 /s], $c = 465$ [J/kg^{o}K] and $\phi\, (F) = \exp\, (F) - 1$ the following results can be obtained.

We notice (Fig. 1) that at low strain rates the temperature rise is strongly affected by the heat flux, whereas for higher strain rates than $\dot{\Psi} = 10^{-2}$ [1/mms] the difference in the solution for adiabatic processes and processes with heat flux vanishes. Concerning the inelastic work (Fig. 2, 3) we state the following. Starting with very small strain rates ($\dot{\psi} = 10^{-6}$) we need at first less work with increasing strain rate according to the rise of the temperature. At higher strain rates, however, the additional viscous stresses lead to a growing demand for mechanical work. Between both regions we find an optimal strain rate minimizing the mechanical work.

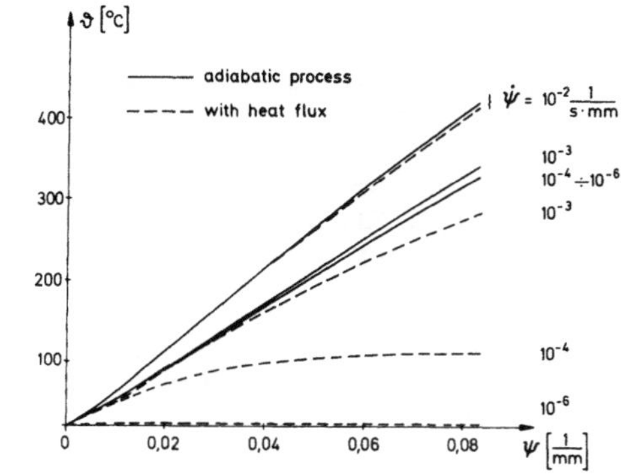

Fig. 1 Temperature rise vs twist in pure shear of a thin-walled tube

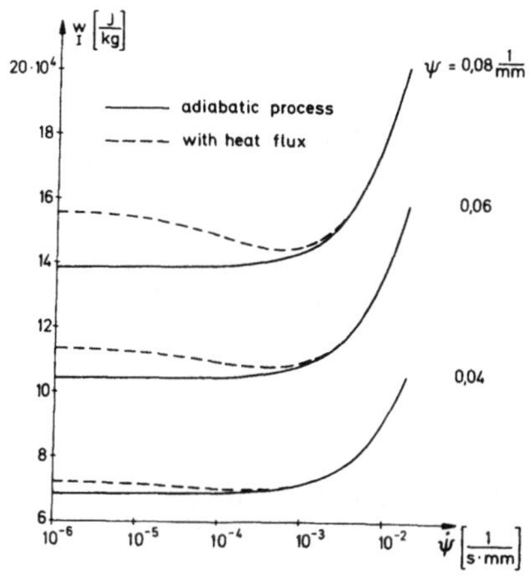

Fig. 2 Specific work in pure shear with simple heat flux

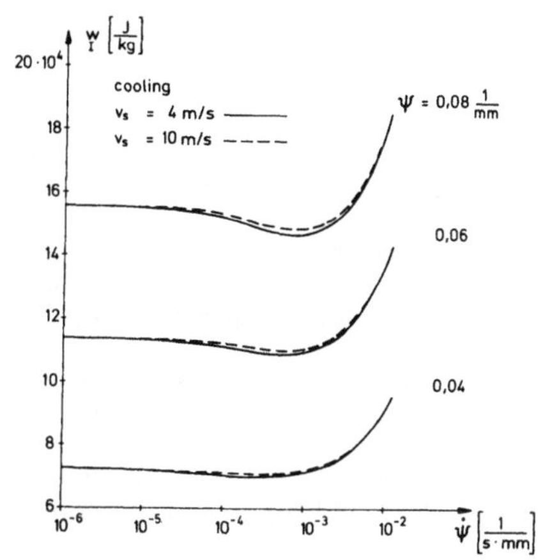

Fig. 3 Specific work in pure shear with cooling

References

1 Lee, E.H.: Elastic-plastic deformations at finite strains; Journ. Appl. Mech. 36 (1969) 1-6

2 Lehmann, Th.: On large elastic-plastic deformations; in: Foundations of Plasticity. Papers contrib. Intern. Symp. on Foundations of Plasticity (Warsaw 1972); ed. by A. Sawczuk, Noordhoff Int. Publ. Leyden 1973

3 Lehmann, Th.: Some aspects of non-isothermic large inelastic deformations; Sol. Mech. Arch. 3 (1978) No. 3

4 Lehmann, Th.: Formänderungen eines klassischen Kontinuums in vierdimensionaler Darstellung; in: Proc. XI. Intern. Congr. Appl. Mech. München 1964, 376-382, Springer-Verlag Berlin/Heidelberg/New York 1966

5 Raniecki, B.; Thermann, K.: Foundations of reversible and irreversible thermo-mechanical processes; Mitteilungen Inst. f. Mechanik, Ruhr-Universität Bochum 2 (1978)

6 Meixner, J.; Reik, H.G.: Thermodynamik der irreversiblen Prozesse; in: Handbuch der Physik, Vol. III/2 (Editor: S. Flügge), Springer-Verlag Berlin/Heidelberg/New York 1959

7 Coleman, B.D.; Gurtin, M.E.: Thermodynamics with internal state variables; Journ. Chem. Phys. 47 (1967) 597-613

8 Farren, W.S.; Taylor, G.I.: The heat developed during plastic extension of metals; Proc. Royal Soc. 107 (1925) 422-451

9 Clarenbrough, L.M.; Hargreaves, M.E.; Loretto, M.H.: The influence of grain size on the stored energy and mechanical properties of copper; Acta Met. 6 (1958) 725-735

10 Schmidt, E.: Einführung in die technische Thermodynamik und in die Grundlagen der chemischen Thermodynamik; 10. Aufl. Springer-Verlag Berlin, 1963

11 Dubbels Taschenbuch für den Maschinenbau, Band 1, Springer-Verlag Berlin, 1966

12 Fritsch, G.; Siegel, R.: Kalt- und Warmfließkurven von Baustählen; Inst. Forsch. Maschinenbau, Karl-Marx-Stadt, 1965

New Trends in Tube and Bar Processing

N. CRISTESCU and S. CLEJA
University of Bucharest, Romania

1. Introduction

In the last years an effort has been done in order to develop a metal working theory, for both cold and hot working processes, based on viscoplastic models. The scope was twofold. First in all existing theories (see Hill [1], Avitzur [2]), which use classic time independent plasticity theories, the working speed is not involved at all. Thus the main task was to estimate the influence of the working speed on the whole process. Secondly, most of the existing theories are approximate and therefore giving generally only an estimate of the forces involved but no information concerning the optimum working regime (i.e. optimum tool geometry, maximum working speed, minimum force, maximum reduction etc.). Therefore the second task was to describe more accurately the process in order to be able to find for all respect the optimum working regimes.

It is well known that in industrial practice the working speeds, mainly in drawing of wires and tubes, are quite high. Thus wires are currently drawn with speeds raising up to 50 m/sec. while in drawing of tubes with floating plug, speeds of several meters per second (often surpassing 10 m/sec.) are usual. Such drawing speeds, combined with the geometry of the product, may lead to very high rates of strains. Thus in wire drawing, mainly if the wire is very thin, the rates of strains are of the order 10^3-10^5 sec^{-1} while in tube drawing the rates of strains are of the order $10^2 sec^{-1}$. These ranges for the rates of strains are well above the ranges where classic plasticity theories can be used. It is well known that these theories have been founded on the basis of quasistatic tests performed with rates of strains of the order 10^{-5}-10^{-3} sec^{-1}. In the last decades very many experimental data, concerning the properties of metals at very high rates of strains, have been published. A great deal of such data were put together in survey papers or books (see Polukhin et al. [3], Lindholm and Bessey [4]). These experimental data, concerning the mechanical properties of various metals for broad ranges of rates of strains and temperatures, can be successfully used in order to describe a metal working process involving very high rates of strains.

This was the reason why in order to describe a high speed working process one has started from a quasilinear viscoplastic constitutive equation of the form

$$\underset{\sim}{D} = \Phi \overset{\circ}{\underset{\sim}{T'}} + \Psi \underset{\sim}{T'} \qquad (1.1)$$

where $\underset{\sim}{D}$ is the strain rate deviator tensor, $\underset{\sim}{T'}$ the stress tensor deviator, $\overset{\circ}{\underset{\sim}{T}}$ is an objective time derivative, while the coefficient factors Φ and Ψ depend on invariants (see Cristescu [5],[6]) and on the plastic work.

Since in all considered problems only stationary processes were considered, it will be assumed that there is a certain domain where the whole viscoplastic deformation takes place and where the first term from the right hand side in (1.1) is negligible. Generally it is expected that this term may be significant at the fast transition at the entry and exit of the die (see fig.2.1). In the first stage it will be assumed that this term contributes little to the overall process of deformation. Thus it is only the last term describing the non-instantaneous response that was maintained and the constitutive equation (1.1) was written in the Bingham form

$$\underset{\sim}{D} = \frac{1}{2\eta}\left\langle 1 - \frac{\sigma(\bar{\varepsilon})}{\sqrt{3}\sqrt{II_{T'}}}\right\rangle \underset{\sim}{T'} , \qquad \mathrm{tr}\, \underset{\sim}{D} = 0 \tag{1.2}$$

where

$$\langle Z \rangle = \begin{cases} 0 & \text{if} \quad Z \leq 0 \\ Z & \text{if} \quad Z > 0 \end{cases} \tag{1.3}$$

η is a viscoplastic coefficient, i.e. a measure of the "rate dependency" of the material. It was found that η is decreasing when the rate of strain is increasing but for certain intervals of variation of the rate of strain η can be considered to be constant. $\sigma(\bar{\varepsilon})$ is describing the work hardening properties of the material and $\bar{\varepsilon}$ is the equivalent strain. To make simpler the calculations for each $\bar{\varepsilon}$, $\sigma(\bar{\varepsilon})$ was replaced by a mean yield stress σ_m. Finally $II_{T'}$ is the second invariant of the stress deviator. The constitutive equation (1.2) has some significant advantages: it is the simplest rate dependent constitutive equation which is objective (the stress fluxes are not involved) and the speed effect, obtained for instance for the drawing stress appears as an additional term, to the one which would be obtained with a classical plasticity theory.
Model (1.2) was used mainly in cold working processes. It was experimentally found (Teodosiu, Soós, Radu [7]) that for hot working processes a non-linear viscoplastic constitutive equation would be more appropriate.

In obtaining the solution described above a basic role played a variational principle for kinematic admissible velocity fields. Let V be an open domain of viscoplastic deformation and Γ its boundary with $\Gamma = \Gamma_T + \Gamma_v$; on Γ_T is prescribed the stress and on Γ_v the velocity. Let $\mathcal{D} = \{ \underset{\sim}{v} \in \underset{\sim}{C}^1(V) \mid \operatorname{div} \underset{\sim}{v} = 0 \text{ in } V, \text{ prescribed values on } \Gamma_v \}$ be the family of all kinematic admissible velocity fields. From the kinematic admissible velocity fields it is the real field which minimizes the functional

$$J(\underset{\sim}{v}) = 2\int_V \left(\eta \, II_D + \frac{\sigma_m}{\sqrt{3}} \sqrt{II_D} \right) dv - \int_V \underset{\sim}{X} \cdot \underset{\sim}{v} \, dv - \int_{\Gamma_T} \underset{\sim}{t} \cdot \underset{\sim}{v} \, d\sigma \tag{1.4}$$

(see [8]), where II_D is the second invariant of the rate of strain tensor, $\underset{\sim}{X}$ are the body forces and $\underset{\sim}{t}$ is the prescribed stress vector on Γ_T. Let $\mathcal{D}_\gamma \subset \mathcal{D}$ a family of kinematic admissible velocity fields depending on a parameter γ, then the minimum of $J(\underset{\sim}{v})$ on $\mathcal{D}_\gamma$ is bigger than the absolute minimum (computed for $\mathcal{D}_\gamma$). The field $\underset{\sim}{v} \in \mathcal{D}_\gamma$ which realizes this minimum will be considered an approximate solution for the considered problem. If the domain V contains a subdomain where a rigid motion takes place, then the family $\mathcal{D} = \{ \underset{\sim}{v} \mid (\exists)$ a surface $\Sigma \subset V$, $\underset{\sim}{v} \in \underset{\sim}{C}^1(V - \Sigma)$, $\operatorname{div} \underset{\sim}{v} = 0$ in $V - \Sigma$, $[\underset{\sim}{v} \cdot \underset{\sim}{n}] = 0$ on Σ and $[\underset{\sim}{v} \cdot \underset{\sim}{\tau}] \neq 0 \}$ is the set of kinematic admissible velocity fields and the term $\int_\Sigma [\underset{\sim}{t} \cdot \underset{\sim}{v}] \, d\sigma$ is to be added to the functional (1.4); here Σ is an a priori unknown rigid-viscoplastic interface, $\underset{\sim}{n}$ is the unit vector of the normal to Σ and $\underset{\sim}{\tau}$ the unit vector of a tangent to Σ . The jump $[\underset{\sim}{t} \cdot \underset{\sim}{v}]$ on Σ is obtained from the dynamic compatibility condition.

2. <u>Wire and bar drawing</u>

A viscoplastic constitutive equation was first used in the theory of wire and bar drawing (Cristescu [9][10]). Excellent experimental data have been published by Wistreich [11], though a single drawing speed was used. Assuming axial symmetry, the velocity field considered is

$$v = v_r = -v_f r_f^2 \frac{\cos\theta}{r^2}, \quad v_\theta = v_\varphi = 0 \qquad (2.1)$$

which was used within the framework of classic plasticity theory by Avitzur [2], Pugh [12] and Renne [13]. The viscoplastic region II (see fig.2.1 where the notations are self-explanatory) is bordered by the die and

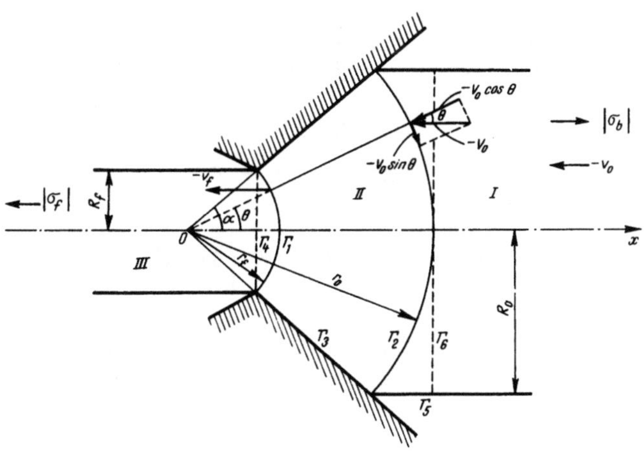

Fig.2.1 Geometry in bar drawing

two discontinuity surfaces Γ_1 and Γ_2 of spherical shape. In domain I and III rigid motion takes place. Across the surfaces Γ_1 and Γ_2 the dynamic jump conditions are used, while the normal component of the velocity is continuous.

For a Bingham constitutive equation the internal dissipative power is found by integrating

$$T'_{ij}D_{ij} = \frac{2\sigma_m}{\sqrt{3}}\sqrt{II_D} + 4\eta\, II_D \tag{2.2}$$

over the entire viscoplastic domain. The friction losses are estimated by assuming a friction law of the form

$$|\underset{\sim}{t}_\tau| = m\sqrt{II_T}, \tag{2.3}$$

which generalizes for the viscoplastic model the friction law $|\underset{\sim}{t}_\tau| = m\,\tau_Y$ used in classical plasticity. From the theorem of total power the drawing stress is obtained as

$$\frac{\sigma_f}{\sigma_Y} = \frac{\sin^2\alpha}{2\cos\alpha - \cos^2\alpha - \frac{2}{3}\cos^3\alpha - \frac{1}{3}}\left\{\frac{\sigma_b}{\sigma_Y}\left[1 + \right.\right.$$

$$\left. + 2\,\frac{-\cos\alpha + \frac{1}{3}\cos^3\alpha + \frac{2}{3}}{\sin^2\alpha}\right] + 2\,f(\alpha)\ln\frac{R_o}{R_f} +$$

$$+ \frac{2}{3}\,\frac{\eta v_f}{\sigma_Y R_f}\,\frac{1}{\sin\alpha}\left[1 - \left(\frac{R_f}{R_o}\right)^3\right]\left[1 - \cos\alpha + \right.$$

$$\left. + \frac{11}{3}(1 - \cos^3\alpha)\right] + \frac{2}{\sqrt{3}}\,m\,\mathrm{cotg}\,\alpha\left\{\ln\frac{R_o}{R_f} + \right.$$

$$\left.\left. + \frac{\eta v_f}{\sigma_Y R_f}\,\frac{\sin\alpha}{\sqrt{3}}\sqrt{11\cos^2\alpha + 1}\left[1 - \left(\frac{R_f}{R_o}\right)^3\right]\right\}\right\} \tag{2.4}$$

where

$$-f(\alpha) = \frac{\sqrt{11}}{2\sqrt{3}}\,\frac{1}{\sin^2\alpha}\left[(\cos\alpha)\sqrt{\cos^2\alpha + \frac{1}{11}} + \right.$$

$$+ \frac{1}{11}\ln(\cos\alpha + \sqrt{\cos^2 + \frac{1}{11}}) - \sqrt{\frac{12}{11}} -$$

$$\left. - \frac{1}{11}\ln(1 + \sqrt{\frac{12}{11}})\right]\,. \tag{2.5}$$

Here σ_Y stands for the actual yield stress of the work-hardened material.

Similar formulae were obtained also for a Coulomb friction law and hydrodynamic lubrication (Cristescu and Suliciu [14] Ch.IX). A comparison with the experimental data by Wistreich [11] obtained for copper wires drawn

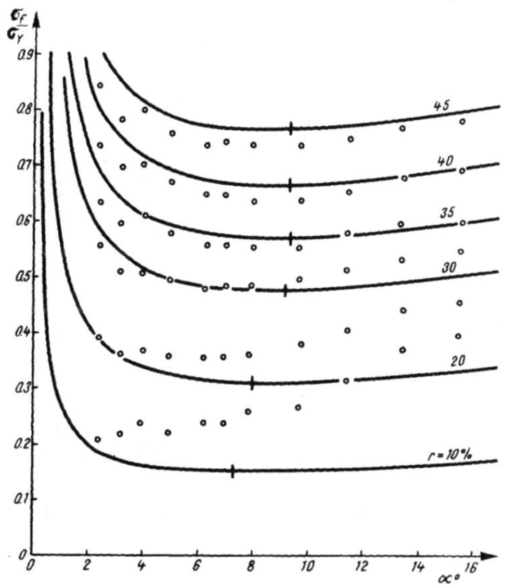

Fig.2.2 Relative drawing stress versus die angle for copper wires as given by formula (2.4) (full lines) and the experimental data by Wistreich [11] (circles)

with 33 mm/sec through various dies is shown on fig.2.2 while a comparison with the experimental data by Cristescu [15] obtained for steel and copper wires drawn with various reductions through dies of 7° and 8° is given on fig.2.3 for three drawing speeds ranging from 5 mm/min to 1100 mm/min. Taking into account a possible bulging effect for relatively small reductions and/or large α , these data are in good agreement with the

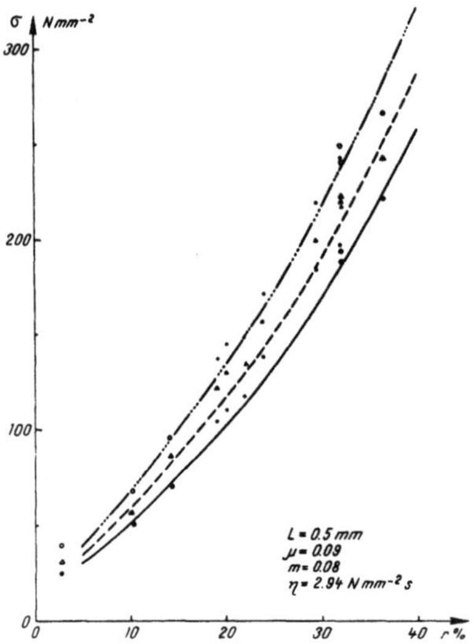

Fig.2.3 Drawing stress versus reduction for copper wires as prescribed by formula (2.4) compared with experimental data from [15]

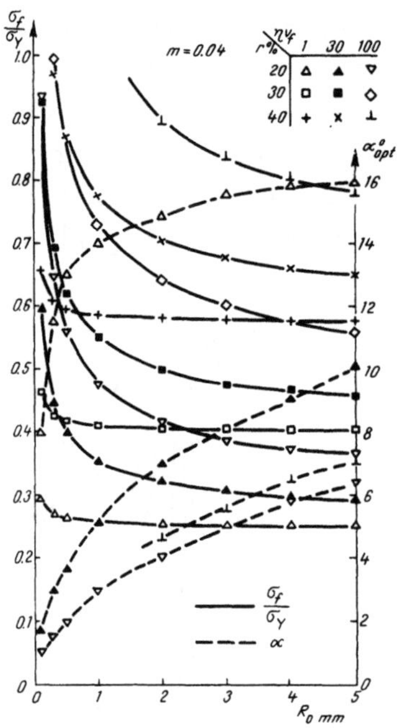

Fig.2.4 The variation of the relative drawing stress and optimum die angle with initial radius of the steel wire for three reductions in area and three speed factor values

given formulae. These formulae allow a full analysis of the drawing process, i.e. an exact estimation of the speed effect, the finding of the maximum possible reduction and also of the optimum die angle (which depends on mechanical properties of the wire, speed, wire diameter, friction etc.). For instance on fig.2.4 is given (see [16]) for steel wires the variation of the relative drawing stress (full lines) and of the optimum die angle (interrupted lines) with the initial radius of the wire, for three reductions in area and three values of the speed factor ηv_f which describes the speed effect. It is evident that the speed effect is remarkable mainly for thin wires and for higher reductions. Die angle is also highly dependent on: the drawing speed (decreases with increasing speed), the wire diameter (α increases with R_o) and friction (increases with friction).

T h e s m a l l p a r a m e t e r m e t h o d was recently applied to the wire drawing problem by Camenschi, Cristescu and Şandru [17]. The same method was first used for the strip drawing problem (Şandru and Camenschi [18]). Starting from the local balance laws, these are integrated using the perturbation series method with respect to a small parameter $B_g=(\sigma_Y R_f)/(\sqrt{3}\,\eta\, v_f)$. In the velocity field both components v_r and v_θ are considered. Both discontinuity surfaces Γ_1 and Γ_2 are now determined from the condition that the normal component of the velocity be continuous across these surfaces. It is found that both surfaces are close to concentric spheres. Since it was found that the Reynolds number $(\varrho v_f R_f)/\eta$ is very small, inertia terms in the equation of motion are neglected while from the dynamic jump conditions yields the continuity of the stress vector at the crossing of discontinuity surfaces. The drawing force, and possibly the back-pull force are determined from the resultant stress on the discontinuity surfaces. A friction law of the form (2.3) was used. The relative drawing stress is obtained as

$$\frac{\sigma_f}{\sigma_m} = \frac{\sigma_b}{\sigma_m} + \frac{2}{\sqrt{3}\,B_g}\left\{\left[1 - \left(\frac{R_f}{R_o}\right)^3\right] F(\alpha,m) + \right.$$

$$\left. + B_g \ln\left(\frac{R_o}{R_f}\right)^3 G(\alpha,m)\right\} \tag{2.6}$$

with the notations given in the Appendix. The first term in the bracket describes the viscous flow while the second one the contribution of plastic deformation. Formula (2.6) was compared with the formula (2.4) obtained with another method. An excellent agreement was found, the two formulae giving practically the same numerical values.

3. Tube drawing

T u b e s i n k i n g with fixed mandrel was the first problem considered with a viscoplastic constitutive equation (Cristescu [19][20]). It is a direct generalization of the bar drawing theory since the same centered velocity field (2.1) was assumed. The drawing stress and the maximum reduction are given as function of drawing speed, wall thickness and friction. The case when the mandrel is ultrasonically activated was also examined; the basic assumptions are the following (Cristescu and Suliciu [14] Ch.IX): the apparent decrease of the yield stress with the intensity of the ultrasonic field is described by an exponential function, the friction coefficient decreases with increasing intensity and the decrease of the viscosity coefficient in the Bingham model is given by a certain empirical formula. The results obtained are compared with experimental data (Cristescu and Drăgan [21]).

The free drawing (sinking) of tubes in the case when along the die the wall thickness remains constant was considered also for a Bingham model by Predeleanu [22]. Numerical examples are also given.

Tube d r a w i n g w i t h f l o a t i n g p l u g was considered for the geometry shown on fig. 3.1. Axial symmetry is assumed. In the viscoplastic domain II a kinematic admissible velocity field of the form (Osakada, Limb, Mellor [23])

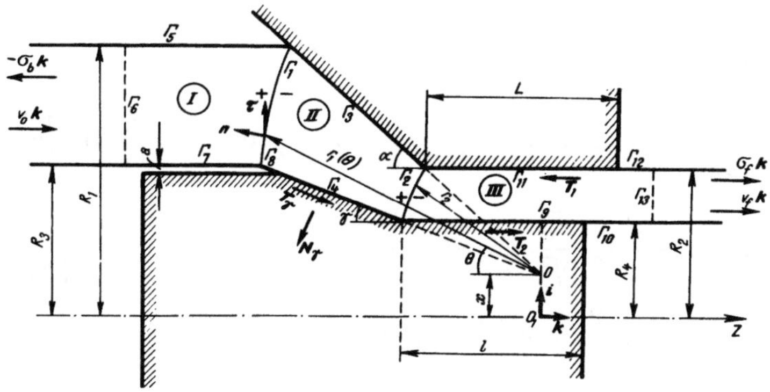

Fig.3.1 Geometry of tube drawing with floating plug

$$v_r = v_f \frac{r_2(r_2 \sin\theta + x)}{r(r \sin\theta + x)} \cos\theta \, , \quad \begin{cases} \theta \in (\gamma, \alpha) \text{ and} \\ r \in (r_2, r_1(\theta)) \end{cases} \tag{3.1}$$

$$v_\theta = v_\varphi = 0$$

is assumed. This implies for Γ_2 the equation r_2=const. while Γ_1 yields from $[\underset{\sim}{v}\cdot\underset{\sim}{n}] = 0$ as

$$r_1(\theta) = \frac{1}{\sin\theta}\left\{\sqrt{\left[(r_2\sin\theta + x)^2 - R_4^2\right]\frac{R_1^2 - R_3^2}{R_2^2 - R_4^2} + R_3^2} - x\right\} \tag{3.2}$$

The shape of the floating plug is defined by the angle γ and the land ℓ. These will be computed assuming that the floating plug is in an ultimate advanced position. Angle γ is obtained by minimizing the functional (1.4) on the subset of kinematic fields (3.1), where V is the domain I and II. Γ_T is the portion of the boundary Γ_5 to Γ_8 where the stress vector is prescribed while Γ_3, Γ_4 that portion where friction law of the form (2.3) is prescribed. The variational principle holds also if the friction law is (2.3), and the added term $\int_\Sigma \underset{\sim}{t}\cdot\underset{\sim}{v}\, d\sigma$ is computed on $\Sigma = \Gamma_1$. Thus the functional becomes

$$J(\underset{\sim}{v}) = 2\pi v_f \Big\{ \sigma_b\Big(\frac{R_2^2 - R_4^2}{2} - I_7\Big) +$$

$$+ \frac{2\sigma_m r_2}{\sqrt{6}} I_1(\gamma,\alpha;\ r_2, r_1(\theta)) + m_3 r_2 \cos\alpha\, R_2 \frac{\sigma_m}{\sqrt{3}} \ln\frac{r_1(\alpha)}{r_2} +$$

$$+ m_4 r_2 \cos\gamma\, R_4 \frac{\sigma_m}{\sqrt{3}} \ln\frac{r_1(\gamma) - \varepsilon}{r_2} + \tag{3.3}$$

$$+ v_f \Big[\sqrt{2}\,\eta\, m_3 r_2^2 \cos\alpha\, R_2\, I_4(\alpha;\ r_2, r_1(\alpha)) +$$

$$+ \sqrt{2}\,\eta\, m_4 r_2^2 \cos\gamma\, R_4\, I_4(\gamma;\ r_2, r_1(\gamma) - \varepsilon) +$$

$$+ \eta\, r_2^2\, I_2(\gamma,\alpha; r_2, r_1(\theta)) - \rho \frac{R_2^2 - R_4^2}{R_1^2 - R_3^2} v_f\, I_8\Big]\Big\}$$

where some notations are given in the Appendix.
In (3.3) the discontinuity surface Γ_2 was disregarded since this surface is part of Γ_v; we recall that on Γ_1^+ stresses are prescribed while on Γ_2^- the velocity. On the subset of kinematic fields (3.1), $J(\underset{\sim}{v})$ becomes a function of γ for which a minimum is computed. With this value of γ, ℓ and the drawing stress σ_f is further computed. ℓ is obtained writing a global equilibrium condition for all forces acting on the floating plug, projected on the symmetry axis. Thus for ℓ we get

$$\ell = \frac{\sqrt{3}\, m_4}{\sigma_Y m_2 R_4}\Big(\frac{\sin\gamma}{\mu} - \cos\gamma\Big) I_5(\gamma; r_2, r_1(\gamma) - \varepsilon) \tag{3.4}$$

where μ is a friction coefficient in a Coulomb type friction law, which is obtained from an experimentally obtained relationship between μ and m.
The drawing stress is computed from the principle of total power extended over the entire rigid-viscoplastic domains I, II and III:

$$\sigma_f = \frac{2\sigma_Y}{\sqrt{3}} \frac{m_2 \ell R_4 + m_1 L R_2}{R_2^2 - R_4^2} +$$

$$+ \left\{ \frac{R_2^2 - R_4^2}{2\left[r_2(\cos\gamma - \cos\alpha) + x(\alpha - \gamma)\right]} \left[r_2\left(-\cos\theta + \frac{\cos^3\theta}{3}\right)\Big|_\gamma^\alpha + \right.\right.$$

$$\left.\left. + \frac{x}{2}\left(\theta - \frac{\sin 2\theta}{2}\right)\Big|_\gamma^\alpha \right] + \frac{R_2^2 - R_4^2}{2} \right\}^{-1} \left\{ \frac{\min J(\underset{\sim}{v})}{2\pi v_f} + \right.$$

$$\left. + \eta v_f r_2^2 I_2(\gamma, \alpha;\ r_2, r_1(\theta)) \right\} . \qquad (3.5)$$

Formulae (3.3)-(3.5) fully describe the process: for any tube geometry, i.e. tube radius, wall thickness (R_3/R_1), diameter reduction (R_3/R_4) and reduction in area

$$r\% = 100 \left[1 - \frac{R_2^2 - R_4^2}{R_1^2 - R_3^2} \right]$$

and mechanical properties of the material (plastic and rate dependent) one can compute the drawing force for any drawing speed and also to find the most appropriate geometry for the tools (i.e. α, γ, ℓ) for that particular drawing process. For instance on figure 3.2 is plotted the variation of the relative drawing stress with the die angle for three values of the speed factor ηv_f and two sets of friction factors. By dotted or interrupted lines are shown the optimum plug angle for various regimes. Thus the drawing stress is increasing significantly with the drawing speed and with friction. A significant increase of the drawing stress with drawing speed was found experimentally by Bisk and Shveikin [24] . With increasing α the drawing stress decreases as well as the difference $\alpha - \gamma$. If γ is kept constant, using a variant of the formula (3.5) a minimum of the

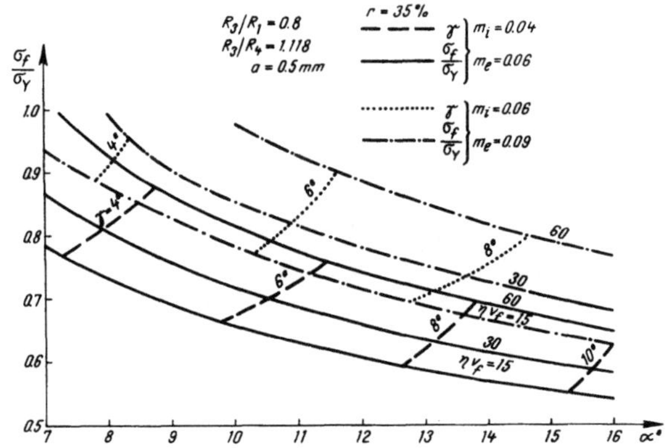

Fig.3.2 Variation of relative drawing stress and optimum plug angle with die angle for three values of the speed factor and two sets of friction factors

drawing stress is obtained when α is varied (see also Bramley and Smith [25]). On fig.3.3 is shown by full lines the optimum γ angle for various die angles, using three values for the speed factor and by dotted lines the maximal ℓ . Both γ and ℓ decrease with increasing speed and increase with increasing α . It was found that for higher speeds and/or small die angles ℓ may become zero. Similar conclusion was obtained for various reductions in area, diameter reductions and wall thicknesses. Limit processes when the floating plug cannot be maintained in equilibrium any longer, or the tube may break, are computed. Drawing stress is increasing very much when the wall thickness decreases, mainly for high reductions in area and high diameter reductions. Some conclusions of importance for industrial practices which yield from formulae (3.3)-(3.5) were discussed by Cristescu et al [26].

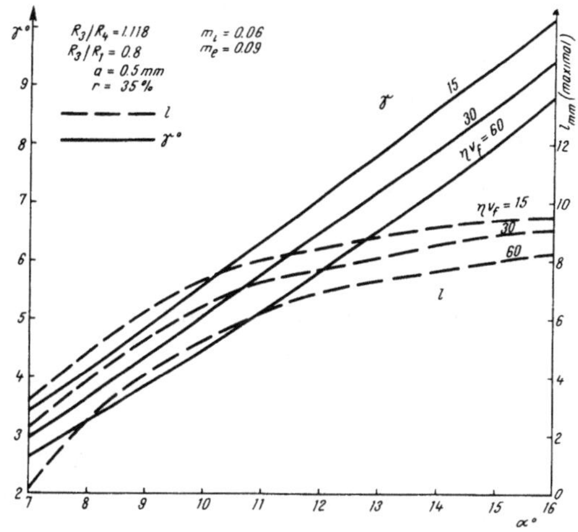

Fig.3.3 Optimum geometry of the plug for various die angles and three speed factor values

4. Conclusions

It was shown that several stationary metal working processes involving very high rates of strains, can be described theoretically using a viscoplastic constitutive equation. Thus it was possible to describe the influence of the working speed on all aspects of the working process. Since relatively simple final formulae were obtained for the drawing stress, one can derive quite easily the optimum working regimes (highest possible speed, smallest drawing force, maximum reduction in area per pass, minimum thickness of the wall of the tube, maximum diameter reduction etc.) but also one can find the optimum geometry of the tools which would lead to superior performances in the considered working process.

The Bingham model was used in the first stage since it is very simple. Improvement of the model is contemplated. Other problems or aspects of the already considered problems are further investigated. These include: a more careful analysis of the discontinuity surfaces bordering the viscoplastic region, the consideration of the inertial terms and of thermal effects, other friction laws, other aspects of the tube processing, extrusion and hydrostatic extrusion, extrusion of composites etc.

References

1 Hill, R.: The Mathematical Theory of Plasticity. Oxford: Clarendon Press 1950.

2 Avitzur, B.: Metal Forming. Processes and Analysis. New York: Mc Graw Hill 1968.

3 Polukhin, P.I.;Gun,G.I.;Galkin,A.M.: The resistence to plastic deformation of metals and alloys (in Russian). Moscow: Metallurgya 1976.

4 Lindholm, U.S.;Bessey, R.L.: A survey of rate dependent strength properties of metals. Report AFML-TR-69-119 1969

5 Cristescu, N.: On the propagation of elastic-plastic waves in metallic rods. Bul.Acad.Pol.Sci.XI,4 (1963) 129-133.

6 Cristescu,N.: Dynamic Plasticity. Amsterdam: North Holland Publ.Comp. 1967.

7 Teodosiu, C.;Soós, E.;Radu, G.: (private communications) 1978

8 Mandel, J.: Introduction a la mécanique des milieux continus déformables. Warsaw: PWN-Editions Scientifiques de Pologne 1974.

9 Cristescu, N.: Plastic flow through conical converging dies, using a viscoplastic constitutive equation. Int.J.Mech.Sci. 17 (1975) 425-433.

10 Cristescu, N.: Drawing Through Conical Dies - An Analysis Compared with Experiments. Int.J.Mech.Sci. 18 (1976) 45-49.

11 Wistreich, J.G.: Investigation of the Mechanics of Wire Drawing. Proc.Inst.Mech.Engrs. 169 (1955) 654-665.

12 Pugh, H.Ll.D.: Mechanical Behaviour of Materials under Pressure. vol.2. Amsterdam: Elsevier 1970.

13 Renne, I.P.: Approximate value (upper bound) of the stress in plastic flow through conical dies (in Russian). Kuznecino-stampovocinoe proizvodstva, 4, (1965) 7-9.

14 Cristescu, N.;Suliciu, I.: Vîscoplasticitate. Bucureşti: Ed.Tehnică 1976

15 Cristescu, N.: Speed influence in wire drawing. Rev. Roum.de Mécanique Appliquée, 22 (1977) 391-399.

16 Cristescu, N.: On the optimum die in fast wire drawing (to be published)

17 Camenschi, G,;Cristescu, N.;Şandru, N.: High speed wire drawing. Archives of Mechanics (to be published)

18 Sandru, N.;Camenschi, G.: Viscoplastic flow through inclined planes with application to strip drawing (to be published).

19 Cristescu, N.: A theory of tube sinking. Mech.Res. Communications, 2 (1975) 233-238

20 Cristescu, N.: Asupra tragerii ţevilor cu diverse viteze. Metalurgia XXVIII (1976) 535-544

21 Cristescu, N.;Drăgan, O.: Speed influence in cold tube drawing in ultrasonic field (to be published)

22 Predeleanu, M.: (private communications) 1978

23 Osakada, K.;Limb, M.;Mellor, P.B.: Hydrostatic extrusion of composite rods with hard cores. Int.J. mech.Sci. 15 (1973) 291-307

24 Bisk, M.B.;Shveikin, V.V.: Tube drawing with floating plug (in Russian). Moscow: Metallurgizdat 1963.

25 Bramley, A.N.;Smith, D.J.: Tube drawing with a floating plug. Metals Technology. (1976) 322-331

26 Cristescu, N.;Cleja, S.;Cuida, O.;Cristescu, C.: Optimizarea geometriei dopului flotant (to be published)

Appendix

Expressions used in various given formulae.

In (2.6):

$$F(\alpha,m) = \frac{\cos^3\frac{\alpha}{2}(2\gamma + \sqrt{3}\sin 2\alpha)}{\gamma\sin\frac{\alpha}{2}(1+2\cos\alpha) + \sqrt{3}\cos\alpha\cos\frac{\alpha}{2}} ,$$

$$G(\alpha,m) = \frac{\frac{m}{3}\cos\alpha\sin^2\alpha - \sin\alpha\left(\gamma\sin\alpha - \frac{2}{\sqrt{3}}\cos\alpha\right) I_1}{(1-\cos\alpha)\left[\sin 2\alpha - \frac{\gamma}{\sqrt{3}}(1-\cos\alpha)(1+2\cos\alpha)\right]} -$$

$$- \frac{\dfrac{\gamma\left[\sqrt{3}\,\sin 2\alpha + 6\gamma\cos 2\alpha - 4\gamma\right] I_2 + 4\gamma^2 I_3}{\sqrt{3}\,\sin 2\alpha + 6\gamma\sin^2\alpha}}{(1-\cos\alpha)\left[\sin 2\alpha - \dfrac{\gamma}{\sqrt{3}}(1-\cos\alpha)(1+2\cos\alpha)\right]}$$

where

$$I_k = \int_0^{\alpha} \frac{\sin^{2k-1}\theta \; d\theta}{\sqrt{1 - \sin^2\theta\,(\lambda_1 - \lambda_2\sin^2\theta)}}\;, \quad k = 1,2,3 \;.$$

In §3:

$$I_1(\gamma,\alpha;r_2,r_1(\theta)) = \int_{\gamma}^{\alpha}\int_{r_2}^{r_1(\theta)} \frac{\sqrt{A\,r^2 + Br + C}}{r(r\sin\theta + x)}\,dr\,d\theta\;,$$

$$I_2(\gamma,\alpha;r_2,r_1(\theta)) = \int_{\gamma}^{\alpha}\int_{r_2}^{r_1(\;)} \frac{A\,r^2 + Br + C}{r^3(r\sin\theta + x)^3}\,dr\,d\theta\;,$$

$$I_4(\alpha;r_2,r_1(\alpha)) = \int_{r_2}^{r_1(\alpha)} \left.\frac{\sqrt{Ar^2 + Br + C}}{r^3(r\sin\theta + x)^2}\right|_{\theta=\alpha} dr\;,$$

$$I_5(\gamma;r_2,r_1(\gamma)-\varepsilon) = \int_{r_2}^{r_1(\gamma)-\varepsilon} \left.\sqrt{II_{T'}}\right|_{\theta=\gamma} (r\sin\gamma + x)\,dr\;,$$

$$I_7 v_f = \int_{\gamma}^{\alpha} \frac{r_1\sin\theta - r_1'\cos\theta}{\sqrt{(r_1')^2 + r_1^2}}\Big[v_o(r_1\sin\theta - r_1'\cos\theta) +$$

$$+ v_r r_1'\Big](r_1\sin\theta + x)\,d\theta\;,$$

$$I_8 v_f^2 = \int_{\gamma}^{\alpha} \frac{v_r r_1'(r_1' \sin\theta + r_1 \cos\theta)}{(r_1')^2 + r_1^2} \Big[v_o(r_1 \sin\theta -$$

$$- r_1' \cos\theta) + v_r r_1' \Big] (r_1 \sin\theta + x) \, d\theta \, .$$

Analysis of Large Deformation and Stress in Metal Forming Processes by the Finite Element Method

Y. YAMADA, A. S. WIFI AND T. HIRAKAWA

Institute of Industrial Science
University of Tokyo, Japan

Summary

In the analysis of metal forming processes featuring large deformations that may be accompanied by a considerable spin of the principal axes as well as by the coupling between tractions and velocity components in the specification of boundary conditions, a careful formulation of the procedure to be chosen for elastic-plastic analysis is crucial. Paper adopts the updated formulation and summarizes first the elastic-plastic constitutive relations incorporating the effect of the change of geometric configuration on the stress equilibrium. This leads to the geometric stiffness matrix, and the similar effect of the change of configuration on the equilibrium of the surface traction rates yields the load correction matrix which will be discussed in some detail for the case of hydrostatic pressure loading. Verificative examples of the formulation include the elastic-plastic deformation of a thick cylinder under internal pressure, the upsetting of a cylindrical block, and the stretch forming of a metal cup. These examples have revealed the importance of the choice of the step size in incremental computations, and of the correction for other potential sources of error inherent, especially to the analysis of large deformations in metal forming processes.

Introduction

The introduction of the finite element method to the plasticity problem dates back to late 1960's when the elastic-plastic constitutive equation was incorporated ingeniously in the standard solution routine that had been enjoying the success in the solution of the elasticity problem. At the earlier stage the application was confined to the contained elastic-plastic deformation where the plastic strain is of the order of 0.1%, i.e. the strain at the elastic limit of metals. The extension to the large strain regime had been a dream, but several breakthroughs were accomplished in quite recent years. Thus it has become possible to attain the steady state of metal flow by the computer simulation at the termination of transient or unsteady elastic-plastic flow, through successive deformation stages.

Rational formulations and solution procedures with proper error corrections have been mandatory for the realization of the dream or the attainment of correct solutions at the end of hundreds of incremental computation steps. The stress rate should be chosen properly in the constitutive relation, the geometric stiffness and load correction matrices which may have nonsymmetric form be handled adequately in the solution routine, and furthermore possible sources of numerical error be carefully eliminated. This paper summarizes the efforts the authors exerted to the development of computational techniques, and experiences they met with in the application of the techniques to the area of metal forming plasticity.

Constitutive Equation for Elastic-Plastic Deformation

The basic theory for problems with material and geometric nonlinearities has been discussed at length in [1, 2], and only a brief outline is given hereunder. We adopt the updated Lagrangian formulation which describes the basic equations taking the current state as the reference configuration for the incipient behavior from it.

The actual mode of incremental deformation is governed by the principle of virtual work rate which is conveniently written as the following in terms of the nominal stress rate $\dot{s}_{ij}$ connected with the virtual velocity gradient $\delta v_{j,i}$ and the nominal traction rate $\dot{F}_{oi}$ defined per unit area of loading surface S_F in the reference state

$$\int_V \dot{s}_{ij}\delta v_{j,i}dV = \int_{S_F} \dot{F}_{oi}\delta v_i dS \tag{1}$$

As detailed in [1], this can be transformed to

$$\int_V (\delta\dot{\varepsilon}_{ij}D_{ijk\ell}\dot{\varepsilon}_{k\ell}+\delta v_{k,i}\sigma_{ij}v_{k,j} - 2\delta\dot{\varepsilon}_{ik}\sigma_{ij}\dot{\varepsilon}_{jk})dV = \int_{S_F} \dot{F}_{oi}\delta v_i dS \tag{2}$$

where $\dot{\varepsilon}_{ij}$ denotes the Euler strain rate

$$2\dot{\varepsilon}_{ij} = v_{i,j} + v_{j,i} \tag{3}$$

and we have chosen the Jaumann rate of Kirchhoff stress $\overset{*}{t}_{ij}$ in the constitutive relation as

$$\overset{*}{t}_{ij} = D_{ijk\ell}\dot{\varepsilon}_{k\ell} \tag{4}$$

In the present study no discrimination is made between $\overset{*}{t}_{ij}$ and the Jaumann rate $\overset{*}{\sigma}_{ij}$ of Euler stress, since they differ only by terms of the order of volume change which has vanishingly small effect on the solution of large strain problems of metal plasticity.

Derivation of the elastic-plastic modulus $D_{ijk\ell}$ in Eq.(4) has been given in [1] and it can be easily shown that $D_{ijk\ell}$ assumes for the isotropic Mises material the form

$$D_{ijk\ell} = 2G(\delta_{ik}\delta_{j\ell} + \frac{\nu}{1-2\nu}\delta_{ij}\delta_{k\ell} - \frac{\alpha}{\beta}\sigma'_{ij}\sigma'_{k\ell}) \tag{5}$$

where

$$\beta = \frac{9G}{2(H'+3G)\bar{\sigma}^2}$$

and $\alpha = 0$ for elastic loading and unloading

$\quad\;\; = 1$ for continuation of plastic state

$\bar{\sigma}$ is the effective stress, G the shear modulus, ν Poisson's ratio, H' the plastic strain-hardening rate, and σ'_{ij} represents the deviatoric part of the Euler stress σ_{ij}.

In the virtual work equation (2), the first term in the volume integral contributes to the conventional stiffness matrix [k] in the finite element formulation and remaining terms are the sources of the geometric stiffness matrix $[k_G]$.

Load Correction Matrices

Corrections concerned in this section are applied to the right hand side of Eq.(1) or (2) under the circumstances that the traction rate is specified in terms of the true rate $\dot{F}_i$ rather than the nominal rate $\dot{F}_{oi}$. An obvious example requiring such corrections is the hydrostatic loading. It acts naturally in the direction normal to the surface which is in turn changing its area under loading. The contact surface or the work-tool interface in material processing is another example which needs corrections. In this case there may be no 'prescribed' traction, but its rate is to be determined from the solution for the nodal displacement rate as the reaction which is coupled with the velocity field along the interface.

In order to derive the specific form of corrections, we note the following relation between F_{oi} and F_i

$$F_{oi} dS_o = F_i dS \tag{6}$$

dS_o and dS are the areas before and after deformation. Differentiation of Eq.(6) yields

$$\dot{F}_{oi} dS_o = \dot{F}_i dS + F_i d\dot{S} \tag{7}$$

The surface traction F_i is connected with the internal stress field σ_{ij} or σ_{ji} by

$$F_i = \sigma_{ji} n_j \tag{8}$$

Thus we have by differentiation

$$\dot{F}_i = \dot{\sigma}_{ji} n_j + \sigma_{ji} \dot{n}_j \tag{9}$$

To have a convenient expression for $d\dot{S}$ appeared in Eq.(7) we use the following relation between the area dS and its projection dS_j

$$dS_j = n_j dS \tag{10}$$

where n_j represents the direction cosine of the normal to the deformed surface. Differentiating Eq.(10) and rearranging it yields

$$n_j d\dot{S} = d\dot{S}_j - \dot{n}_j dS \tag{11}$$

Introduction of Eqs.(8), (9) into Eq.(7) furnishes, in view of Eq.(11)

$$\begin{aligned} \dot{F}_{oi} dS_o &= (\dot{\sigma}_{ji} n_j + \sigma_{ji} \dot{n}_j) dS + \sigma_{ji} (d\dot{S}_j - \dot{n}_j dS) \\ &= \dot{\sigma}_{ji} n_j dS + \sigma_{ji} d\dot{S}_j \end{aligned} \tag{12}$$

As detailed in [2, 3], the following relation holds for the rate $d\dot{S}_j$ of dS_j in the updated formulation

$$d\dot{S}_j = \dot{\varepsilon}_{mm} n_j dS - v_{k,j} n_k dS \tag{13}$$

where $\dot{\varepsilon}_{mm}$ represents the rate of volumetric strain or minus of the changing rate $\dot{\rho}/\rho$ of density. Thus we have from Eqs.(12) and (13)

$$\dot{F}_{oi} dS_o = \dot{\sigma}_{ji} n_j dS + (\sigma_{ji} \dot{\varepsilon}_{mm} n_j - \sigma_{ji} v_{k,j} n_k) dS \tag{14}$$

Expressions with an explicit inclusion of the Jaumann rate of Kirchhoff stress $\overset{*}{t}$ or the similar rate of Euler stress $\overset{*}{\sigma}$ are obtainable by the use of following conversions

$$\dot{\sigma}_{ji} = \overset{*}{t}_{ji} + \sigma_{jk} \dot{\omega}_{ki} + \sigma_{ik} \dot{\omega}_{kj} - \sigma_{ji} \dot{\varepsilon}_{mm} \tag{15}$$

$$\overset{*}{\sigma}_{ji} = \overset{*}{t}_{ji} - \sigma_{ji} \dot{\varepsilon}_{mm} \tag{16}$$

where the rotation is defined by

$$2\dot{\omega}_{ki} = v_{i,k} - v_{k,i} \tag{17}$$

It should be noted that the anti-clockwise rotation is taken positive following the definition adopted in [1-3]. Introducing Eqs.(15) and (16) to Eq.(14) leads to

$$\dot{F}_{oi} dS_o = \overset{*}{t}_{ji} n_j dS + (\sigma_{jk} \dot{\omega}_{ki} n_j + \sigma_{ik} \dot{\omega}_{kj} n_j - \sigma_{ji} v_{k,j} n_k) dS \tag{18}$$

$$\dot{F}_{oi} dS_o = \overset{*}{\sigma}_{ji} n_j dS + (\sigma_{jk} \dot{\omega}_{ki} n_j + \sigma_{ik} \dot{\omega}_{kj} n_j + \sigma_{ji} \dot{\varepsilon}_{mm} n_j - \sigma_{ji} v_{k,j} n_k) dS \tag{19}$$

Second terms in parentheses on the right hand side of Eqs.(14), (18) and (19) are the sources of load correction matrices. These matrices have been effectively incorporated in the analysis of steady metal flow in extrusion [3]; the problem against which a breakthrough was done first by Lee, Mallett and Yang [4].

Considering the future application of the formulation to

the analysis of the hydraulic bulging of sheet metals, we derive the specific form of the load correction matrix for the case of hydrostatic loading in the following. For that purpose, we first note that the right hand side of Eq.(1) or (2) is written for the hydrostatic tensile loading p(>0), in view of Eq.(14) as

$$\int_{S_F} \dot{F}_{oi}\delta v_i dS = \int_{S_F} \dot{p}\delta_{ji} n_j \delta v_i dS + \int_{S_F} p(\delta_{ji}\dot{\varepsilon}_{mm} n_j - \delta_{ji} v_{k,j} n_k)\delta v_i dS \tag{20}$$

where δ_{ji} denotes the Kronecker delta.

The second surface integral in the right hand side of Eq.(20) which is associated with the load correction matrix can be written in matrix form as

$$\int \lfloor \delta v \rfloor [p^c]\{\dot{e}\} dS \tag{21}$$

where

$$[p^c] = p\left[\begin{array}{ccc|ccc|ccc} 0 & n_2 & n_3 & 0 & -n_1 & 0 & 0 & 0 & -n_1 \\ -n_2 & 0 & 0 & n_1 & 0 & n_3 & 0 & 0 & -n_2 \\ -n_3 & 0 & 0 & 0 & -n_3 & 0 & n_1 & n_2 & 0 \end{array}\right] \tag{22}$$

and

$$\{\dot{e}\}^T = \lfloor \dot{e} \rfloor = \lfloor \dot{e}_{11} \;\; \dot{e}_{12} \;\; \dot{e}_{13} \mid \dot{e}_{21} \;\; \dot{e}_{22} \;\; \dot{e}_{23} \mid \dot{e}_{31} \;\; \dot{e}_{32} \;\; \dot{e}_{33} \rfloor \tag{23}$$

$$\dot{e}_{ji} = \partial v_i/\partial x_j \tag{24}$$

For the axisymmetric case shown in Fig. 1 where the directions 1, 2, 3 stand for r, θ, z coordinates, $\delta v_2=0$ and $n_2=0$, and thus Eqs.(21) and (22) assume the form

$$\int \lfloor \delta v_r \;\; \delta v_z \rfloor [p^c]\{\dot{e}\} dS \tag{25}$$

$$\{\dot{e}\}^T = \lfloor \dot{e} \rfloor = \lfloor \partial v_r/\partial r \;\; \partial v_z/\partial z \;\; \partial v_r/\partial z \;\; \partial v_z/\partial r \;\; v_r/r \rfloor \tag{26}$$

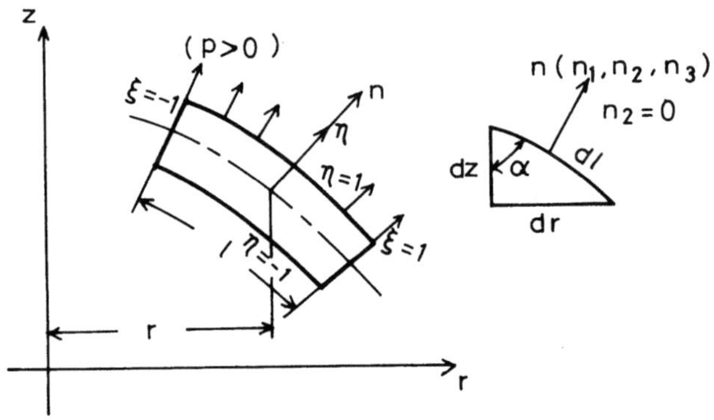

Fig. 1 Axisymmetric sheet element under hydrostatic surface loading

with

$$[p^c] = p\begin{bmatrix} 0 & -n_1 & 0 & n_3 & -n_1 \\ -n_3 & 0 & n_1 & 0 & -n_3 \end{bmatrix} \tag{27}$$

Now assume parametric representation of geometric and field quantities by ξ,η coordinates and let the velocity field $\{v(\xi,\eta)\}$ be related to the nodal velocity vector $\{\dot{u}\}$ by

$$\begin{aligned} v_r(\xi,\eta) &= N_1(\xi,\eta)\dot{u}_{r1} + \cdots + N_i(\xi,\eta)\dot{u}_{ri} + \cdots + N_n(\xi,\eta)\dot{u}_{rn} \\ v_z(\xi,\eta) &= N_1(\xi,\eta)\dot{u}_{z1} + \cdots + N_i(\xi,\eta)\dot{u}_{zi} + \cdots + N_n(\xi,\eta)\dot{u}_{zn} \end{aligned} \tag{28}$$

which can be written compactly as

$$\begin{Bmatrix} v_r(\xi,\eta) \\ v_z(\xi,\eta) \end{Bmatrix} = [N(\xi,\eta)]\begin{Bmatrix} \dot{u}_r \\ \dot{u}_z \end{Bmatrix} \tag{29}$$

Identical interpolation functions $N_1(\xi,\eta)$ are assumed for the virtual velocity field $\{\delta v(\xi,\eta)\}$.

An expression of $\{\dot{e}\}$ for the axisymmetric case is obtainable by introducing Eq.(29) to Eq.(26). We write the resulting equation in the form

$$\{\dot{e}(\xi,\eta)\} = [B_e(\xi,\eta)]\begin{Bmatrix} \dot{u}_r \\ \dot{u}_z \end{Bmatrix} \tag{30}$$

It can be shown, by an introduction of Eqs.(27) and (30) to Eq.(25) and subsequent straightforward matrix multiplication, that the generic component in the load correction matrix $[k_C]$ is written as

$$[k_{Cji}] = \int_{S_F} \begin{bmatrix} N_j n_1 \frac{N_i}{r} & N_j (n_1 \frac{\partial N_i}{\partial z} - n_3 \frac{\partial N_i}{\partial r}) \\ N_j (n_3 \frac{\partial N_i}{\partial r} - n_1 \frac{\partial N_i}{\partial z} + n_3 \frac{N_i}{r}) & 0 \end{bmatrix} p dS \tag{31}$$

Integration extends over the surface S_F where the hydrostatic loading p is applied. It should be noted that the reversal of sign in the derivation of Eq.(31) is due to the shift of the term for load correction to the left hand side so that the element stiffness equation is described as follows

$$([k]+[k_G]+[k_C])\{\dot{u}\} = \{\dot{F}\} \tag{32}$$

where [k] and $[k_G]$ are the conventional and geometric stiffness matrix respectively.

A convenient expression for $[k_{Cji}]$ of Eq.(31), generic component of the load correction matrix $[k_C]$, is obtainable for the isoparametric element shown in Fig.1. To this end, we note that the surface area dS of the axisymmetric element is

$$dS = 2\pi r d\ell \tag{33}$$

$d\ell$ is the meridional length of the element and we have

$$\begin{aligned} n_1 d\ell &= -dz = -(\partial z/\partial \xi)d\xi \\ n_3 d\ell &= dr = (\partial r/d\xi)d\xi \end{aligned} \tag{34}$$

Substituting Eqs.(33) and (34), we have

$$[k_{Cji}] = \int_{S_F} \begin{bmatrix} -N_j \frac{\partial z}{\partial \xi} \frac{N_i}{r} & -N_j (\frac{\partial z}{\partial \xi} \frac{\partial N_i}{\partial z} + \frac{\partial r}{\partial \xi} \frac{\partial N_i}{\partial r}) \\ N_j (\frac{\partial r}{\partial \xi} \frac{\partial N_i}{\partial r} + \frac{\partial z}{\partial \xi} \frac{\partial N_i}{\partial z} + \frac{\partial r}{\partial \xi} \frac{N_i}{r}) & 0 \end{bmatrix} \cdot 2\pi r p d\xi \tag{35}$$

$\partial N_i/\partial r$ and $\partial N_i/\partial z$ are determined by the following well-known formula

$$\begin{Bmatrix} \dfrac{\partial N_i}{\partial r} \\ \\ \dfrac{\partial N_i}{\partial z} \end{Bmatrix} = [J]^{-1} \begin{Bmatrix} \dfrac{\partial N_i}{\partial \xi} \\ \\ \dfrac{\partial N_i}{\partial \eta} \end{Bmatrix}, \quad [J] = \begin{bmatrix} \dfrac{\partial r}{\partial \xi} & \dfrac{\partial z}{\partial \xi} \\ \\ \dfrac{\partial r}{\partial \eta} & \dfrac{\partial z}{\partial \eta} \end{bmatrix} \qquad (36)$$

A Generalized Self Correcting Approach

Considering that several hundreds of computation stages or increments are usually required to get a final solution for large deformation problems, some comments are in place on the sources of 'lack of correctness' and furthermore the remedies we have tested to eliminate the accumulation of errors.

For the elastic-plastic incremental solution, it is often observed that the response of finite elements is too stiff, particularly for low hardening materials when the computation approaches to the fully plastic state where the incompressible plastic straining dominates.

The over-stiff behavior of the element is dependent on the element type and the integration scheme of the element stiffness, specifically the number of Gaussian points in numerical integration [5-7]. According to the result of our numerical experiments, the nine-nodes complete Lagrangian quadrilateral element with 2×2 Gaussian integration points gives the best result. However, the eight-nodes 'serendipity' quadrilateral element is also found effective, provided 2×2 integration scheme is used and the shape of the element is not an irregular one. Quadrilateral elements composed of four triangles have been shown to be useful in plane strain problems[5]. Unfortunately this element is not suitable to the problem where the bending effect is one of the factors of primary importance as in the sheet metal forming.

A correction for the departure of the stress state from the exact yield condition and/or from the actual effective stress-strain curve is also important in the elastic-plastic analysis.

In the present study, a modification to the so called r_{min} technique [8] is adopted to compensate the departure of the computed stress state from the yield condition.

Let $\{\Delta\sigma\}$ be the computed stress increment in a generic elastic element, and $r\{\Delta\sigma\}$ be the increment just enough to bring the element to yielding. Then the jump in stress increment due to the violation of yield condition is $\{\sigma_r\}=(1-r)\{\Delta\sigma\}$ for the element in the transition region from the elastic to plastic state. The apparent load vector $\{f_2\}$ is calculated from $\{\sigma_r\}$ by

$$\{f_2\} = \int_V [B_\varepsilon]^T \{\sigma_r\} dV \tag{37}$$

where $[B_\varepsilon]$ is the matrix relating the strain rate $\{\dot{\varepsilon}\}$ to the nodal velocity vector $\{\dot{u}\}$. $\{f_2\}$ is treated similarly to the load vector $\{f_1\}$ which is used in the standard finite element analysis routine for the check of nodal force equilibrium

$$\{f_1\} = \{F\} - \int_V [B_\varepsilon]^T \{\sigma\} dV \tag{38}$$

In Eq.(38), $\{F\}$ is the prescribed nodal force and $\{\sigma\}$ the current element stress. $\{f_1\}$ and $\{f_2\}$ are added to the prescribed nodal load rate $\{\dot{F}\}$ to form the corrected load rate vector $\{\dot{F}\}$corr. as follows

$$\{\dot{F}\}\text{corr.} = \{\dot{F}\} + \{f_1\} + \{f_2\} \tag{39}$$

The effect of the correction $\{f_2\}$ is to bring or raise the stress of the element in transition to the level corresponding to $\{\sigma+\Delta\sigma\}$ rather than $\{\sigma+r\Delta\sigma\}$ after each step of computation.

The improvement by the corrections can be seen in Fig.2 which compares the elastic-plastic solutions for a thick cylinder under internal pressure. The material is assumed almost incompressible (Poisson's ratio ν=0.49999) and non-hardening (H'=0). The prescribed pressure value of p=7.2 kgf/mm^2 (70.6 MPa) corresponds to the deformation stage where the plastic region expands to half the thickness of the cylinder (c/a=0.5).

Solutions for the radial displacement u_a of inner the wall and the effective stress $\bar{\sigma}$ at point A justify the correction by $\{f_2\}$,

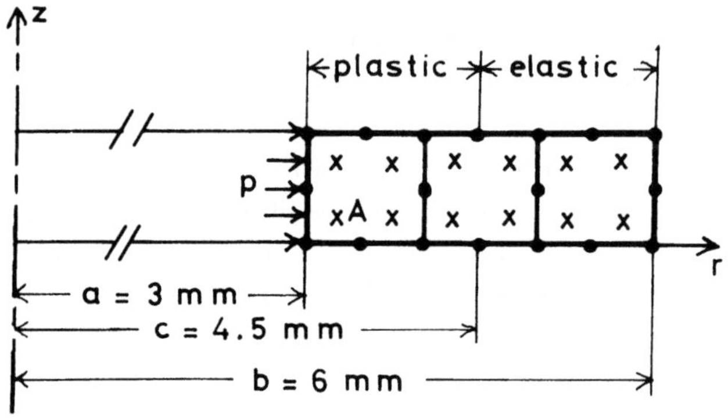

$E = 3 \times 10^4$, $\sigma_y = 10$ in kgf/mm^2

$H' = 0.0$, $p = 7.2$ in kgf/mm^2

$\upsilon = 0.49999$

$c/a = 1.5$ ($p=7.2078$ kgf/mm^2 , theoretically for $c/a = 1.5$)

1 kgf/mm^2 = 9.80 N/mm^2 (MPa)

	(a)	(b)	(c)	exact
u_a $\times 10^3$mm	1.88450	1.89310	1.92060	1.94855
$\bar{\sigma}$ at A kgf/mm^2	10.0968	10.0847	10.0129	10.0

(a) simple incremental solution

(b) with correction $\{f_1\}$ for equilibrium

(c) with correction $\{f_1\} + \{f_2\}$ for equilibrium and deviation from yield condition

Fig.2 Example of error assessment in elastic-plastic solution ; thick cylinder under internal pressure

along with the conventional one $\{f_1\}$. Furthermore, it should be emphasized that the Gaussian integration of order 2×2 is applied to the eight nodes rectangular element adopted in this verificative example.

Numerical Examples

The first example is the upsetting of a cylindrical block shown in Fig.3. Sticking, i.e. no slip condition, is assumed along the tool-work interface. Due to the symmetry of the problem, a quarter part of the block is analysed by using the mesh division depicted in the figure. It is worth mentioning that this is the bench mark test problem being promoted by the Joint Examination Program of the Validity of Various Numerical Methods for the Analysis of Metal Forming Processes and outcomes from the inter-

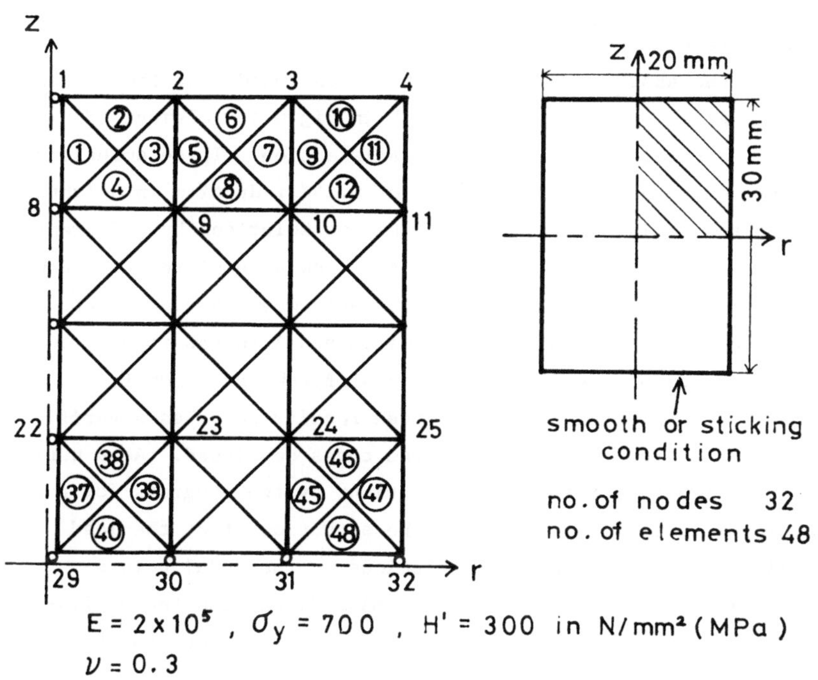

Fig. 3 Upsetting of a cylindrical block

national cooperation. The Program will be reported on by Kudo in this Proceedings. Two nonlinear analysis programs MAGNAP and ELPLASA being developed parallelly in the author's laboratory participated in the Joint Examination Program.

Fig.4 illustrates the computed results for the upsetting load as a function of the reduction in height of the block. It is observed that the solution is quite sensitive to the increment size chosen in the computation. In the present example the increment of 0.50% per each reduction step leads to an erroneous overestimate of the upsetting load, while the increment of 0.25% is considered to be an appropriate choice. This is confirmed by making an inspection of the effective stress-plastic strain relation in element 40 of Fig.3. As can be seen from Fig.5, the computed stress values deviate considerably from the input material data when the increment size is 0.50%, in constrast to an almost satisfactory correlation with the input data in case of the increment of 0.25%.

From the above observations, it is concluded that the stress solutions given by the curves C and D in Fig.6 for element 48 of Fig.3 are accurate to the author's satisfaction. These solutions are obtained by using the increment of 0.25% reduction in height per step and it is remarkable that the two solutions by different programs, C by ELPLASA and D by MAGNAP, show a complete coincidence. The solutions are deteriorated considerably when the increment size is doubled as illustrated by the curves A in the figure, but the coincidence of predictions by two different programs is heldover. It is considered that these results serve as a verification of the programs as well as the codes developed. As a note, a slight difference between the solutions by two programs for the upsetting load in Fig.4 is due to the situation that the load is estimated from the computed reactions at the nodal points along tool-work interface in case of MAGNAP, whereas the estimation is based on the stresses in the elements adjacent to the interface in case of ELPLASA.

The discrepancies of the upsetting load due to the increment

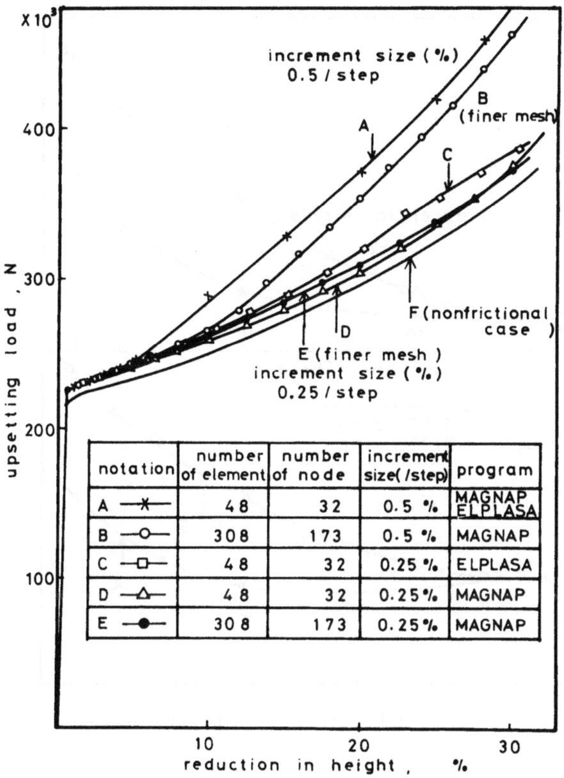

Fig. 4 Comparison of computed upsetting load

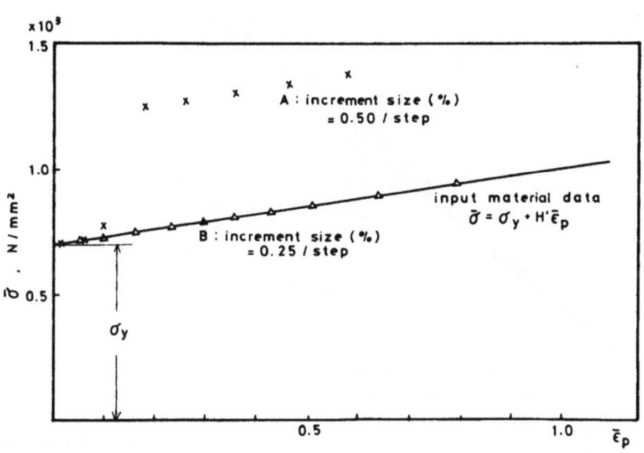

Fig. 5 Effective stress ($\bar{\sigma}$) - plastic strain ($\bar{\varepsilon}^p$) relation in element 40 of Fig. 3

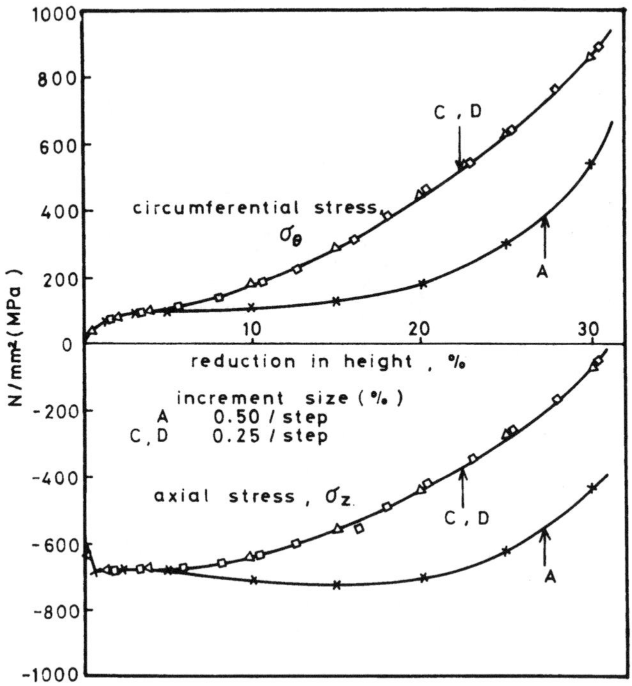

Fig. 6 Comparison of computed stress in element 48 of Fig. 3

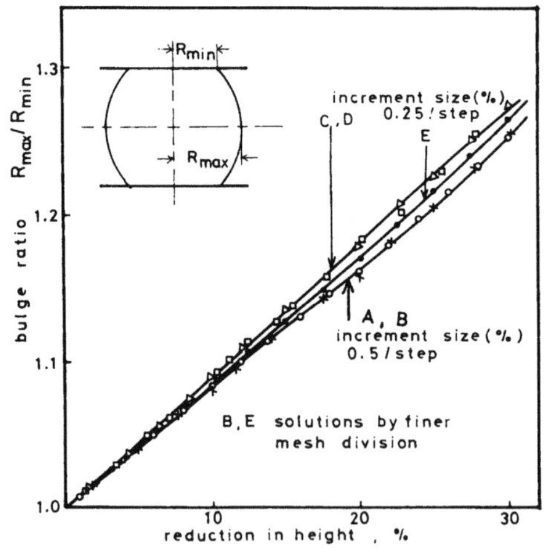

Fig. 7 Bulge ratio at different stages of reduction in height

size are reflected in the bulge ratio curves in Fig.7 which show an erroneous underestimate for the amount of bulging when the step size is not small enough. In comparison to the remarkable effect of the step size on the solutions, advantages attainable by means of the refinement of mesh size are not significant as can be deduced from the comparison of curves B and E obtained by the use of finer mesh with corresponding curves A and C, D in Figs.4 and 7.

The second numerical example in the present paper is the stretch forming of a cylindrical cup shown in Fig.8. The finite element solution of this problem has been given by Wifi [9]. The total Lagrangian formulation was adopted in that solution and use was made of the six nodes linear strain triangular element. The identical problem is studied in the present paper by applying the updated Lagrangian formulation which incorporates the eight node isoparametric quadrilateral element.

The deformed shape and the incident of necking being predicted from the solutions of two formulations are compared in Fig.8. Although the method of fixation of the flange part of the blank is slightly different, the deformed shape and the extent of contact zone of the sheet with the hemispherical headed punch are almost identical. A distinct necking formation is observed at the stage of almost the same punch travel U_p=30 mm where a sudden decrease is marked in the punch load.

Concluding Remarks

Present paper demonstrates the versatility and usefulness of the elastic-plastic analysis based on the updated Lagrangian formulation. By careful considerations on the pertinent factors due to geometric nonlinearities, the formulation enables the accurate computation of the deformation and stress as well as the prediction of the defect in large deformation-large strain problems such as those associated with the metal forming plasticity. A number of computer programs for the nonlinear analysis have been under active development and some of them are in the public domain.

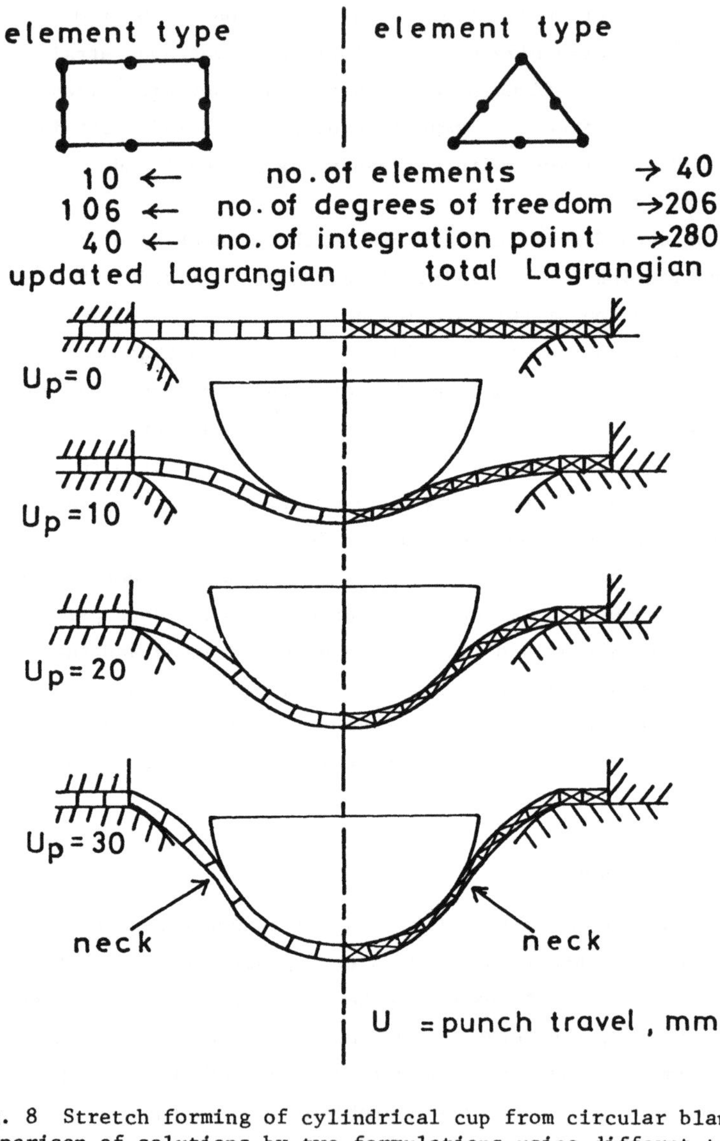

Fig. 8 Stretch forming of cylindrical cup from circular blank; comparison of solutions by two formulations using differnt type of finite element

Although the examples shown in this paper deal merely with the axisymmetric deformation, it is presumed that the extension of the work to more general three dimensional cases is not an intractable task in view of the advent of larger and faster hardware as well as more efficient routines for the solution of large system of equations. On the finite element side, ingenious incorporation of singularities in the nonlinear analysis routines and also, elaboration of the attributes of elements themselves are quite essential.

Acknowledgement

A part of this work was carried out under the sponsorship by the Grant in Aid for Scientific Research, Ministry of Education, Japan. This support is gratefully acknowleged, and thanks go further to Miss Fumie Terasaka for typing the manuscript.

References

1. Yamada, Y., Hirakawa, T. and Wifi, A.S., Analysis of large deformation and bifurcation in plasticity problems by the finite element method, Finite Element in Nonlinear Mechanics, Tapir Publishers, Trondheim, Norway, 1978.
2. Yamada, Y., Constitutive modelling of inelastic behavior and numerical solution of nonlinear problems by the finite element method, Computers and Structures, vol.8 (1978), 533-543.
3. Yamada, Y. and Hirakawa, T., Large deformation and instability analysis in metal forming processes, a bound volume of ASME, Applications of Numerical Methods to Forming Processes, Winter Annual Meeting, San Francisco, December, 1978.
4. Lee, E.H., Mallett, R.L. and Yang, W.H., Stress and deformation analysis of the metal extrusion process, Computer Methods in Applied Mechanics and Engineering, vol.10 (1977), 339-353.
5. Zienkiewicz, O.C. and Hinton, E., Reduced integration, function smoothing and non-conformity in finite element analysis (with special reference to thick plates), J. of the Franklin Institute, vol.302 (1976), 443-460.

6. Zienkiewicz, O.C., The Finite Element Method, Third Edition, McGraw-Hill (UK), 1977, chap.11.

7. Nagtegaal, J.C., Parks, D.M. and Rice, J.R., On numerically accurate finite element solutions in the fully plastic range, Computer Methods in Applied Mechanics and Engineering, vol.4 (1974), 153-177.

8. Yamada, Y., Yoshimura, N. and Sakurai, T., Plastic stress-strain matrix and its application for the solution of elastic-plastic problems by the finite element method, Int. J. of Mechanical Sciences, vol.10 (1968), 343-354.

9. Wifi, A.S., An incremental complete solution of stretch forming and deep drawing of a circular blank using hemispherical punch, Int. J. of Mechanical Sciences, vol.18 (1976), 23-31.

Stress and Deformation Analysis of Metal-Forming Processes

E. H. LEE, R. L. MALLETT
Stanford University

R. M. McMEEKING
University of Illinois-Urbana

Summary

A complete stress-analysis of a metal-forming process is necessary in order to be able to predict the onset of metal-forming defects such as the generation of internal or surface cracks and of high residual stresses. Even though plastic strains dominate elastic strains, it is necessary to carry out an elastic-plastic analysis in order to evaluate stresses throughout the material since rigid-plastic theory can only predict stresses in the regions exhibiting significant current plastic flow. A computer program to evaluate complete stress distributions has been developed and applied to the extrusion process. Examples of stress and deformation distributions are presented for plane strain extrusion of a slab through rigid frictionless curved dies. The program has also been generalized to include axi-symmetric extrusion and an example is given. The stress fields obtained exhibit features which are consistent with the known development of extrusion defects.

The Challenge

A successful metal-forming process must produce a sound product free from defects such as internal and surface cracks, local concentrations of plastic strain due to flow instabilities and high residual stresses. These can all lead to loss of strength in either steady or fatigue-type loading, loss of dimensional accuracy and of dimensional stability during subsequent machining operations. A successful metal-forming analysis must be capable of predicting the occurrence of such defects and this requires an evaluation of the complete history of stress and deformation distributions throughout the work-piece. A fracture criterion, for example, may require the knowledge of the history of stress at a material element throughout its deformation until the crack occurs.

Metal-forming processes commonly involve finite strains, as, for example, in extrusion processes which produce substantial reduction in cross-sectional area and thus strains of order unity. Elastic strains are essentially limited to the order of 10^{-3} (the yield stress divided by the elastic modulus) but these cannot be neglected, and thus plastic-rigid theory cannot be utilized in this project, because elastic strains play an essential role in determining the stress distribution. In principle, stress distributions cannot be determined in the rigid regions of plastic-rigid solutions since the stress is independent of the strain there, and hence the compatibility equations cannot be expressed in terms of stress, which prevents evaluation of the stress field.

Because of the severe nonlinearity of the plasticity laws and the large strains occurring in metal-forming processes, it is mandatory to adopt numerical methods for stress and deformation evaluations. We adopt the finite-element method because it results in a convenient computational procedure, particularly for problems involving curved boundaries as arise in the flow of metal over die surfaces. In order to evaluate stresses accurately, calculations must involve the determination of strains accurate relative to elastic strains which, as already mentioned, are of the order 10^{-3}. Thus computed strains must be accurate to within about 10^{-5} , while the strains themselves are of unit order of magnitude. This illustrates the stringent requirements on the computing procedure to achieve satisfactory stress evaluation which could be used to assess the success of a forming process.

The preceding considerations should not be taken to imply that plastic-rigid theory cannot generate useful information about metal forming processes. Limit analysis can be used to seek tool geometries which call for minimum power usage, for example.

However the prediction of important metal forming defects, such as central burst or chevroning when an internal crack appears across the axis in extrusion, falls beyond its scope [1].

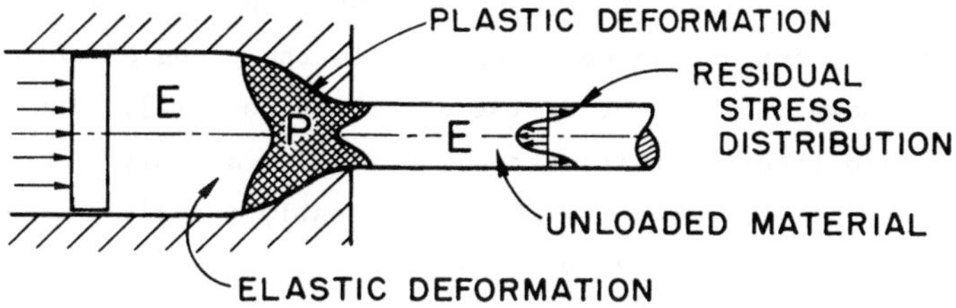

Figure 1. Extrusion configuration.

Figure 1 shows a schematic diagram of an extrusion process. A driven piston forces the billet through the die to produce the extrudate. Plastic flow is confined to the vicinity of the die: the material which has not yet entered the die region is elastic, the driving stresses being below yield, and material unloads as it emerges from the die and hence is also deforming purely elastically. The non-homogeneous plastic deformation in the die generates residual stresses in the extrudate as depicted. For plastic strain components of order unity, variations of the order of elastic strain $(\sim 10^{-3})$ suffice to produce residual stresses of the order of the yield stress.

Figure 1 indicates that most of the work-piece is deforming purely elastically, so that elastic-plastic theory is essential for the evaluation of stresses in most of the work-piece including the important residual stresses in the extrudate. Plastic-rigid theory determines the stresses only in the material currently exhibiting appreciable plastic flow, and the cross-hatched plastic region is significantly more extensive than is defined by this requirement, since it includes the contained plastic regions in which plastic strains are constrained by supporting elastic material to be of the order of elastic strain. They must therefore be considered rigid according to plastic-rigid theory. To cite specific examples of configurations of flowing regions one must turn to perfectly-plastic-rigid slip-line theory of deformation in plane strain which contains most of the rigorous solutions in more than one dimension. For extrusion with short dies, the circumstance in which the central burst or chevroning defect is most likely, the region of current appreciable plastic flow narrows to a point on the symmetry axis in the plane of deformation as shown in Fig. 40, p. 181 of Hill's book [2]. Thus plastic-rigid theory permits stress evaluation at only one point on the axis of symmetry which is clearly totally inadequate for expressing a fracture criterion.

The Method of Analysis.

As already stated, the finite-element method is adopted to formulate a computational scheme for evaluation of the deformation and stress distributions. We consider metal forming by cold working and so use a rate independent plasticity law with work hardening. In particular we select an elastic-plastic incremental or flow type constitutive relation with isotropic hardening which can reproduce a measured stress-strain curve. An incremental elastic-plastic law is a linear relation between stress-rate and strain-rate, the coefficients being functions of stress or strain history. For numerical analysis it is convenient to work in terms of stress and strain increments by multiplying the rates by the time increment Δt , and using the constitutive and strain-displacement relations, the equilibrium equations and boundary conditions to determine the increments of stress and displacement in the time interval from t to $(t + \Delta t)$. At time t the solution is known from numerical integration over the previous time increments from the initial undisturbed state, and this solution provides the coefficients in the equations to determine increments in the new time step. Requirements on the desired accuracy of the solution can be used to set limits on the magnitude of the time step Δt at each time in order to ensure the accuracy of the solution based on linear differential analysis.

For many metals, including the aluminum alloy which we shall consider, the tangent modulus during plastic flow is of the same order of magnitude as the stress. In this circumstance it was shown by Rice [3] that in the material stress-rate, which appears in the incremental law, the convected component is in general of the same order as the partial derivative of stress at a fixed position. Thus it is essential to retain convected terms in the analysis, including convected terms due to rotation. This requirement can be automatically satisfied by using Lagrange coordinates, that is choosing time and the coordinates in a fixed reference state as the independent variables for analysis. Then a partial time derivative, corresponding to fixed referential coordinates and hence to a particular material point, includes all convected terms associated with a material derivative. The finite element relations used were obtained from a variational principle in referential coordinates developed by Hill [4]. On the other hand, the plasticity laws are most easily expressed in their classical form in terms of the current geometry of the body. For example, the Cauchy stress or "true stress" determines the yield limit, so that introduction of a nominal stress based on the undeformed geometry would have to involve inclusion of deformation variables in the yield condition. To overcome the two conflicting requirements just described, the artifice is adopted, [4], of taking the configuration at time t to be the reference state for evaluation of stress and displacement increments in the time interval $(t, t + \Delta t)$. Thus the reference state is also the current configuration for that time step and both the classical laws of plasticity and the reference-state-based variational principle can be easily utilized. This approach is termed the sequentially updated Lagrangian scheme, and McMeeking and Rice developed such a finite element formulation [5] which has been modified somewhat for application to metal-forming problems [6,7].

When appreciable plastic flow has occurred, the plastic strains commonly dominate elastic strains as do their increments. Since plastic flow is essentially incompressible, it follows that during the plastic flow following the initial loading process, total deformation increments are essentially incompressible. For most finite-element configurations, the geometrical constraint associated with incompressibility commandeers virtually all the degrees of freedom to satisfy this requirement, leaving no flexibility for the displacement increments to satisfy the variational principle. This difficulty was elucidated by Nagtegaal, Parks and Rice [8] who suggested methods of overcoming the problem. We have adopted the simplest approach, which has been shown to be effective in plane-strain deformation problems, of using groups of four constant-strain triangular elements in the form of quadrilaterals divided up by their diagonals. It is shown in [8] that if the nodal displacement increments associated with three of these triangles correspond to incompressible deformation, then the deformation of the fourth will automatically be so. This reduces the number of constraint relations by the factor 3/4, which provides sufficient flexibility for the variational principle to generate a satisfactory solution. We were able to show that this approach gives satisfactory results for extrusion problems in plane strain, and a trial application to axially-symmetrical extrusion also proved to be successful.

The variational principle in terms of the unsymmetric nominal stress [4] was developed for either prescribed velocity or prescribed nominal tractions on the boundaries. For convenience of mathematical formulation, the principle was expressed in terms of the Kirchhoff stress, which is the Cauchy or "true" stress based on the current configuration multiplied by the material density ratio ρ_o/ρ, where ρ_o is the density of the undeformed material. Also to yield a convenient analytical structure, the Jaumann stress-rate was adopted, which is a material derivative which includes convection of the average rotation of the material element in addition to convection due to translation. For metal forming problems, the sliding of the deforming work-piece along curved tool surfaces poses more general boundary conditions than the prescribed velocities or prescribed tractions for which the variational principle was developed. However relatively minor changes can accommodate such generalizations [6]. For example, in the case of a frictionless curved die controlling plane-strain deformation, the stream-line at the surface is an envelope of principal-stress directions because the shear traction is zero. The tangent at each material point moving along the envelope thus rotates with the average angular velocity of the material element there, so that the shear component of the Jaumann stress rate, which would contribute to the boundary work terms in the variational principle, remains zero. However, the normal component of the surface traction does contribute to the rate of external work done by the traction rate, since as the particle moves, the direction of the normal traction changes and a non-zero rate component arises parallel to the die tangent. The resulting non-zero rate of work done by the rate of change of nominal traction contributes to the composite stiffness matrix of the finite-element formulation without changing its structure and hence without changing the method of solution [6].

Some Examples

Stress and deformation distributions were evaluated by means of the finite-element program for a slab of aluminum alloy extruded in plane strain to a thickness reduction of 25% through dies having continuous curvature [6,7]. Details of the computational aspects of the solutions and the stress-strain relation are given in the references. The configurations of the work-piece and dies before and after deformation are shown in Figure 2. The quadrilateral finite elements are depicted in both configurations, and each quadrilateral was divided by the diagonals into four constant-strain triangles.

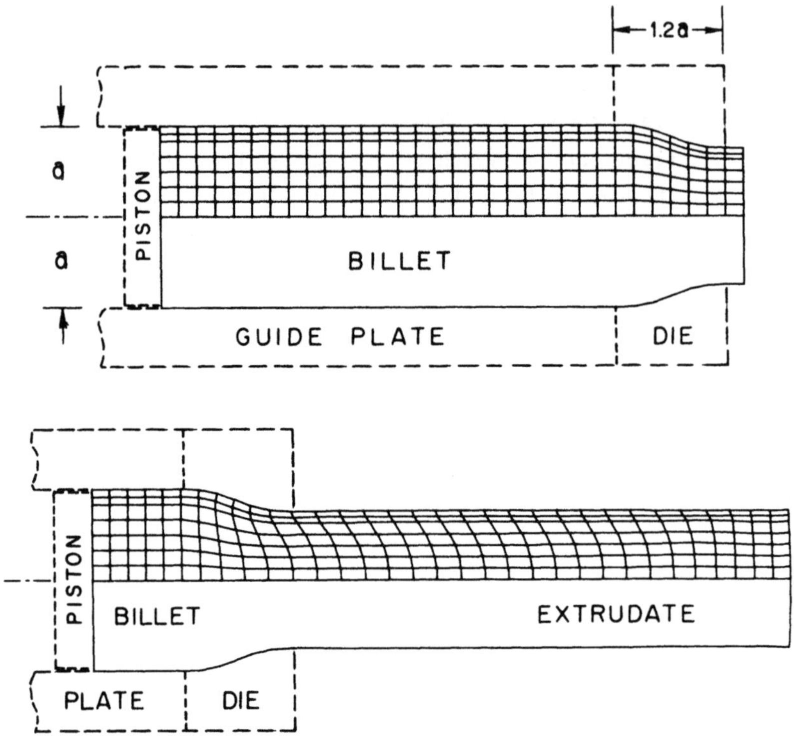

Figure 2. Undeformed and deformed plane-strain finite-element mesh.

Smaller elements are chosen near the surface to evaluate more accurately the high shear strain gradient there, for at the surface itself the shear stress was zero throughout the motion and hence so also was the shear strain. The finite-element configuration could be considered scribed on the undeformed billet in the upper part of Figure 2, and the mesh deforms with the material to generate the configuration in the lower part of the figure in conformity with the Lagrange or convected type variables adopted. The extruded material appears to emerge from the die as a rigid body but elastic recovery of lateral strain does occur in the solution and increases the thickness of the extrudate but too slightly to be apparent on the scale of Figure 2.

The deformed finite-element mesh exhibits a transient deformation pattern in the extrudate just inside the die and adjacent to it due to continuing plastic flow, and also near the free end due to the non-homogeneous undeformed mesh pattern there. Between these two regions a steady strain distribution associated with the onset of a steady-state extrusion solution appears to have developed. The appearance of the steady state is evident in the growth of the total driving force to a value which remains constant [6], and also in characteristics of the stress fields to be described later. The attainment of a steady state indicates that the quadrilateral elements with diagonals, known as crossed triangles, did adequately accommodate the incompressibility constraint since otherwise the driving force would not have settled to a steady value [8].
The fact that cross-sections initially normal to the axis of symmetry become curves in the extrudate, as shown in Figure 2, arises due to shear strain caused by the non-homogeneous straining in the die. This is known as parasitic shear since it does not contribute to the section reduction, but causes additional energy absorption in plastic flow known as "lost work".

In order to illustrate properties of the stress distribution, values of stress components at the centers of the quadrilateral elements are plotted as each element moves through the process. Figure 3 depicts the variation of the longitudinal stress σ_{xx}, normalized by the initial yield stress Y. The material point represented by each curve at steady state is denoted by its ordinate in the undeformed billet normalized with respect to the half-width a, shown in Figure 2. Figure 3 shows that the billet material is in essentially uniform longitudinal compression before entering the die. Elements near the surface are subject to high longitudinal compression in the die which changes to a tensile peak as the material emerges from the die. Elements near the plane of symmetry are subject to tensile longitudinal stress in the die which changes to compression on emergence. In the steady region of the extrudate

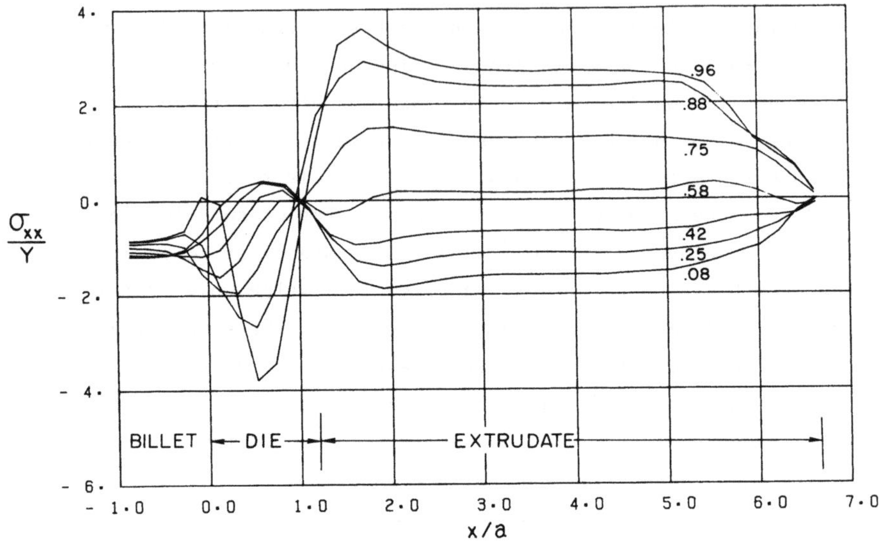

Figure 3. Longitudinal variation of normalized longitudinal stress at 7 lateral stations for frictionless plane strain extrusion with 25% thickness reduction.

the stress distribution across sections does not vary with the longitudinal coordinate x . This represents the residual stress distribution for steady-state operation of the process. It comprises tensile stress near the surfaces about twice the magnitude of the compressive stress near the plane of symmetry. The tensile stress peak near the surfaces just outside the die indicates the potential for surface cracks to be initiated there, as does the tensile stress in the die near the central axis.

It is interesting to observe that the only other available stress analyses for the extrusion process are based on rigid-ideally-plastic analysis using slip-line-field theory. As already discussed, this determines stresses only in regions of appreciable plastic flow which covers a region in Figure 3 as wide as the die length near the work-piece surfaces, but reducing to zero, or nearly zero width near the central axis (see Fig. 40, p. 181 of [2]). Moreover the influence of work hardening is then neglected.

The distribution of longitudinal residual stresses depicted in Figure 3 can be used to check the accuracy of the solution. Since no surface tractions are applied to the extrudate, the resultant longitudinal force should be of zero magnitude. Numerical integration determines an average stress equal to about 2% of the maximum stress. This is a cogent test of the accuracy of the method, since the variational principle used is for the velocity field, and stresses are evaluated from the resulting strain history. Thus the achievement of equilibrium confirms the validity of the basic structure of the method. Another test of the accuracy of the solution arises from the fact that in the region of axially uniform residual stresses the equilibrium equations require that the shear stress σ_{xy} must be zero. Examination of the stresses show this to be the case to a satisfactory degree of accuracy.

Axially-symmetric extrusion through a frictionless die having a continuously turning tangent and the same 25% area reduction as considered in plane strain yields the distribution of longitudinal stress, σ_{zz}, shown in Figure 4. Axisymmetric ring finite elements with quadrilateral cross-sections were combined to form a network similar to the plane-strain case shown in Figure 2. Each quadrilateral ring element was constructed from four ring elements with triangular cross sections and displacement patterns which were linear in the radial and axial coordinates. Comparison of Figure 4 for the axially symmetric case with the analogous Figure 3 for plane strain reveals that the stress magnitudes adjacent to the axis of symmetry are considerably enhanced in the axi-symmetric case. The reason is clear when one considers the equilibrium conditions across a section of the extrudate in the two cases:

$$\int_o^b \sigma_{xx}\, dy = 0 \; ; \quad 2\,\pi \int_o^b \sigma_{zz}\, r\, dr = 0$$

The weighting factor, r, in the axially symmetric case requires enhanced stresses near the axis to achieve equilibrium.

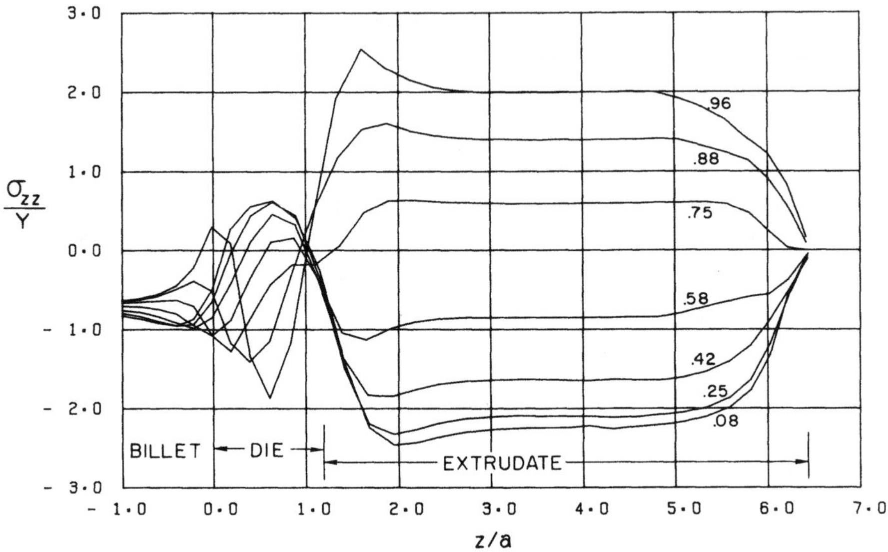

Figure 4. Axial variation of normalized axial stress at 7 radial stations for frictionless axisymmetric extrusion with 25% area reduction.

Consquently, as evident in Figure 4, relatively larger tensile stress occurs near the axis in the die, and hence increased tendency to initiate central-burst cracks.

For lack of solutions, even with the plastic-rigid model, assessments of axially-symmetric cases have often been based on plane-strain slip-line-field solutions. The comparison between Figures 3 and 4, particularly when combined with the limited information supplied by plastic-rigid slip-line theory, indicate the shortcomings of such an approach.

Conclusion

The examples presented indicate that the finite-element elastic-plastic approach offers the feasibility of making rational analytical assessments of design choices for a forming process which will ensure a defect free product.

Acknowledgements

Thanks are due to the U. S. National Science Foundation's Materials Research Laboratory Program through the Center for Materials Research at Stanford University for support of the work involved in preparation of this paper. The Figures are reproduced from the ASME publication, Reference [7].

References

1 Lee, E. H.; McMeeking, R. M.: Concerning analysis of central burst in metal forming, J. of Eng. for Indust. Trans. of ASME, 100, (1978) 386-387.

2 Hill, R.: The mathematical theory of plasticity, Clarendon Press, Oxford (1950).

3 Rice, J. R.: A note on the "small strain" formulation for elastic-plastic problems, Tech. Rep. N00014-67-A-0191-0038/8, Div. of Eng. Brown University (1970).

4 Hill, R.: Some basic principles in the mechanics of solids without a natural time, J. Mech. Phys. Solids, 7. (1959) 209-225.

5 McMeeking, R. M.; Rice, J. R.: Finite-element formulations for problems of large elastic-plastic deformation, Int. J. Solids Structures, 11, (1975) 601-616.

6 Lee, E. H.; Mallett, R. L.; Yang, W. H.: Stress and deformation analysis of the metal extrusion process, Comp. Meth. in Appl. Mech. Engg. 10, (1977) 339-353.

7 Lee, E. H.; Mallett, R. L.; McMeeking, R. M.: Stress and deformation analysis of metal forming processes, Numerical Modeling of Manufacturing Processes, PVP-PB-025, Ed. by R. F. Jones, Jr., H. Armen, and J. T. Fong,(1977) 19-33, ASME.

8 Nagtegaal, J. C.; Parks, D. M.; Rice, J. R.: On numerically accurate finite element solutions in the fully plastic range, Comp. Meth. in Appl. Mech. Engg., 4, (1974) 153-177.

Inhomogeneity of Deformation and Tool Design

T.Z. BLAZYNSKI
Department of Mechanical Engineering,
University of Leeds, U.K.

Summary

The nature and effects of inhomogeneous, or redundant, deformation, occurring in metal-forming, are discussed. Methods of assessment of the levels of redundancy are outlined, and the relationship with the geometry of the pass is indicated. Two methods of design of tool profiles are presented.

Introduction

Inhomogeneity of deformation, associated with the presence of macro-shears in a metal-forming process, is more often considered from the point of view of its effect on the mechanics of the operation than from that of its influence on the soundness of the flow and, therefore, on the quality of the final product. Whereas, undoubtedly, the measure of agreement obtained between a mathematical model and physical reality depends on the ability of the latter to account for the inhomogeneity present, both the physical behaviour and properties of the material processed are affected, to a far greater degree, by inhomogeneity, while, at the same time, their practical significance is far greater. It is for this reason that the relationship between the characteristics of the metal and the incidence and level of inhomogeneity, or redundancy, acquires a particular significance in metal forming studies.

Specific Effects of Inhomogeneity

Internal macro-shearing of the metal along and across a working forming pass - as distinct from the mechanisms of micro effects such as deformation along the slip planes etc.- is totally redundant to the main objective of the operation, i.e. change of shape. The presence of these additional, often large, strains increases the severity of the process by affecting: (i) the general level of straining, (ii) properties of the material, and (iii) build-up of residual stresses.

In cold forming processes, the induced shear strains will increase the levels of stress-strain curves, while reducing the strain to fracture, and -by providing additional constraints to the flow- may also increase the build-up of residual stresses. Conditions for predisposition to selective cracking can thus be created. Further, large grain recrystallisation - a feature of some processes- when physically combined with the appearance of redundant straining can easily induce complete failure of the material.

Distortions of the internal structure of the metal, caused by the inhomogeneity of deformation, are responsible for the appearance of many material faults, especially in the rotary hot working processes where, for instance, central cavitation - or internal cracking- occurs in 2-roll tube piercing systems, and wall cracking is present in tube rolling. Hot extrusion of solid core, sheathed, bimetallic rods, as well as the extrusion of conventional solid rods, provide further examples of the incidence of material failure by cracking. A link between the development of faults and high level of redundancy has been clearly established in a number of conventional processes in which concentrations of inhomogeneous macroshears are found to coincide with the zones of crack and lap formation. In all these cases, whether hot or cold, the appearance of inhomogeneity either limits the processed materials to those of better quality only, or, alternatively, reduces the amount of the possible deformation.

Irrespective of its mechanical properties, any material capable of being plastically formed will, of course, suffer the same homogeneous strain when worked in the pass of a given geometry. The incidence and level of redundancy, however, will depend on both the geometry of the pass and the stress-strain characteristics of the metal. It is clear therefore that providing the geometry of the tool system is correctly established, the redundant effects can be controlled by the removal of the unnecessary straining constraints imposed on the flow. A reduction in the longitudinal warping, circumferential shearing and -in rotary processes-twisting, will allow the deformation of the metal to proceed along, almost, optimal internal flow lines.

Measure of Redundancy

The assessment of the levels of redundancy is made by means of the redundancy factor which relates theoretical, or ideal, conditions, to those which obtain in physical reality. A stress redundancy factor would thus relate theoretically computed drawing stresses, extrusion pressures etc. to those measured experimentally, whereas a strain redundancy factor would provide a link between the generalized homogeneous strain - associated with the desired change of shape- and the total generalized strain actually imposed, including the inhomogeneous component. Since the latter factor does not depend on any arbitrarily chosen theory, and, further, gives a direct indication of the mean yield stress operating in the system considered, it is preferable to the former. Thus the strain redundancy factor, ϕ, (>1) is defined as:

$$\phi = \varepsilon_T/\varepsilon_H \tag{1}$$

where ε is the generalized strain, and subscripts, T, and H, refer to the total and homogeneous strains respectively. In optimal conditions ϕ would be one. If the approximate value of ϕ is known and ε_H is calculated from the expression

$$d\varepsilon_H = \frac{\sqrt{2}}{3}[(d\varepsilon_x-d\varepsilon_y)^2+(d\varepsilon_y-d\varepsilon_z)^2+(d\varepsilon_z-d\varepsilon_x)^2]^{\frac{1}{2}} \tag{2}$$

where subscripts x, y and z refer to homogeneous strains in three mutually perpendicular directions, the value of ε_T and, therefore, that of the operative yield stress is assessable. Conversely, if the magnitudes of the respective components of inhomogeneous strains - in a given operation- are known ε_T can be calculated from the following general expression:

$$\varepsilon_T = \sqrt{\frac{2}{3}}\ [\varepsilon_x^2+\varepsilon_y^2+\varepsilon_z^2+\tfrac{1}{2}(\varepsilon_L^2+\varepsilon_C^2+\varepsilon_t^2)]^{\frac{1}{2}} \tag{3}$$

where subscripts L, C and t refer to the longitudinal, circumferential and due to twist shears respectively. Some of the redundant effects may not be present in the process considered and equation (3) will simplify accordingly.

The "efficiency" of the pass, in the context of its level of redundancy, can thus be made easily and can then be used to c o m p a r e the effect of tool geometry on the flow. To avoid constant reliance on experimental data, implied by the outlined procedures, the method of determination of ϕ, for a given alloy, should be related to the geometry of the pass, and to the generalized homogeneous strain only - as both these quantities are known a priori. Such an approach was, in fact, adopted by, among others, Wistreich in wire drawing [1], Hill and Green in sheet drawing [2], and Caddell and Atkins in rod drawing and extrusion [3]. By considering geometrical configuration of the pass, definitions of geometry factor Δ have been derived. These relate the basic properties of the systems considered to the levels of inhomogeneity. It is known from the experimental evidence that in all these cases the relationship between ε_T, ε_H and die semi-angle α is of the form:

$$\varepsilon_T = K_1\varepsilon_H + K_2\sin\alpha \tag{4}$$

or

$$\phi = K_1 + K_2\sin\alpha/\varepsilon_H \tag{5}$$

also

$$\phi = C_1 + C_2\Delta \tag{6}$$

The equivalence of equations (5) and (6) depends on the definition of Δ. According to Caddell and Atkins, for instance,

$$\Delta = (D+d)\sin\alpha/(D-d) \tag{7}$$

where D and d are the initial and final rod diameters, and Δ represents the ratio of die diameter, at mid-points of the face, to the length of contact between the tool and workpiece.

A recent investigation, carried out by the author, indicates that the relationship between ε_T and ε_H in the three basic cold tube drawing processes of sink, stationary plug and moving mandrel, is linear (Fig. 1). Consequently, the discussed method of treatment can be extended to these three processes.

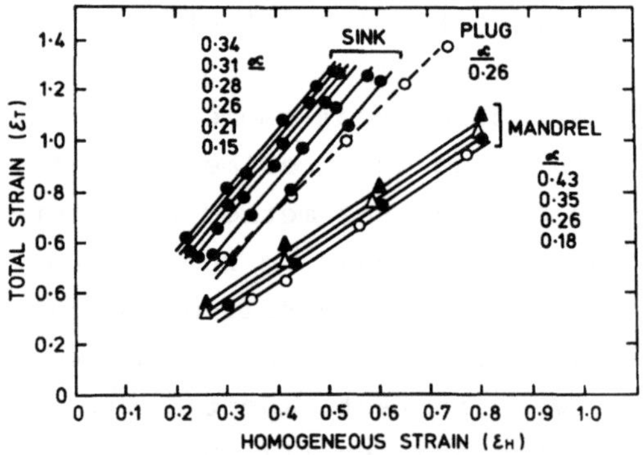

Fig. 1. Variation of total generalized strain with homogeneous strain in cold drawing of steel tubes.

To account for the different geometry of the workpiece, the definition of Δ has to be modified. In tube drawing, Δ can be defined as the ratio of the cross-sectional area of the tube at mid-points of the die to the area of contact between the die and the tube.

The derivation of the appropriate relationship is conveniently illustrated with reference to plane strain mandrel drawing through a die of a semiangle α. ϕ is determined from equations (4) and (6) in the following way. Let A_M and A_D be the cross-sectional areas of the tube and die, and D_o, D, h_o and h be the initial and final tube diameters and wall-thicknesses respectively, then in a conical die, or in one that approximates closely to this basic profile,

$$\Delta = \frac{A_M}{A_D} = \frac{1}{2}\,\frac{(h_o+h)\,[(D_o+D)-(h_o+h)]}{(D_o+D)\,(h_o-h)}\,\sin\alpha$$

But, $(D_o+D) \gg (h_o+h)$, and, hence

$$\Delta \simeq \frac{1}{2}\,\frac{(h_o+h)}{(h_o-h)}\,\sin\alpha \tag{8}$$

The generalized homogeneous strain - in plane strain conditions- is given by [4] as

$$\varepsilon_H = \frac{2}{\sqrt{3}}\,\ell n\,(h_o/h) \tag{9}$$

and substituting into equation (8),

$$\Delta = \frac{1}{2}\,\frac{1}{\tanh(\sqrt{3}\;\varepsilon_H/2)}\,\sin\alpha \simeq \frac{1}{\sqrt{3}}\,\frac{\sin\alpha}{\varepsilon_H} \tag{10}$$

Comparing equations (5) and (6) one notices that:

$$K_1 = C_1,\text{ and } K_2 = \frac{1}{\sqrt{3}}\,C_2 \tag{10a}$$

For the given metal the constants K_1 and K_2 are determined experimentally. Equation (10) applies also to the process of plug drawing, but a similar reasoning shows that in tube sinking:

$$K_1 = C_1,\ K_2 = 4a\ C_2/\sqrt{3}\ \text{and}\ a = \pi\ h/D_o \tag{10b}$$

These relationships are valid if it is assumed that no changes in wall-thickness occur. In consequence, the "constants" K_1 and K_2 reflect not only the material properties, but also the initial dimensions of the sunk tube.

The following m e a n values of ϕ can be established for low carbon tubing:

Sink: $\phi = 2.4 + 0.02D\Delta/h$ (11)

Plug: $\phi = 2.0 - 0.264\Delta$ (12)

Mandrel: $\phi = 1.35 - 0.334\Delta$ (12a)

Fig. 2 indicates the variation of the actual values of ϕ about the mean curves representing equations (11) to (12a).

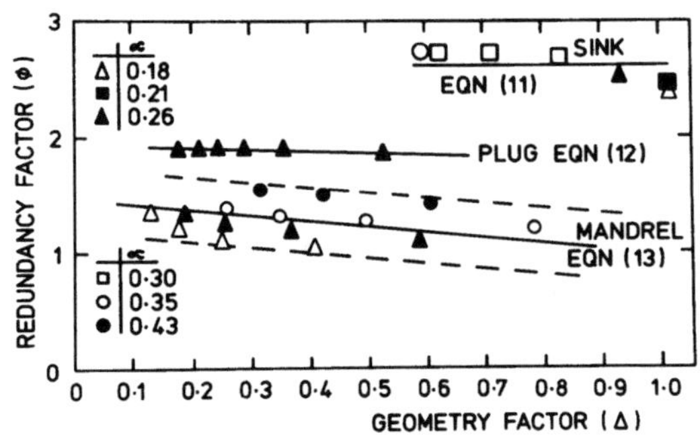

Fig. 2. Relationship between factor ϕ and factor Δ in steel tube drawing.

Since the equations of the type discussed above are not yet applicable to a number of processes, experimental determination of redundant strains is often necessary. Cold worked, strain-hardening alloys can be investigated by means of Hill and Tupper principle of superposition of stress-strain curves [5]. Extrusion and drawing lend themselves to the use of square or circular grids, imprinted by the photographic or electrochemical techniques [4,6], whereas, severe rotary processes call for the use of model materials incorporating inserts or laminates [4] of the types illustrated in Fig. 3.

Fig. 3. Examples of the insert technique applied to forging and rotary piercing and elongating:

Wherever possible circular grids - imprinted on the actual metal- are preferable since in isotropic materials the minor and major axes of the resulting ellipses indicate automatically the directions of principal strains.

The most convenient procedure consists in printing a grid on the surface of the already deformed element and, then, in deforming it further by imposing a single increment of strain - corresponding to the diameter of the original circle of the grid. Incremental summation of deformations occurring along the flow lines, across any transverse section of the specimen, will provide a reliable guide to the pattern of flow and the incidence of redundancy.

In the case of inserts and laminates, model non-metallic and metallic materials have to be used such as mixtures of waxes, clays, Plasticine, or lead, pure aluminium and copper alloys. The problem of similitude of properties of the model and prototype materials arises and consequently careful matching of stress- strain curves - at appropriate strain rates - and of strain-hardening functions of the respective materials is required [4].

Concepts of Tool Design

Imposition of homogeneity of the deformation - associated with either the total removal of, or a reduction in the shearing constraints on the flow - is of primary importance. This is directly linked with the lessening of the severity of the operation and consequently with ensuring that the individual flow lines are smooth and do not suffer unnecessary discontinuities. In any forming process the pattern of the deformation depends either on the order in which the magnitudes of the successive ratios of generalised strain increments vary along the pass, or on the variation in the strain rate itself. If therefore a constant rate of either of these parameters can be maintained throughout the operation, the likelihood of the incidence of flow discontinuities will be reduced. Which of the two parameters is likely to be of greater importance in any given case, will be dictated by the characteristics of the worked material and the mechanics of the process.

The possible bases of tool design can therefore be associated with two, pre-arranged, modes of predominantly homogeneous flow that consist in maintaining a constant rate of either the ratios of successive generalized homogeneous strains increments (CRHS), or the m e a n strain rate (CMSR) along the pass.

The CRHS concept can be defined, with reference to equation (2), and by considering equidistant transverse sections of the pass, as follows:

$$\frac{\varepsilon_{Hn} - \varepsilon_{H(n-1)}}{\varepsilon_{H(n-1)} - \varepsilon_{H(n-2)}} = s \qquad (13)$$

where, n, is the considered section and, s, is the rate of deformation.

In any specific physical situation, equation (2), when integrated, can be represented as:

$$\varepsilon_H = \ln (Z) \tag{2a}$$

where function Z reflects the physical dimensions of the specimen. At the entry to the working zone of the pass Z = 1, and consequently it follows from equations (2a) and (13) that:

$$Z_n/Z_{n-1} = (Z_1)^{s^{(n-1)}} \tag{14}$$

With an assumed, but c o n s t a n t, value of, s, the solution of equation (14) provides the required outline of the pass geometry and, therefore, of tool profile. The rate, s, can take any value; with s = 1 corresponding to a uniform rate of deformation (U). Rates smaller than one will give decelerated flow (D), and those larger than one accelerated deformations (A). Equation (14) can not predict conditions in section (1) of the pass and so a physically appropriate set of dimensions has to be assumed.

The mean strain rate ($\dot{\varepsilon}_m$) is referred to the physical boundary and is defined as [4]:

$$\dot{\varepsilon}_m = \frac{1}{x} \int \dot{\varepsilon} dx \tag{15}$$

and $$\dot{\varepsilon} = d\varepsilon_H/dt \tag{16}$$

where, t, is the time, and, x, is the axial distance.

Although equation (15) can be evaluated for any given process - in terms of a constant value of strain rate- and therefore the CMSR concept can be applied, it will be found that quite arbitrarily chosen values of, x, will give a variation in diameters, or wall-thicknesses that satisfies the required condition. Optimisation of tool profile does not necessarily follow since the variation can be fairly random. If, however, the CMSR concept is operated in conjunction with the CRHS one, and a rate of deformation, s, is selected first, the resulting tool profile will be smooth and acceptable. In either situation, in a 2-tool system the profile of one of the tools - most likely that subjected to the higher rate of wear - has to be assumed and, then, used as datum for the profiling of the other. Fig. 4 shows the difference in tool profiles - designed on the basis of the two concepts - for two operations that are representative of the extremes of metal forming, i.e. rotary, on the one hand, and extrusion/ drawing processes on the other. Representative plots

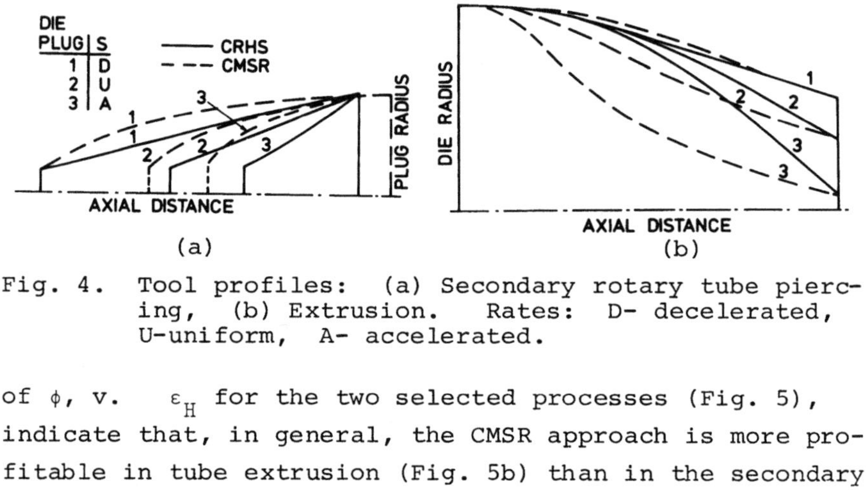

Fig. 4. Tool profiles: (a) Secondary rotary tube piercing, (b) Extrusion. Rates: D- decelerated, U-uniform, A- accelerated.

of ϕ, v. ε_H for the two selected processes (Fig. 5), indicate that, in general, the CMSR approach is more profitable in tube extrusion (Fig. 5b) than in the secondary rotary piercing (Fig. 5a). Although in both cases redundancy reduces with the increase in the homogeneous strain, the effect of the rate of deformation on ϕ varies with the concept employed and the process considered. The choice between the concepts depends, basically, on the degree of wear of the tool produced, on the one hand, by the complexity or otherwise of the basic profile, and, on the other, by the increase in the throughput of the material - resulting from the reduction in inhomogeneity- and reflected in the increased loading. The CMSR tool profiles are often smoother, but they are also responsible for a considerable increase in the operational loads, torques, etc. Both concepts give, however, much lower levels of redundancy than the standard industrial tools.

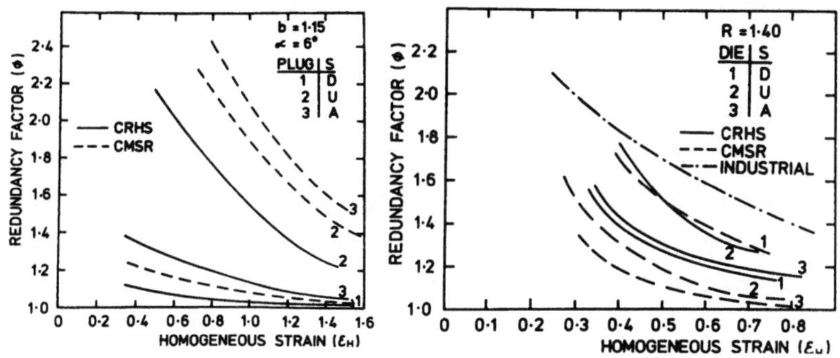

(a) (b)

Fig. 5. Influence of tool profile on redundancy: (a) Secondary piercing of wax tubes (Tube dia/gorge dia ratio b = 1.15), (b) Extrusion of lead tubes. (Extrusion ratio R = 1.40)

The incidence and level of inhomogeneity are far greater in the rotary processes than in the drawing and similar operations and the "efficiency" of tools is therefore of greater importance. The CRHS-type tools are found to give lower load requirements, while reducing substantially the levels of redundancy. Table 1 shows the influence

Table 1. Variation of ϕ with Process, Strain and Rate of Deformation in Wax Tubing. (CRHS Tools).

Rotary Process	Ranges of Parameters				
	ε_H	ϕ			
		U	D	A	Indust.
Piercing	1.4-1.7	3.0-4.3	2.4-3.5	2.6-4.0	4.2-4.8
Secondary Piercing	0.4-1.6	1.1-1.0	2.1-1.2	1.3-1.1	-
Assel	0.2-1.0	3.7-1.8	4.7-2.5	1.8-1.2	-
P.E.	1.5-3.0	2.8-3.2	2.7-3.1	2.4-2.9	4.0-5.2

Feed angle - 6°, Rates ÷ U-Uniform, D-Decelerated, A-Accelerated.

of the rate, s, on ϕ for 3-roll primary and secondary piercing, Assel elongating, and combined piercing-elongating (P.E.) [4, 7] processes. Irrespective of the rate, redundancy increases with ε_H in the primary piercing and P.E. processes, but is reduced in the secondary piercing and Assel elongation.

Conclusions

The use of tools that facilitate the flow of metal through a metal-forming pass can reduce the levels of inhomogeneity by as much as 25 to 50 per cent in comparison with the effects of ordinary industrial tools. The improvement is more pronounced in the rotary processes than in the drawing, and similar operations. In the latter group of processes, however, it becomes possible to predict, fairly accurately, the values of redundancy factors that are likely to obtain in any given tool configuration.

References

1 Wistreich, J.G.: The fundamentals of wire drawing. Met. Rev. 3 (1958) 97.

2 Hill, R.; Green, A.P.: Calculations on the influence of friction and die geometry in sheet drawing. J. Mech. Phys. Solids 1 (1953) 31.

3 Caddell, R.M.; Atkins, A.C.: The influence of redundant work when drawing rods through conical dies. Trans. ASME No. 67-WA/Proc. 11 (1967).

4 Blazynski, T.Z.: Metal Forming: Tool Profiles and Flow. London: Macmillan Press Ltd. 1976.

5 Hill, R.; Tupper, S.J.: A new theory of plastic deformation in wire drawing. J. Iron Steel Inst. 159 (1948) 353.

6 McClean, W.; McCandless, A.J.; Bahrani, A.S.: A strain measurement system for sheet metal forming. Sheet Metal Ind. 55 (1978) 653.

7 Blazynski, T.Z.: An assessment of a combined rotary piercing and elongating process for tube making. J. Inst. Metals 95 (1967) 97.

A New Approach for Generalized Three-Dimensional Extrusion of Sections from Round Billets by Conformal Transformation

D.Y. YANG, M.U. KIM and C.H. LEE
Korea Advanced Institute of Science, Seoul, KOREA

Summary

An analytical expression for the continuous die surface is found by utilizing external conformal mapping of sections and by introducing a new concept of "profilization" in the three-dimensional extrusion of sections from round billets. A kinematically admissible velocity field is derived utilizing the incompressibility condition with the aid of conformal transformation and thus a formulation for the upper-bound pressure is obtained. Optimized configurations of the die surface for some chosen sections are obtained by minimizing the extrusion power with respect to chosen design variables.

Introduction

The extrusion of sections from round billets through continuous dies has many advantages, especially from the viewpoint of billet procurement. The lubricated cold extrusion through continuous dies enables the improvement of strength and surface quality as compared with the conventional hot extrusion.

In determining the intermediate contours between a billet and a general non-circular section, only graphical methods or some experimental methods using analogies have been reported. Beyer (1) surveyed these methods in his recent paper. These graphical and experimental methods lack theoretical back-up in that the plastic deformation energy is not taken into consideration at all. Especially, the die length is excluded from the design variables, not to speak of intermediate die profile. Therefore, some practical method of die design which embodies the effect of the whole deformation energy is required, together with systematic theoretical analysis.

As for the three-dimensional extrusion of sections from similar billets, only a few results have been reported [2] [3] [4] [5]. Concerning the three-dimensional extrusion of sections from round billets, there has been no systematic analysis except the work of Nagpal and Altan [6], but their extrusion model is limited to elliptical sections only. Since Nagpal and Altan formulated the upper-bound solution for the extrusion of elliptic sections from round billets, no analytical method has been reported for the extrusion of other general shapes from round billets. The difficulty of this problem lies in obtaining the analytical expression of the die surface and also in the theoretical formulation for the internal metal flow bounded by the complex three-dimensional surface. Therefore, some simplifying assumptions are necessary for the analytical attempt. In this paper, a theory related to a suggested design method will be developed for the three-dimensional extrusion of arbitrarily-shaped sections from lubricated round billets through continuous dies.

Method of Die Surface Construction and Generalized Mapping Coefficients

In order to initiate the mathematical analysis, the die surface should be so determined as to permit analytic expressions of the surface and also continuous changes of the cross-sectional shape. As a consistent means of a smooth geometrical change of the shape, Conformal transformation of the outside of a section onto the outside of a unit circle(external mapping) is introduced, as shown in Fig.1. Conformal transformations of the doubly-connected region onto a ring-shaped region may be preferable but it seems practically very difficult to find a mapping function for a general doubly-connected region [7]. Regarding external mapping of a section, it is found that the shape of a conformally-mapped circumference corresponding to a certain circular circumference lying considerably farther outside of a unit circle decays to almost a circle, with a slight deviation(Fig.1). The conformally-mapped circumference obtained by external mapping can thus be used in determining the shape of the intermediate die contour when a round billet is extruded into a section through a continuous die cavity.

The external mapping function can be approximated by a finite number of terms;

$$W(\tau) = \sum_{k=1}^{M} B_k \, \tau^{1-(k-1)N_s} \qquad (1)$$

$$\tau = re^{i\omega}$$

where N_s is the number of axes of symmetry. A continuous placement of circumferential contours in between the entrance and exit planes can be achieved by putting the variable r as a continuous function of z, r(z). For the time being, let r(z) be any continuous function of z satisfying the following boundary conditions;

$$r(z) = R_c \text{ at } z = 0, \; r(z) = 1 \text{ at } z = L \qquad (2)$$

where R_c represents the value of r at which the corresponding circumferential contour on the mapped plane is almost a circle with a prescribed deviation, and L is the die length. The surface formed from external mapping in this manner is expressed by the following two equations;

$$x = \sum_{k=1}^{M} \bar{b}_k(z) \cos[1-(k-1)N_s]\omega \quad (3a)$$

$$y = \sum_{k=1}^{M} \bar{b}_k(z) \sin[1-(k-1)N_s]\omega \quad (3b)$$

where $\bar{b}_k(z) = B_k[r(z)]^{1-(k-1)N_s}$

As in Fig.1(c), the shapes of entrance and exit are prescribed by the billet and the product section. Usually, the surface defined by Eqs.(3a, 3b) does not fit in with the billet, because R_C was not determined to fit in with the billet but only to attain circular congruency. So, the surface contours should be scaled up (or down) continuously along the axial direction while permitting the consideration of the die profile. An "effective" die profile is defined such that a cross-section at any distance from the entrance plane is transformed onto a circle having the same area, and the radius of this circle is expressed by a continuous profile function of z, h(z) (Fig.1(d, e)). This procedure of scaling the surface contours of Fig.1(b) to the effective die profile is now called "profilization". h(z) can be any continuous, monotonically decreasing function of z satisfying the following boundary conditions;

$$h(z) = 1 \text{ at } z = 0, \; h(z) = \lambda \text{ at } z = L \quad (4)$$

where λ is defined such that $1-\lambda^2$ equals fractional reduction of area. The following profile function of a biquadratic polynomial is chosen as a continuous profile function,

$$h(z) = 1+\left[CL^2-\frac{3(1-\lambda)}{L^2}\right]z^2+\left[\frac{2(1-\lambda)}{L^3}-2CL\right]z^3+Cz^4 \quad (5)$$

where C is a geometric variable determining the position of the inflection point of this biquadratic polynomial. This profile function was effectively used in the Battelle Memorial Institute [6] in various lubricated metalforming processes. In lubricated axisymmetric extrusion, this die profile seems more effective than any other known die profile, because it produces no shear energy on the elastic-plastic boundaries. Another interesting characteristic is that elastic-plastic boundaries render theoretically flat in the axisymmetric upper-bound formulation [8][9]. The biquadratic polynomial has two optimizable geometric variables (C and L) by which the geometric shape of die surface minimizing the extrusion power can be determined.

Through this study, it is found to be recommendable that r(z) be in a similar form to h(z). Then, r(z) satisfying the boundary conditions of Eq.(22) is given as follows;

$$r(z)=R_c+\left(CL^2-\frac{3(R_C-1)}{L^2}\right)z^2+\left(\frac{2(R_C-1)}{L^3}-2CL\right)z^3+Cz^4 \tag{6}$$

Now, new profilized coefficients for the profilized die boundary are as follows;

$$b_k(z) = \frac{\bar{b}_k(z)\ h(z)}{\sqrt{\sum_{k=1}^{M}[1-(k-1)N_s]\,[\bar{b}_k(z)]^2}} \tag{7}$$

Then, the profilized die boundary is expressed by the following two equations;

$$x = \sum_{k=1}^{M} b_k(z)\ \cos[1-(k-1)N_s]\omega \tag{8a}$$

$$y = \sum_{k=1}^{M} b_k(z)\ \sin[1-(k-1)N_s]\omega \tag{8b}$$

When the sectional shape is complex, it will be difficult to set up any direct mathematical formulation of the three-dimensional extrusion. The analysis for the internal metal flow can be treated more readily once the complex cross-section can be transformed onto a unit circle, for example by internal conformal transformation (Fig.2). Considering the smooth change of cross-sectional shape coefficients of an internal mapping function should vary continuously. Therefore, let us introduce a "generalized" mapping function in which each of the mapping coefficients is permitted to be some continuous function of z as follows;

$$W(\tau) = \sum_{k=1}^{M} a_k(z)\, \tau^k = x + iy \tag{9}$$

where $\tau = \rho e^{i\omega}$

$$x = \sum_{k=1}^{M} a_k(z)\, \rho^k \cos k\omega \tag{10a}$$

$$y = \sum_{k=1}^{M} a_k(z)\, \rho^k \sin k\omega \tag{10b}$$

It is almost impossible to find exact "generalized" mapping coefficients that exactly fit in with the given die surface. Nevertheless, we may find approximate generalized mapping coefficients serviceable to our purpose. We can choose N+1 cross-sections which lie at equal intervals, including the entrance and exit cross-sections. We find N+1 internal mapping functions corresponding to N+1 cross-sections and then interpolating polynomial for each coefficient can be obtained either by solving N+1 linear equations simultaneously, or by the Lagrangian method.

For such sections as have deep re-entrant contours(e,g.,

gears or splines), internal mapping does not render sufficient congruency unless a large number of terms are taken. Another disadvantage of using internal conformal mapping functions lies in using approximate coefficients by interpolating polynomials. These drawbacks can be eliminated by defining a special internal transformation(Fig.2), expressed by the following equations;

$$x = \sum_{k=1}^{M} b_k(z)\ \rho \cos[1-(k-1)N_s]\omega \tag{11a}$$

$$y = \sum_{k=1}^{M} b_k(z)\ \rho \sin[1-(k-1)N_s]\omega \tag{11b}$$

Derivation of a Kinematically Admissible Velocity Field

In formulating an upper-bound solution, a kinematically admissible velocity field is to be found for a given extrusion model. Any kinematically admissible velocity field should satisfy a continuity equation and two velocity boundary conditions [10];

i) Continuity equation or incompressibility should be satisfied.

ii) Velocity components normal to the elastic-plastic boundaries should be continuous across the boundaries.

iii) Velocity components normal to the die surface should vanish.

The continuity equation in incompressible three-dimensional flow is given by,

$$\frac{\partial V_x}{\partial x} + \frac{\partial V_y}{\partial y} + \frac{\partial V_z}{\partial z} = 0 \tag{12}$$

Velocity components satisfying Eq.(12) should also satisfy the two velocity boundary conditions, ii) and iii). The difficulty of the mathematical treatment in three-dimensional problem demands some simplifying assumptions as in other papers of the authors [5] [11]. The velocity boundary condition ii) can be readily satisfied by assuming elastic-plastic boundaries to be flat planes(as entrance and exit planes) and again by assuming a uniform axial velocity at each cross-section;

$$V_z = F(z) \tag{13}$$

Then, F(z) is uniqely determined simply by the cross-sectional area bounded by the die surface, from volume constancy. When we use the internal conformal transformation of Eq.(9), an expression for F(z) is given by

$$F(z) = \frac{V_0}{\sum_{k=1}^{M} k\,[a_k(z)]^2} \tag{14}$$

Alternative expression for F(z) can also be obtained by using Eqs.(3a, 3b). Substituting Eq.(13) into Eq. (12)

$$\frac{\partial V_x}{\partial x} + \frac{\partial V_y}{\partial y} = -F'(z) \tag{15}$$

where F'(z) is the first derivative of F(z) with respect to z. The general solution of Eq. (17) can be obtained as follows,

$$V_x = -\frac{xF'(z)}{2} + \frac{\partial G}{\partial y} \tag{16a}$$

$$V_y = -\frac{yF'(z)}{2} - \frac{\partial G}{\partial x} \tag{16b}$$

where G is a continuous and differentiable function, G(x, y, z).

Coordinates (x,y) can be related to the curvilinear Coordinates (ρ,ω) by Eqs. (10a, 10b). G(x, y, z) should be determined to satisfy the velocity boundary condition iii).

The mathematical expression for the velocity boundary condition iii) is

$$\vec{V} \cdot \nabla H(x, y, z) = 0 \tag{17}$$

where H(x, y, z) = 0 represents the die surface. When we employ Eqs. (10a, 10b) for the expression of the die surface, it can be obtained by putting $\rho = \rho_0$ (taken to be the unity), and can be expressed by the following parametric form;

$$x = \sum_{k=1}^{M} \rho_0^k a_k(\zeta) \cos k\omega \tag{18a}$$

$$y = \sum_{k=1}^{M} \rho_0^k a_k(\zeta) \sin k\omega \tag{18b}$$

$$z = \zeta \tag{18c}$$

Substituting the velocity components of Eqs. (13), (16a) and (16b) into Eq. (17),

$$\left[-\frac{xF'(z)}{2} + \frac{\partial G}{\partial y}\right]_{\rho=\rho_0} \cdot \frac{\partial H}{\partial x} + \left[-\frac{yF'(z)}{2} - \frac{\partial G}{\partial x}\right]_{\rho=\rho_0} \cdot \frac{\partial H}{\partial y}$$

$$+ F(z) \cdot \frac{\partial H}{\partial z} = 0 \tag{19}$$

The components of the normal vector of the die surface (i.e, $\frac{\partial H}{\partial x}$, $\frac{\partial H}{\partial y}$ and $\frac{\partial H}{\partial z}$) can be related to Eqs. (18a, 18b, 18c) by means of differential geometry;

$$\frac{\partial H}{\partial x} : \frac{\partial H}{\partial y} : \frac{\partial H}{\partial z} = \frac{\partial y}{\partial \omega} : \left(-\frac{\partial x}{\partial \omega}\right) : \left(\frac{\partial x}{\partial \omega}\frac{\partial y}{\partial \zeta} - \frac{\partial y}{\partial \omega}\frac{\partial x}{\partial \zeta}\right) \tag{20}$$

Substituting the relationship (20) into Eq. (19), and using the chain rule, we obtain

$$\left.\frac{\partial G}{\partial \omega}\right|_{\rho=\rho_o} = \frac{F'(\zeta)}{2}\left[x\frac{\partial y}{\partial \omega} - y\frac{\partial x}{\partial \omega}\right] - F(\zeta)\left[\frac{\partial x}{\partial \omega}\frac{\partial y}{\partial \zeta} - \frac{\partial y}{\partial \omega}\frac{\partial x}{\partial \zeta}\right] \tag{21}$$

The partial derivatives of Eq. (21) can be easily found by differentiating Eq. (18a, 18b) with respect to ω and ζ. Utilizing Eq. (14) and some trigonometric relations, Eq. (21) finally becomes,

$$\left.\frac{\partial G}{\partial \omega}\right|_{\rho=\rho_o} = \frac{F'(\zeta)}{2} \sum_{\substack{i=1 \\ (i\neq j)}}^{M}\sum_{j=1}^{M} j\,\rho_o^{i+j} a_i(\zeta)a_j(\zeta)\cos(i-j)\omega$$
$$+ F(\zeta) \sum_{\substack{i=1 \\ (i\neq j)}}^{M}\sum_{j=1}^{M} i\,\rho_o^{i+j} a_i(\zeta)a_j'(\zeta)\cos(i-j)\omega \tag{22}$$

Integrating $\left.\frac{\partial G}{\partial \omega}\right|_{\rho=\rho_o}$ with respect to ω,

$$G(\rho_o,\omega,\zeta) = \frac{F'(\zeta)}{2} \sum_{\substack{i=1 \\ (i\neq j)}}^{M}\sum_{j=1}^{M} \frac{j}{i-j}\rho_o^{i+j} a_i(\zeta)a_j(\zeta)\sin(i-j)\omega$$
$$+ F(\zeta) \sum_{\substack{i=1 \\ (i\neq j)}}^{M}\sum_{j=1}^{M} \frac{i}{i-j}\rho_o^{i+j} a_i(\zeta)a_j'(\zeta)\sin(i-j)\omega \tag{23}$$

Eq.(23) is derived from the velocity boundary condition on the die surface. One of the natural extensions of $G(\rho_o,\omega,\zeta)$ as a function of ρ to describe the internal flow is simply by replacing ρ_o by ρ defined in the conformal transformation. We now return to the transformation Eqs.(10a, 10b).

$$G(\rho,\omega,z) = \frac{F'(z)}{2} \sum_{\substack{i=1 \\ (i\neq j)}}^{M}\sum_{j=1}^{M} \frac{j}{i-j}\rho^{i+j} a_i(z)a_j(z)\sin(i-j)\omega$$
$$+F(z) \sum_{\substack{i=1 \\ (i\neq j)}}^{M}\sum_{j=1}^{M} \frac{i}{i-j}\rho^{i+j} a_i(z)a_j'(z)\sin(i-j)\omega \tag{24}$$

The velocity field of Eqs.(13), (16a) and (16b) is now completely determined and kinematically admissible. From the velocity field of Eqs.(13), (16a) and (16b), the strain rate components can be easily determined. Referring to Fig.4, the material is divided into three zones. In the undeformed zone I, the rigid body motion in the axial direction is given by V_o. In the zone III, the rigid body motion of the product is given by $V_f = V_o/\lambda^2$. The derived velocity field is of course consistent with these boundary velocities. The deforming zone II is bounded by Γ_1 a Γ_2 and the die surface Γ_3. The velocity discontinuities do not exist on the entrance and exit planes(Γ_1 and Γ_2) for the chosen die profile of Eq.(5).

Upper-bound

Introducing the upper-bound theorem the expression of Eq.(25) is a minimum for the actual velocity distribution [12], i.e.

$$J^* = \dot{W}_i + \dot{W}_f \tag{25}$$

where J* is an upper-bound on the energy consumption. There is no energy dissipation by shear on the elastic-plastic boundaries, since no velocity discontinuity exists on either of them for the chosen die profile.
In Eq.(25), the internal power of defomation, $\dot{W}_i$, is calculated from the derived strain rate field

$$\dot{W}_i = \int_V \bar{\sigma}\,\dot{\bar{\epsilon}}\,dV \tag{26}$$

where

$$\dot{\bar{\epsilon}} = \frac{2}{\sqrt{3}}\left[(\dot{\epsilon}_x^2 + \dot{\epsilon}_y^2 + \dot{\epsilon}_z^2)/2 + \dot{\epsilon}_{xy}^2 + \dot{\epsilon}_{yz}^2 + \dot{\epsilon}_{zx}^2\right]^{\frac{1}{2}}$$

In the above equation, $\bar{\sigma}$ is the flow stress for work-hardening material and is assumed to be a function of only the total strain, $\bar{\epsilon}$. The total strain ($\bar{\epsilon}$)

can be obtained by integrating the strain rate($\dot{\bar{\epsilon}}$) with respect to time along the strain path.
The frictional power dissipated along the die surface Γ_3 , $\dot{W}_f$, is given by

$$\dot{W}_f = \int_{S_3} \frac{m\bar{\sigma}}{\sqrt{3}} |\Delta V_3| \, dS_3 \tag{27}$$

where m is a friction factor between the working material and the die surface. ΔV_3 is the velocity at the die sion pressure is determined to be

$$P_{avg} = \frac{J^*}{\pi V_o} \tag{28}$$

Results and Discussion

The derived upper-bound formulation can be applied to the extrusion of arbitrarily-shaped sections from round billets. Several types of sections are chosen for computation. The formulation is not given in the closed form in any case. So, computation is carried out numerically. As the working material, annealed commercially pure aluminum (Al 1100-0) is chosen whose flow stress is given by the following stress-strain relationship;

$$\sigma = 15.5 \, \bar{\epsilon}^{0.295} \ \text{kg/mm}^2 \tag{29}$$

Instead of integrating the strain rate along the strain path, the mean strain rate on each chosen plane in the numerical calculation is integrated numerically to obtain the mean total strain on each cross-section.

The length, l_i , from the entrance plane to an inflection point is direct and more useful. l_i can be expressed in terms of C as

$$l_i = \frac{L}{2} \left[1-K+ \sqrt{\left(\frac{1}{3} + K^2 \right)} \right] \tag{30}$$

where $K = (1-\lambda)/CL^4$

In the calculations, the radius of the billet is set equal to unity. Therefore, the die length L is treated dimensionless in all calculations which follow.

In extruding sections from round billets, the following sections are chosen for computation;

1. Four-leaf section with curvilinear boundary (R.A.=60%)
2. Square sections of three sizes (R.A.=40, 60 and 80%)
3. Rectangular sections of two aspect ratios (R.A.=80%, A/B=2.0 and 1.25)
4. Gear section (R.A.=60%, Module of gear=3 Number of teeth=20)

The aspect ratio in rectangular sections, A/B(ratio of longer side to a shorter one, unity in a square), can be considered to be a simple measure of shape complexity. Eight terms are used for external and internal mapping functions in most of the cases. The generalized mapping functions defined by the interpolating polynomials were used for the first three types of sections. For the gear section, a special internal mapping function defined by Eqs.(11a, 11b) was employed instead.

As proposed in this study, the geometrical change of the die contour along the axial direction can be controlled by two geometric variables, C and L.

C_{opt} and L_{opt} are the values of C and L at which the total extrusion power is a minimum, and by these values the optimal die configurations can thus be determined.

Fig.5(A) shows the variation of extrusion power with die length for four-leaf section(R.A.=60%, m=0.1) when C is set as 0.14, and Fig.5(B) shows the pressure variation with C when L is fixed at 0.182 (L_{opt}, determined in Fig.5(A)). Fig.6 and Fig.7 show the variation of extrusion pressure with respect to C and L, for extrusion of rectangular section(R.A.=80%, A/B=2.0, m=0.1) and for extrusion of gear section(R.A.=60%, Module=3, m=0.1), respectively.

Optimal values of C and corresponding pressures for the chosen sections are shown in Table 1 together with l_i/L_{opt} relative position of an inflection point at optimum.

As expected, the pressure increases with increasing reduction of area and slightly with increasing aspect ratio. But, the effect of sectional shape on pressure is found to be very small. The optimal die length becomes considerably longer with increasing reduction of area, but becomes slightly larger with incressing aspect ratio. The relative position of an inflection point(l_i/L_{opt}) moves downstream toward the exit with increasing reduction of area. All in all, it is confirmed that the optimal configuration of the die does not vary appreciably for moderate changes of the sectional shape when the reduction of area is kept constant. Optimized configurations of the die surface for the sections are shown in Fig.8 through Fig.14.

Conclusions

1. A new method of analysis is proposed for the three-cimensional extrusion of arbitrarily-shaped sections from round billets through continuous dies. External mapping of a section is effectively utilized to construct a continuous die surface and thus to design an optimal die using a biquadratic polynomial as overall die profile. The new concept of "profilization" is found to be very useful in constructing a die surface to an "effective" die profile defined in this study.

2. A kinematically admissible velocity field is derived utilizing the incompressibility, by two types of internal mapping both assuming the uniform distribution of axial velocity. From the derived velocity field, an upper-bound solution is formulated to find the average extrusion pressure.

3. Computation is carried out for some chosen sections of various shapes in extruding annealed, commercially pure aluminum. In lubricated extrusion, the optimal extrusion pressure is mainly a function of the reduction of area. For the given reduction of area, it is found that moderate variations under such geometric conditions as the optimal position of an inflection point and cross-sectional shape of the product do not influence the extrusion pressure appreciably. Using an approximate optimization, the optimal configurations of the die surface are found for the chosen sections under the given assumptions.

References

1. Beyer, K.: Ein Allgemeines Schema fuer das Kalibrien von gezogenen Sonderprofilen. Neue Huette 21 (1976) 389-394.

2. Juneja, B.L.; Prakash, R.: An Analysis for Drawing and Extrusion of Polygonal Sections. Int. J. MTDR 15 (1975) 1-30.

3. Gun, G.Y.; Polukhin: Method of Conformal Transformations in Three-Dimensional Problems of the Theory of Rolling and Drawing-I. Izvest. V.U.Z. Chernaya Met. 10 (1971) 63-66.

4. Nagpal, V.: On the Solution of Three-Dimensional Metal-Forming Processes. J.Eng. for Ind. Trans.ASME 99 (1977) 624-629.

5. Yang, D.Y.; Lee, C.H.: Analysis of Three-Dimensional Extrusion of Sections Through Curved Dies by Conformal Transformation. to be published in Int. J. Mech. Sci. (1978)

6. Nagpal, V. and Altan, T.: Analysis of the Three-Dimensional Metal Flow in Extrusion of Shapes with the Use of Dual Stream Functions. Proc. 3rd NAMRC, Pittsburgh, Pa. (1975) 26-40.

7. Kantorovich, I.V. and V.I. Krylov: Approximate methods of Higher Analysis, Interscience, NewYork(1964)

8. Nagpal, V, and Underwood, E.E. Analysis of Axisymmetric Flow Through Curved Dies Using a Generalized Upperbound Approach. Proc. 2nd NAMRC, Madison, Wisconsin (1974) 225-238.

9. Nagpal, V.: General Kinematically Admissible Velocity Fields for Some Axisymmetric Metal Forming Problems. ASME Paper No. 73-WA/Prod-4 (1973)

10. Avitzur, B.: Metal Forming: Processes and Analysis, McGraw-Hill, NewYork (1978)

11. Yang. D. Y.; Kim, M. U.: Lee, C. H.: An Analysis for Extrusion of Helical Snapes from Round Billets. to be publisned in Int. J. Mech. Sci. (1978)

12. Drucker, D. C.: Coulomb Friction, Plasticity and Limit Loads. J. Appl. Mech. 21 (1954) 71-74.

Table 1

sectional shape	R. A. (%)	L_{opt}	C_{opt}	l_i/L_{opt}	P_{avg} (kg/mm^2)
Four-leaf	60	1.82	0.145	0.6945	15.7
Square	40	1.35	0.192	0.6618	7.5
Square	60	1.832	0.14	0.6948	13.7
Square	80	2.454	0.11	0.7276	29.3
Rectangle A/B=1.25	80	2.467	0.115	0.731	32.5
Rectangle A/B=2.0	80	2.49	0.12	0.7349	33.3
Gear No.of teeth=20 Module=3	60	1.827	0.143	0.6955	15.6

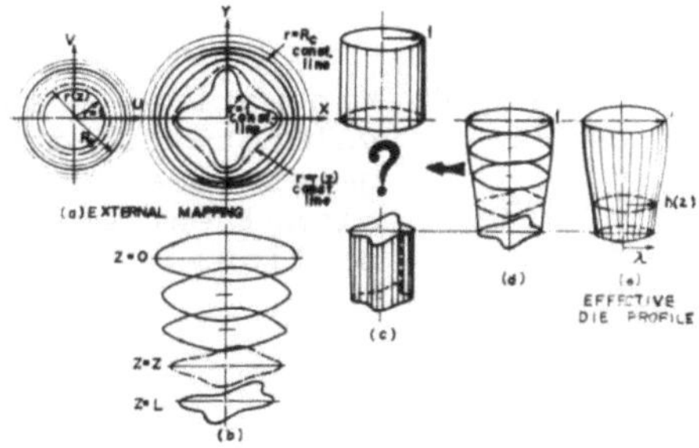

Fig.1. Construction of die surface and profilization in generalized three-dimensional extrusion of sections

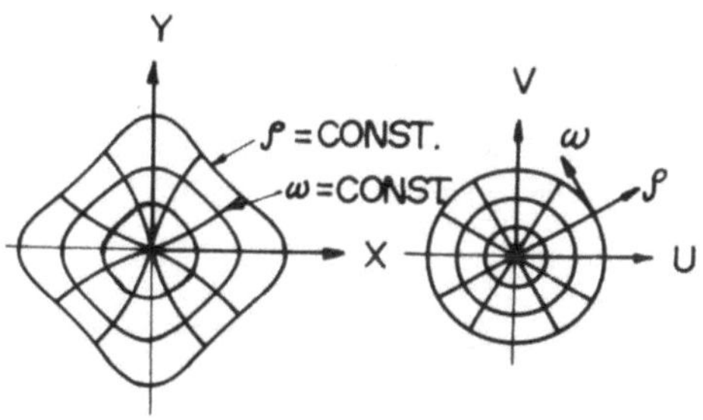

Fig.2. Internal conformal transformation of a unit circle onto an intermediate cross-section

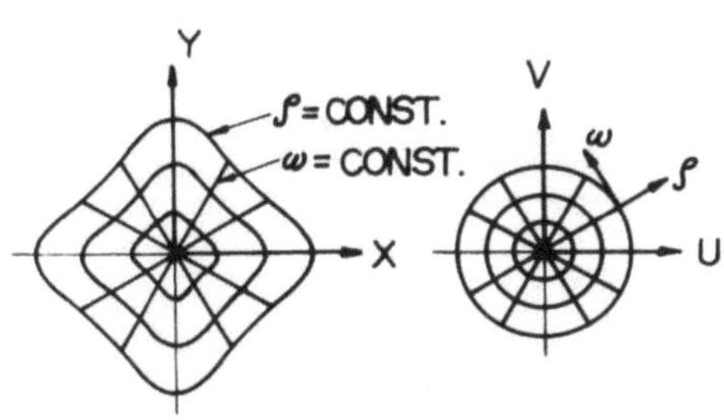

Fig.3. Special transformation of a unit circle onto an intermediate cross-section as internal mapping

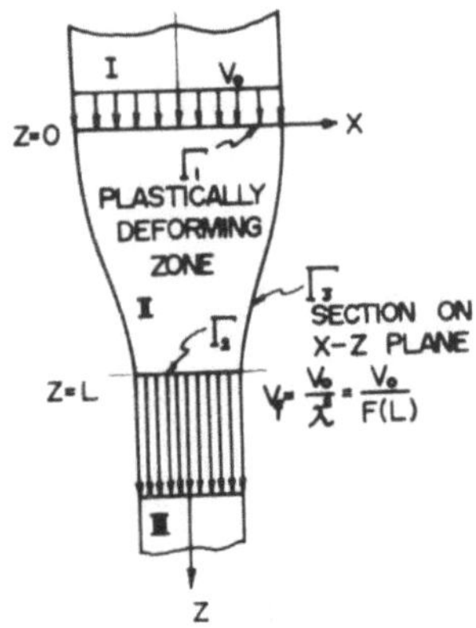

Fig.4. Kinematically admissible velocity field for generalized three-dimensional extrusion of sections

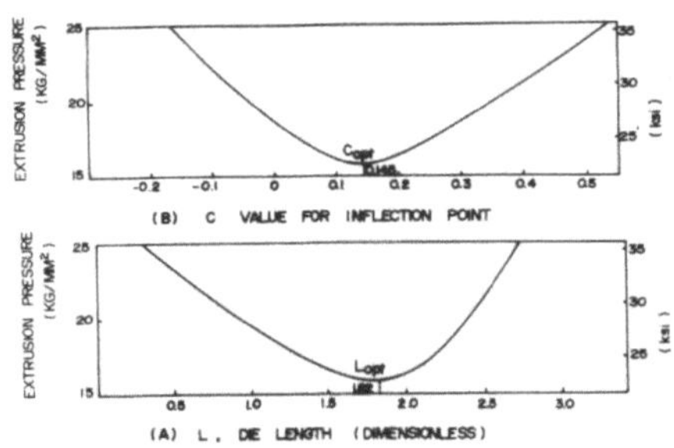

Fig.5. (A) Effect of die length on extrusion pressure for a four-leaf section(Reduction of area=60%, m=0.1, C=0.14)
(B) Effect of position of an inflection point on extrusion pressure for a four-leaf section (L=1.82)

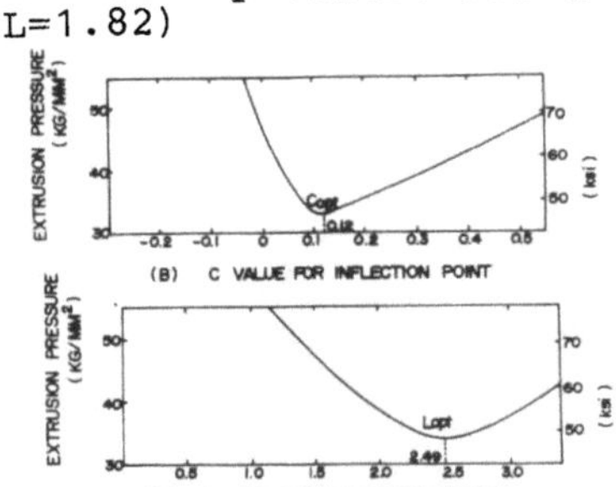

Fig.6. (A) Effect of die length on extrusion pressure for a rectangular section(R.A.=80%, m=0.1, A/B=2.0, C=0.1)
(B) Effect of position of an inflection point on extrusion pressure for a rectangular section (L=2.49)

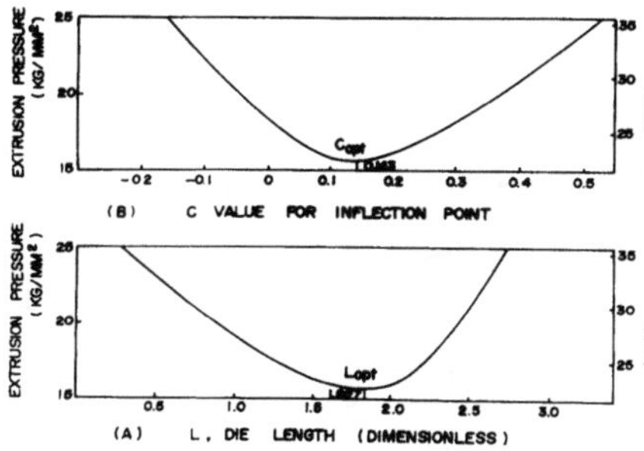

Fig.7. (A) Effect of die length on extrusion pressure for a gear section(R.A.=60%, m=0.1, Module of gear=3, no. of teeth=20, C=0.14)
(B) Effect of position oI an inflection point on extrusion pressure for a gear section(L=1.827)

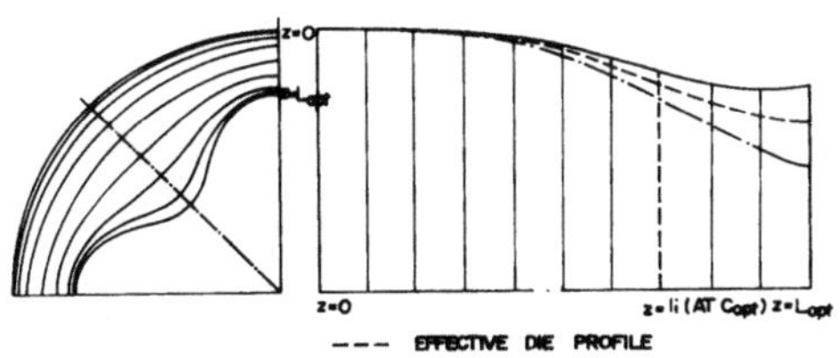

Fig.8. Optimal configuration of the die surface for four-leaf section(R.A.=60%, m=0.1, L_{opt}=1.82, C_{opt}=0.145)

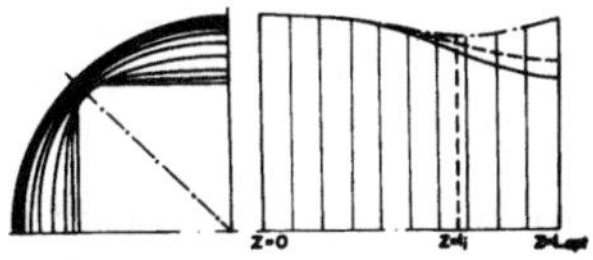

Fig.9.

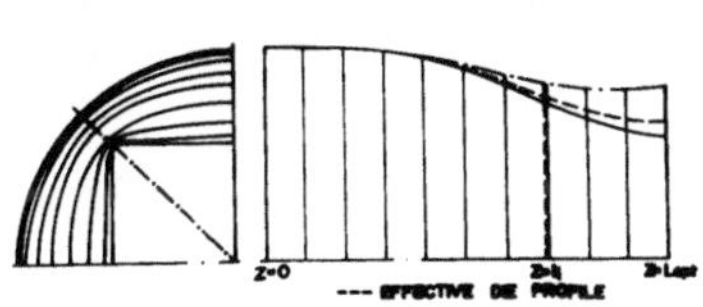

Fig.10.

Fig.9. Optimal configuration of the die surface for a square section(R.A.=40%, m=0.1, L_{opt}=1.35, C_{opt}=0.192)

Fig.10. Optimal configuration of the die surface for a swuare section(R.A.=60%. m=0.1, L_{opt}=1.832, C_{opt}=0.14)

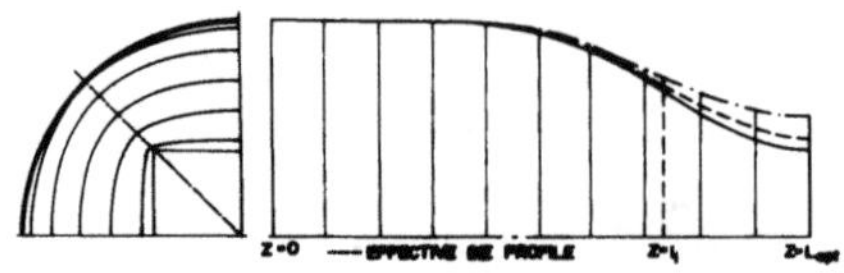

Fig.11.

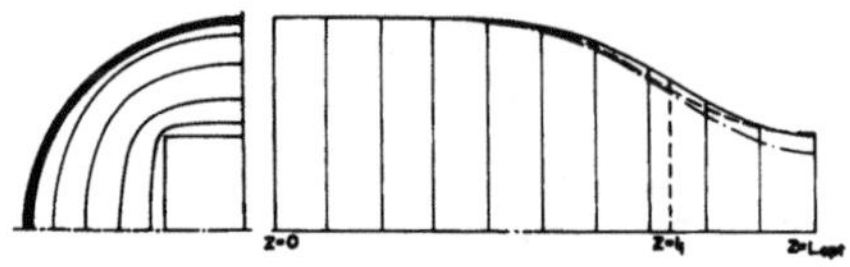

Fig.12.

Fig.11. Optimal configuration of the die surface for a square section(R.A.=80%, m=0.1 L_{opt}=2.454, C_{opt}=0.11)

Fig.12. Optimal configuration of the die surface for a rectangular section(R.A.=80%, A/B=1.25, m=0.1, L_{opt}=2.49, C_{opt}=0.12)

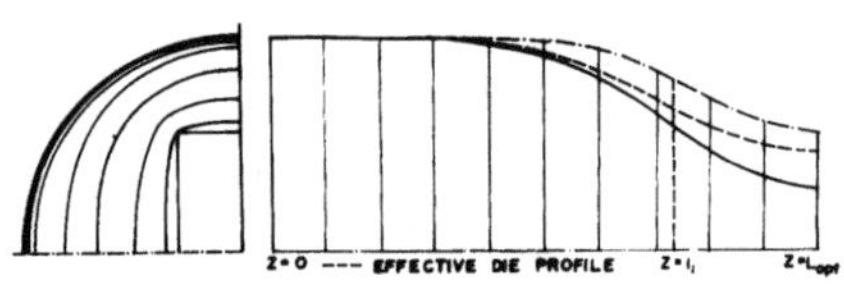

Fig.13

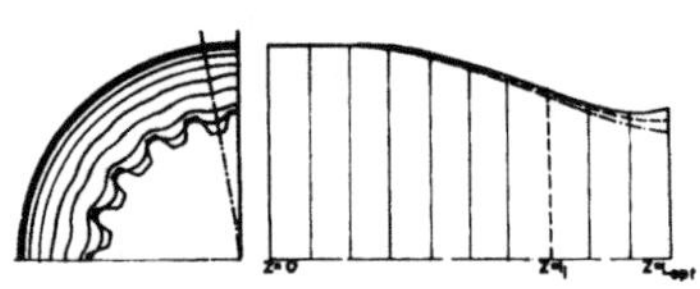

Fig.14

Fig.13. Optimal configuration of the die surface for a rectangular section(R.A.=80%, A/B=2.0, m=0.1, L_{opt}=2.647, C_{opt}=0.115)

Fig.14. Optimal configuration of the die surface for a gear section(R.A.=60%, Module of gear=3, no. pf teeth=20, m=0.1 L_{opt}=1.827, C_{opt}=0.143)

Ideal Metal Forming

(Summary)

O. RICHMOND, H.L. MORRISON and M.L. DEVENPECK

U.S. Steel Corporation Research Laboratory
Monroeville, Pa., USA

Ideal forming, as its name implies, is a theoretical concept for eliminating redundant work during a shape change. It is based upon concepts like frictionless dies not strictly attainable in practice. Yet actual processes which are designed to approach ideal processes can lead not only to maximum efficiency but also to minimum forces and wear on forming dies, and to minimum structural damage in the formed product.

A principle for designing ideal processes, but limited to steady state processes, was proposed some time ago [1]. We used it to obtain specific die profiles for ideal strip drawing [1][2], wire drawing [2][3], and tube drawing [2], and it was described in greater generality by Hill [4]. It requires the attainment of coincidence of streamlines with principal stress lines [1] and it achieves continued orthogonality of material surfaces which are at any time orthogonal to the streamlines [5]. Processes which are designed according to this principle were called streamlined by Richmond [2] and ideal by Hill [4].

A more general principle for designing ideal forming processes, not limited to steady state processes, is proposed here. It unifies the aforementioned steady streamlined processes with such nonsteady processes as simple tension and bending, tube and sphere expansion, tube nosing [6] and forging of incomplete rings [7]. It requires the attainment of continued coincidence of material surfaces which are at any time coincident with principal stress surfaces, and it reduces to the aforementioned streamlining principle for steady state processes. These material surfaces, and coincidentally principal stress surfaces, are also principal strain rate surfaces whether or not coincidence of principal stress and principal strain rate is required by the constitutive model. Usually they can be utilized to form a convenient orthogonal material coordinate system (ξ, η, ζ) with equi-mass elements.

In an ideal flow the principal strain rates, $\dot{\varepsilon}_i$, are related to the Lamé metric coefficients [8] by the simple relations, $\dot{\varepsilon}_i = (\ln H_i)_{,t}$. Thus for the material coordinate system with equi-mass elements the product of the density, ρ, and the Lamé coefficients, i.e. $\rho H_\xi H_\eta H_\zeta$ is constant throughout the flow. The rate of working is

$$\dot{w} = (\sigma_\xi - \sigma_\zeta)(\ln H_\xi)_{,t} + (\sigma_\eta - \sigma_\zeta)(\ln H_\eta)_{,t} - \sigma_\zeta(\ln\rho)_{,t}$$

and the equations of motion are [8] ,

$$\sigma_{\xi,\xi} + (\sigma_\xi - \sigma_\eta)(\ln H_\eta)_{,\xi} + (\sigma_\xi - \sigma_\zeta)(\ln H_\zeta)_{,\xi} = \rho H_\xi (f_\xi - \gamma_\xi)$$

$$\sigma_{\eta,\eta} + (\sigma_\eta - \sigma_\zeta)(\ln H_\zeta)_{,\eta} + (\sigma_\eta - \sigma_\xi)(\ln H_\xi)_{,\eta} = \rho H_\eta (f_\eta - \gamma_\eta)$$

$$\sigma_{\zeta,\zeta} + (\sigma_\zeta - \sigma_\xi)(\ln H_\xi)_{,\zeta} + (\sigma_\zeta - \sigma_\eta)(\ln H_\eta)_{,\zeta} = \rho H_\zeta (f_\zeta - \gamma_\zeta)$$

where f_j and γ_j are the gravity and inertial body forces respectiveley. For streamlined, or steady, ideal flow one of the families of material coordinates, say ξ, is coincident with the streamlines, and the product of the velocity and its orthogonal metric coefficients, i.e. $v_\xi \; H_\eta \; H_\zeta$, is constant throughout the flow.

Ideal flows of a rigid/plastic material with strain and pressure dependence of the yield stress [9] are considered. Many ideal flows are possible for the Tresca-type material where strain and pressure dependence of yielding are neglected, but the class of ideal flows is severely restricted by the inclusion of such effects. Nonetheless even in cases where exact ideal flows cannot be found, it is possible to find nearly ideal flows by perturbing solutions for the simple Tresca model. Similar remarks can be made for the effects of friction at tool interfaces.

For the Tresca-type material, ρ is constant, $\sigma_\xi - \sigma_\eta = 2K$ and either $\dot{\varepsilon}_\zeta = 0$ (biaxial, or simple, flow) or $\sigma_\xi = \sigma_\eta$ (triaxial, or full, flow). For biaxial flow,

$$\sigma_\xi - 2K \ln H_\xi = f(\zeta,t) \text{ and}$$

$$\sigma_\zeta - 2K \ln H_\xi = g(\xi,\eta,t)$$

and for triaxial flow

$$\sigma_\xi - 2K \ln H_\xi = h(t) \; .$$

For steady state, $\ln H_\xi = \ln v_\xi + C$. Thus for plane strain biaxial, or for triaxial, streamlined flow, $\sigma_\xi - 2K \ln v_\xi$ is constant throughout the flow [1][2].

The rate of working is $\dot{w} = 2K\,\dot{\varepsilon}_{\xi}$ for all cases of ideal flow in the Tresca material, which is the same as that for simple tension. Thus there is no redundant work in a nonreversing ideal flow.

Specific new examples of ideal flows are given and compared with available experiments.

References

1 O. Richmond and M.L. Devenpeck: Proc. 4th U.S. Congr. Appl. Mech., 1962, p. 1053.

2 O. Richmond: Mechanics of the Solid State, U. of Toronto Press, 1968, p. 154.

3 O. Richmond and H.L. Morrison: J. Mech. Phys. Solids 15(1967) 195.

4 R. Hill: J. Mech. Phys. Solids 15(1967) 223.

5 R. Hill: J. Mech. Phys. Solids 14(1966) 245.

6 H. Lippmann: J. Mech. Phys. Solids 10(1962) 111.

7 H. Lippmann: Int. J. Mech. Sci. 9(1967) 223.

8 G. Lamé: Leçons sur la Théorie Mathématique de l'Elasticité des Corps, 1852.

9 W.A. Spitzig, R.J. Sober and O. Richmond: Met. Trans. 7A(1976) 1703.

The Application of Singular Perturbation Techniques to the Analysis of Forming Processes for Strain-Hardening Materials

I. F. COLLINS

Department of Mathematics, University of Manchester Institute of Science and Technology, Manchester, U.K.

Summary

Velocity and strain-rate discontinuities are common features of slipline field solutions to plane strain problems for perfectly plastic materials. In a real material exhibiting hardening these discontinuities are replaced by narrow bands across which the gradients of velocity or strain-rate are large but finite. This paper sets up a mathematical description of these bands using concepts of singular perturbation theory and investigates the form of the singularity in the perfectly plastic limit. The analysis suggests a possible singular perturbation procedure by which approximate solutions to forming processes involving a small amount of hardening can be deduced from known perfectly plastic solutions. In this paper discussion will be limited to steady flow problems.

Introduction

The most powerful analytical tool available for the analysis of metal forming processes is the method of characteristics as developed in the theories for plane strain, plane stress and axisymmetric deformations. Whilst this method is capable of coping with problems with quite complicated geometries it is severely limited as far as material behaviour is concerned, since all hardening effects are ignored and the deforming metal is assumed to behave as a perfectly plastic material. It is a straightforward matter to generalise the basic field equations for steady or unsteady flows to the case of an isotropic hardening material as has been done for example by Richmond - see Devenpeck and Weinstein [1]. In plane strain these equations are most conveniently written with reference to the α- and β- sliplines (maximum shear strain-rate trajectories):

$$\frac{\partial p}{\partial s_\alpha} + 2k\frac{\partial \phi}{\partial s_\alpha} - \frac{\partial k}{\partial s_\beta} = 0 \tag{1.1}$$

$$\frac{\partial p}{\partial s_\beta} - 2k\frac{\partial \phi}{\partial s_\beta} - \frac{\partial k}{\partial s_\alpha} = 0 \tag{1.2}$$

$$\frac{\partial u}{\partial s_\alpha} - v\frac{\partial \phi}{\partial s_\alpha} = 0 \tag{1.3}$$

$$\frac{\partial v}{\partial s_\beta} + u\frac{\partial \phi}{\partial s_\beta} = 0 \tag{1.4}$$

$$\frac{Dk}{Dt} \equiv u\frac{\partial k}{\partial s_\alpha} + v\frac{\partial k}{\partial s_\beta} + \frac{\partial k}{\partial t} = h\,(2\dot{\gamma})$$

$$\equiv h\left(\frac{\partial u}{\partial s_\beta} + \frac{\partial v}{\partial s_\alpha} - v\frac{\partial \phi}{\partial s_\beta} + u\frac{\partial \phi}{\partial s_\alpha}\right) \tag{1.5}$$

where ds_α and ds_β denote elements of arclength along the α- and β- curves respectively. The first two equations are the equilibrium conditions (the deformation being assumed to be quasi-static), k is the yield shear stress, p is the mean (hydrostatic) pressure and ϕ is the inclination of the sliplines to a fixed reference direction. If k is a constant, the last term in each of these equations vanishes enabling the equations to be integrated to give the Hencky relations of the perfectly plastic theory. Equations (1.3) and (1.4) express the fact that the extension rates are zero in the local slipline directions and are just Geiringer's equations of the classical theory (u and v are the velocity components in the local α- and β- directions respectively). The final equation is the isotropic hardening law, Dk/Dt is the material rate of change of the yield shear stress, $\dot{\gamma}$ is the shear strain-rate and h is a hardening modulus. If the distribution of yield stress is assumed known the first four equations of this system can be solved to find the plastic stress and velocity fields for a non-homogeneous plastic body. Some particular solutions may be found in [2] and [3], whilst Spencer [4] has developed a systematic perturbation scheme appropriate for a small degree of inhomogeneity. There seems however to have been very few analytical studies of the practically important problem of the deformation of a hardening solid in which the current distribution of yield stress has to be found as part of the solution by making use of the hardening rule. Hill in an unpublished note of 1948 investigated the steady flow of a linearly hardening material through a simple wave region around a convex or concave corner, whilst Halling and Mitchell [5] have discussed an iterative scheme for constructing the solution for an extrusion process in which the velocity fields associated with the known perfectly plastic slipline solution is used to obtain a first order estimate of the distribution of yield stress. This scheme has been critically discussed by Pearce [6] who emphasizes, as indeed do Halling and Mitchell, that this procedure runs into severe difficulties when the perfectly plastic solution possesses singular curves such as velocity discontinuities which are of course very common features of slipline solutions. The main object of this paper is to use elementary concepts of singular perturbation theory to show how such singular curves are "opened up" in the corresponding hardening solution. The need for such a study has been emphasized by Hopkins [7] who also discusses some general aspects of the equations governing steady flows of hardening materials.

There has of course been much interest in recent years in the use of finite element procedures to analyse forming processes for hardening materials [8], although in some special situations it is possible to obtain solutions using rather more simple numerical methods [9]. There are also available a number of experimentally obtained slipline field solutions (c.f. [1] and [10] - [12]) which in particular demonstrate the "opening up" of discontinuities.

Steady Flows - General Discussion

In the study of the deformation of hardening materials it is necessary to distinguish between steady and unsteady processes. For perfectly plastic materials the basic uniqueness theorem [13] holds for steady flows as well as for initial deformation, yield-point, problems. The same is not true for hardening materials however, where there is a uniqueness theorem for the incremental (or rate) problem [14], but not for steady flows. Indeed due to the path dependent nature of the plastic constitutive equations one would not anticipate that such flows would be unique.

Despite the fact that steady flow problems are in this sense not "well-posed" boundary value problems, they are still attractive for theoretical studies since the local time derivative in (1.5) vanishes, which means that the equations have two rather than three independent variables and hence can be analysed using the powerful procedures for quasi-linear systems of equations in two independent variables [15-16]. Some aspects of such an analysis have been given by Pearce [6] and Hopkins [7] who have adopted the "interior derivative" approach to characteristics and have shown that although the system of five equations (1.1)-(1.5) has five real characteristics - the α- and β- sliplines (each repeated twice) and the γ - streamlines - it is impossible to decouple the equations into five ordinary differential equations along these characteristics. In particular it is impossible to rewrite the equilibrium equations (1.1) and (1.2) so as to obtain an equation with all the derivatives in the same direction. If one considers the corresponding unsteady equations with the $\partial k/\partial t$ term still in (1.5) then one has a quasi-linear system in t h r e e independent variables and the best that one can hope to do is to combine them so that all the derivatives are in t w o directions lying in a characteristic surface [16]. The fact that one obtains this structure in the steady flow situation too only serves to further emphasize the ill-posed nature of such problems.

Characteristics can alternatively be viewed as curves across which the normal derivatives can be discontinuous. The relations between the jumps in the normal derivatives of the five independent variables can be found by writing the equations in matrix form as discussed by Hopkins [7] and by calculating the right-eigenvector of the system, c.f. [16]. Here however we prefer to deduce them directly from (1.1)-(1.5). Consider the possible jumps in derivatives across an α-line; the dependent variables themselves are assumed continuous so that all derivatives in the α-direction will be continuous. Hence from (1.1), $\partial k/\partial s_\beta$ must be continuous and so from the hardening rule (1.5) Dk/Dt and $\dot{\gamma}$ must be continuous. Thus not only the velocity but also the strain-rate must be continuous in the steady flow of a hardening material. The discontinuities are hence much weaker than for a perfectly plastic material. Since $\dot{\gamma}$ is continuous, the combination $\partial u/\partial s_\beta - v\ \partial\phi/\partial s_\beta$ is continuous from (1.5) though each need not separably be. It is hence still possible for the slipline curvature $\partial\phi/\partial s_\beta$ to be discontinuous.

If [ ] denotes the jump in a quantity across a curve Γ, then arguing as above and using the remaining equations it is easy to show that the various jump relations can be summarized as follows:

$$\frac{\left[\frac{\partial p}{\partial n}\right]}{\pm 2k} = \frac{\left[\frac{\partial \phi}{\partial n}\right]}{1} = \frac{\left[\frac{\partial u}{\partial n}\right]}{v} = \frac{\left[\frac{\partial v}{\partial n}\right]}{-u} = \frac{\left[\frac{\partial k}{\partial n}\right]}{0} \qquad (2.1)$$

where $\partial/\partial n$ denotes the normal derivative to Γ and the upper/lower signs are taken according as Γ is an α- or β-line. If θ is the anticlockwise inclination of the streamline from the α-line and q is the speed (so $u = q\cos\theta$, $v = q\sin\theta$) it follows from (2.1) that

$$\frac{\partial q}{\partial n} = 0 \quad \text{and} \quad \frac{\partial(\theta+\phi)}{\partial n} = 0 \qquad (2.2)$$

so that the curvature of the streamline is continuous across a slipline - a result which does not hold in the perfectly plastic limit. Corresponding to (2.1) the jump relations across a streamline (γ- characteristic) are

$$\frac{\left[\frac{\partial p}{\partial n}\right]}{2k\sec 2\theta} = \frac{\left[\frac{\partial \phi}{\partial n}\right]}{1} = \frac{\left[\frac{\partial u}{\partial n}\right]}{v} = \frac{\left[\frac{\partial v}{\partial n}\right]}{-u} = \frac{\left[\frac{\partial k}{\partial n}\right]}{-2k\tan 2\theta} \qquad (2.3)$$

It may be noted in passing that since equilibrium demands that the normal derivative of k must be continuous across a characteristic, it must in particular be so across an unloading characteristic. However this derivative (evaluated in the sense of the local streamline) must be non-negative on the plastic side and non-positive on the rigid side and hence must be actually zero at the unloading characteristic.

Steady Flows - Perturbation Procedures

In this section we will discuss possible procedures for constructing approximate solutions for steady flow processes such as drawing, extrusion or strip rolling, involving a small amount of hardening by employing perturbation techniques on the corresponding perfectly plastic solution. From now on we shall assume that the various quantities in equations (1.1)-(1.5) have been non-dimensionalised: the lengths ds_α, ds_β by a typical length scale L (e.g. strip thickness), speeds u and v by a typical speed V (e.g. strip entry speed) and stresses p and k and h by the constant yield stress k_0 of the corresponding perfectly plastic material. The non-dimensional hardening modulus $\lambda = h/k_0$ will be assumed to be constant and much less than unity.

If we were to follow Spencer's [4] perturbation approach we should expand the five dependent variables in power series in λ of the form:

$$f = f^0 + \lambda f^1 + \lambda^2 f^{11} + \ . \ . \ . \qquad (3.1)$$

As can be readily verified, when substituted in (1.1)-(1.5) the zeroth order terms give the equations for a perfectly plastic material, whilst the first order ones give a system of linear equations. The functions f^0, f^1, f^{11} ... etc in these series are functions of (s_α^0, s_β^0) - the spatial co-ordinates in the zeroth order, perfectly plastic solution. There is an intrinsic difficulty in this approach however, for as has already been noted velocity and strain-rate discontinuities in the zeroth order, perfectly plastic solution, must be expanded into regions of finite thickness in the hardening solutions. If such a discontinuity occurred across an α-line say, then all the points on a given β-line in the expanded layer would correspond to a single point, with a single value of s_β^0 in the zeroth order solution. The "regular" expansion procedure of (3.1) must hence be replaced by a "singular" perturbation procedure. In the standard terminology of perturbation theory [17-19] the expansion procedure (3.1) is not "uniformly valid" over the complete domain of the solution.

One way of proceeding would be to employ Lighthill's method of strained co-ordinates which have been widely used in parallel situations in gas dynamics [17-19]. This would involve expanding an independent variable, in particular the "through-thickness variable s_β^0" in a series of the general form (3.1) as well as the dependent variables. However we prefer here to first reformulate the equations and take advantage of the hyperbolic nature of the equations by introducing characteristic co-ordinates α and β as the independent variables. In this we are following Lin [20] and Fox [21] who generalized Lighthill's procedure in a similar way in the corresponding gas dynamical situation. Richmond and Morrison [22] have used this formulation of the perturbation procedure in a plasticity context to deduce the approximate solution to an axisymmetric perfectly plastic problem from the corresponding plane strain solution. Their's was a regular perturbation analysis however and did not involve the expansion of discontinuities.

There are many ways of introducing an (α,β) - parameterization of the two families of characteristics. In classical slipline theory for perfectly plastic materials it is usual to define α and β as the angles turned through along the sliplines from some arbitrary chosen origin [23]. This is only possible however if the characteristic network possesses the Hencky-Prandtl property which of course does not extend to the hardening situation. Whenever possible however in these perturbation analyses we shall choose α and β in this way on the zeroth order perfectly plastic characteristic network.

The arclengths are then given by

$$ds_\alpha = h_\alpha \, d\alpha \quad \text{and} \quad ds_\beta = h_\beta \, d\beta \tag{3.2}$$

where h_α , h_β are the Lamé parameters, which satisfy the equations

$$h_{\alpha,\beta} + \phi_{,\alpha} h_\beta = 0 \quad \text{and} \quad h_{\beta,\alpha} - \phi_{,\beta} h_\alpha = 0 \tag{3.3/3.4}$$

which express the fact that the α- and β- lines cut orthogonally [24]. The Lamé parameters depend on the choice of the (α,β) - parameterization. In the classical theory $d\alpha$ and $d\beta$ are differentials in the angle ϕ so that h_α and h_β are then equal to the radii of curvature of the sliplines.
The basic equations (1.1)-(1.5) are now rewritten:

$$h_\beta (P_{,\alpha} + 2k\, \phi_{,\alpha}) - h_\alpha k_{,\beta} = 0 \tag{3.5}$$

$$h_\alpha (P_{,\beta} - 2k\, \phi_{,\beta}) - h_\beta k_{,\alpha} = 0 \tag{3.6}$$

$$u_{,\alpha} - v\, \phi_{,\alpha} = 0 \tag{3.7}$$

$$v_{,\beta} + u\, \phi_{,\beta} = 0 \tag{3.8}$$

$$h_\beta u k_{,\alpha} + h_\alpha v k_{,\beta} = \lambda\, (h_\alpha u_{,\beta} + h_\beta u_{,\alpha}$$

$$+ h_\beta u\, \phi_{,\alpha} - h_\alpha v\, \phi_{,\beta}) \tag{3.9}$$

The system (3.3) - (3.9) is a set of 7 equations in the 7 unknowns h_α, h_β, ϕ, p, k, u and v.
By expanding each of the 7 dependent variables as power series of the form (3.1), substituting in (3.3) - (3.9) and comparing coefficients of like powers of λ, it is easy to set up a regular perturbation scheme, starting from the zeroth order perfectly plastic solution, in which the first order equations are essentially equivalent to those employed by Halling and Mitchell [5]. However our main aim here is to see how to cope with the velocity and strain-rate discontinuities.

Expansion of Discontinuities

Consider firstly a velocity discontinuity across an α-line in the base perfectly plastic solution, the jump in tangential velocity being of magnitude U. In the new (α,β) - formulation it is easy to see how to cope with zero thickness of the discontinuity layer in the zeroth order solution, one simply puts $h_\beta^{\,0} = 0$ in the series expansion of h_β. The characteristic co-ordinate α can still be chosen as the angle turned through by the base α-line, but β must be defined differently. If the velocity components on the minus side of the discontinuity are $(u^-(\alpha), v^0(\alpha))$ say, then on the plus side they are $(u^-(\alpha) + U$ $v^0(\alpha))$. We will choose β so that the zeroth order tangential velocity component is

$$u^0\ (\alpha,\beta) = u^-(\alpha) + \beta\ , \tag{4.1}$$

so that the boundaries of the image of the discontinuity in the characteristic diagram are $\beta = 0$ and $\beta = U$. In fact any monotonic function of β could be used in (4.1) in place of β but a linear function is chosen for simplicity.
We now look for solutions of (3.3) - (3.9) in the form of perturbation series of the form:

$$\begin{aligned}
h_\alpha &= h_\alpha^{\,0}(\alpha) + \delta(\lambda)\, h_\alpha{}'\ (\alpha,\beta) + \dots \\
h_\beta &= 0 + \delta(\lambda)\, h_\beta{}'\ (\alpha,\beta) + \dots \\
\phi &= \phi^0(\alpha) + \delta(\lambda)\, \phi'\ (\alpha,\beta) + \dots \\
p &= p^0(\alpha) + \delta(\lambda)\, p'\ (\alpha,\beta) + \dots \\
u &= (u^-(\alpha)+\beta) + \delta(\lambda)\, u'\ (\alpha,\beta) + \dots \\
v &= v^0(\alpha) + \delta(\lambda)\, v'\ (\alpha,\beta) + \dots \\
k &= 1 + \nu(\lambda)\, k'\ (\alpha,\beta) + \dots
\end{aligned} \tag{4.2}$$

where δ and ν are functions of λ to be determined. (N.B. $\phi^0(\alpha) = \alpha$). From general experience with singular perturbation series [17-19] , we would anticipate that δ and ν will involve fractional powers of λ or possibly logarithmic functions. They both must of course vanish with λ. These functions are determined by the condition that when substituted into (3.3) - (3.9) they should yield a systematic iterative scheme starting from the zeroth order perfectly plastic solution. There is no a priori reason for the functions of λ occurring in the expansions of h_α - v in (4.2) to be all the same - this becomes necessary however when one substitutes these series into (3.3) - (3.8).

Consider now the hardening rule (3.9): the lowest order term on the right-hand side comes from the first term and is $\lambda h_\alpha^{\,0}$ and is hence $O(\lambda)$. The lowest order term on the left-hand side comes from the second term and is $h_\alpha^{\,0}(\alpha)\ v^0(\alpha)\ \nu(\lambda)\ k_{,\beta}{}'(\alpha,\beta)$ and is hence $O(\nu)$. (An important exception to this is when $v^0(\alpha) = 0$, i.e. when no material is crossing the discontinuity in the base perfectly plastic solution. Such "contact discontinuities" frequently occur in slipline solutions at the boundary of "dead metal zones". In such a case the lowest order term in the convective derivative of the left-hand side of (3.9) is $O(\delta\nu)$.) These lowest order terms only balance therefore if $\nu = \lambda$ (or $\delta\nu = \lambda$ for a contact discontinuity).

If we now consider the equilibrium equation (3.5). We know that $p_{,\alpha}^{\,0} + 2\phi_{,\alpha}^{\,0} = 0$ from Hencky's relations. The lowest order terms in (3.5) are hence $\delta h_\beta{}'\delta(p'_{,\alpha} + 2\phi'_{,\alpha})$ and $h_\alpha^{\,0}\nu k'_{,\beta}$ which balance only if $\delta^2 = \nu$. Thus we must choose

$$\delta = \lambda^{\frac{1}{2}} \quad \text{and} \quad \nu = \lambda \tag{4.3}$$

in the expansions (4.2), though in the case of a contact discontinuity the singularity is stronger and we must choose

$$\delta = \lambda^{1/3} \quad \text{and} \quad \nu = \lambda^{2/3} \tag{4.4}$$

The complete set of first order perturbation equations are:

$$
\begin{aligned}
&v^{o}(\alpha)\, k'_{,\beta}\,(\alpha,\beta) = 1 \\
&h'_{\alpha,\beta} + \phi^{o}_{,\alpha}\, h_{\beta}' = 0 \\
&h'_{\beta,\alpha} - \phi^{o}_{,\beta}\, h_{\alpha}^{o} = 0 \\
&h'_{\beta}(p'_{,\alpha} + 2\phi'_{,\alpha}) - h_{\alpha}^{o}\, k'_{,\beta} = 0 \\
&p'_{,\beta} - 2\phi'_{,\beta} = 0 \\
&u'_{,\alpha} - v^{o}\,\phi'_{,\alpha} - v'\,\phi^{o}_{,\alpha} = 0 \\
&v'_{,\beta} + u^{o}\,\phi'_{,\beta} = 0
\end{aligned}
\tag{4.5}
$$

These equations result from equating the coefficients of $\lambda^{\frac{1}{2}}$. Due to the different order of the first term in the expansion of k one also gets a relation between these first order quantities from the $\lambda^{3/2}$ terms in the hardening relation (3.9):

$$
\begin{aligned}
&h'_{\beta}\,(u^{-} + \beta)\, k'_{,\alpha} + h_{\alpha}^{o}\, v'\, k'_{,\beta} + h'_{\alpha}\, v^{o}\, k'_{,\beta} \\
&= h'_{\alpha} + h_{\alpha}^{o}\, u'_{,\beta} + h'_{\beta}\, v^{o}_{,\alpha} + h'_{\beta}\,(u^{-} + \beta)\phi^{o}_{,\alpha} \\
&\qquad - h^{o}_{\alpha}\, v^{o}\, \phi'_{,\beta}
\end{aligned}
\tag{4.6}
$$

Clearly it will only be possible to obtain general solutions to these equations in particularly simple cases - where the zeroth order slipline is straight for example - but in this paper we will concentrate on investigating the form of the singularities. It follows from above that in the perfectly plastic limit ($\lambda \to o$) the tangential velocity gradient $\partial u/\partial s_{\beta} = u_{,\beta}/h_{\beta} \to \infty$ as $\lambda^{-\frac{1}{2}}$ in general but as $\lambda^{-1/3}$ in the case of a contact discontinuity. The above technique can be easily modified to cope with the expansion of a single point, this point being the intersection of two velocity discontinuities, by putting h_{α}^{o} as well as h_{β}^{o} equal to zero in (4.2), and by putting $u^{o\alpha} = u^{-} + \beta$ and $v^{o} = v^{-} + \alpha$, where u^{-}, v^{-} are constants. The lowest order of the non-vanishing terms in the equations now changes and one finds that one must now put

$$\delta = \nu = \lambda \tag{4.7}$$

which of course is the same as for a regular expansion.
Finally consider a strain-rate discontinuity in which once again $h_\beta^{\,o} = 0$ but where $u^o = u^o(\alpha)$ and does not vary through the discontinuity. The shear strain rate terms in (3.9) are hence $O(\delta)$ and the right-hand side hence $O(\lambda\delta)$. The lowest order convected derivative is $O(\nu)$, so that we must put $\lambda\delta = \nu$. Equilibrium again demands that $\delta^2 = \nu$, so that for a strain-rate discontinuity we require

$$\delta = \lambda \quad \text{and} \quad \nu = \lambda^2 \tag{4.8}$$

The Matching Problem

In the previous section we have derived the form of the "natural perturbation scheme" for various types of "singular region", each region being considered in isolation. When one comes to evaluate complete solutions however one will need to match these various types of solution, by applying continuity conditions at their interfaces. Since we are dealing with an essentially hyperbolic situation we cannot use the term by term matching of inner and outer expansions commonly used in fluid dynamics [17-18]. Such procedures are appropriate for elliptic equations where the dependent variables and their derivatives of all orders are continuous. Instead we must employ the much weaker continuity conditions derived in an earlier section of this paper.
At first sight there is a paradox arising when one considers this matching process due to the differing forms of the perturbation series in the various regions. As a specific example consider ϕ which must be continuous everywhere in a plastic region: yet in an expanded velocity jump the first order

term is $O(\lambda^{\frac{1}{2}})$ whilst in a neighbouring regular region or expanded strain-rate jump it is $O(\lambda)$ which makes it impossible to equate first order terms. (N.B. Since it has been shown that the strain-rate must be continuous it is in fact necessary to regard a velocity discontinuity as being sandwiched between two strain-rate discontinuities). This paradox is resolved however by appreciating that in the preceding analysis we have assumed that the changes from the perfectly plastic solution are caused by the local increase in yield stress due to hardening. However there may well be more important effects induced on the boundary of such a region. For example in a "regular region" hardening induces changes of $O(\lambda)$ but if such a region is bounded by one or more velocity discontinuities, in which hardening produces larger changes of $O(\lambda^{\frac{1}{2}})$, such changes will be transmitted through the entire "regular region" due to the need to preserve continuity and equilibrium. In other words we must impose a perturbation series of the form (4.2) (with $\delta = \lambda^{\frac{1}{2}}$ and $\nu = \lambda$) on such a "regular region". As can easily be verified the first order coefficients of $\lambda^{\frac{1}{2}}$: ϕ', p', etc will then satisfy the equations for a perfectly plastic material, so that to order $\lambda^{\frac{1}{2}}$ the characteristic network in a "regular region" will still form a Hencky-Prandtl net. Hardening in such a region only has to be considered when working to order λ. These considerations suggest that it may be possible to deduce the main effect of a small amount of hardening on a particular process by perturbing the known perfectly plastic solution but retaining only the dominant terms in the series expansions, which will be $O(\lambda^{\frac{1}{2}})$ unless there are contact discontinuities present in which case they will be $O(\lambda^{1/3})$. This possibility however has still to be investigated.

Conclusions

There has been very little analytical work done on the nature of flows of hardening materials, and this paper has been concerned with one of the main areas of difficulty - the "opening up" of velocity and strain-rate discontinuities. Here we have concentrated on the mathematical nature of the singularities involved in steady flows. Rather similar considerations are possible for unsteady, post-initial yield point, problems where one expands in terms of a suitable non-dimensional time variable about the known initial perfectly plastic solution. These considerations suggest a possible perturbation procedure in which the dominant effect of hardening can be deduced, but complete solutions are yet to be investigated.

The result of most practical significance arising from this analysis is that whilst in general hardening induces changes of the order of the hardening modulus λ in the yield stress it produces larger changes, of order $\lambda^{\frac{1}{2}}$, in the geometrical and velocity variables h_α, h_β and ϕ etc. Thus if one considers a particular solution for strip drawing for example, one would surmise that the drawing stress will change by $O(\lambda)$ but that the dominant effect will be that the reduction ratio for which this particular solution is valid will change by $O(\lambda^{\frac{1}{2}})$.

This investigation has been limited to materials exhibiting isotropic strain-hardening, behaviour however it should be possible, using the same device of introducing characteristic co-ordinates and Lamé parameters, to cope similarly with materials exhibiting anisotropic hardening, or strain-rate dependence and also include temperature effects. Indeed it could be argued that since the strain-rates are high and theoretically infinite at a velocity discontinuity in the perfectly plastic limit, their expansion is governed primarily by the rate dependent behaviour of the material, even in a material whose global response would normally be considered to be rate independent. This is a matter for future research.

References

1 Devenpeck, M.L. and Weinstein, A.S.: Experimental investigation of workhardening effects in wedge flattening with relation to nonhardening theory. J. Mech. Phys. Solids 18 (1970) 213-232.

2 Olszak, W. and Rychlewski, J.: Geometrical properties of stress fields in plastically non-homogeneous bodies under conditions of plane strain. Second-order Effects in Elasticity, Plasticity and Fluid Dynamics, Haifa (1962) 269-294.

3 Olszak, W. and Rychlewski, J., Urbanowski, W.: Plasticity under non-homogeneous conditions. Advances in Applied Mechanics, Academic Press, 7 (1962) 131-214.

4 Spencer, A.J.M.: Perturbation methods in plasticity - I Plane strain of non-homogeneous plastic solids. J. Mech. Phys. Solids 9 (1961) 279-288.

5 Halling, J. and Mitchell, L.A.: The effect of variation in yield stress on slipline field solutions. Int. J. Prod. Res. 5 (1967) 249-262.

6 Pearce, J.H.B.: Theory of plane strain for a rigid/work-hardening material. M. Sc. Thesis, University of Manchester (1970).

7 Hopkins, H.G.: Mathematical methods in plasticity theory. Problems in Plasticity, ed. by Sawczuk, A. (1972), 235-260.

8 Zienkiewicz, O.C.: The finite element method (Third Edition) McGraw-Hill Book Co., London (1977).

9 Collins, I.F. and Meguid, S.A.: On the influence of hardening and anisotropy on the plane-strain compression of thin metal strip. J. Appl. Mech. 44 (1977), 271-278.

10 Christopherson, D.G., Oxley, P.L.B. and Palmer, W.B.: Orthogonal cutting of a work-hardening material. Engineering 186 (1958), 113-116.

11 Palmer, W.B. and Oxley, P.L.B.: Mechanics of orthogonal machinery. Proc. Inst. Mech. Engrs 173 (1959), 623-654.

12 Farmer, L.E. and Oxley, P.L.B.: A slipline field for plane-strain extrusion of a strain-hardening material. J. Mech. Phys. Solids 19 (1971) 369-388.

13 Hill, R.: On the state of stress in a plastic-rigid body at the yield point. Phil. Mag. 42 (1951) 868-875.

14 Hill, R.: On the problem of uniqueness in the theory of a rigid-plastic solid - I. J. Mech. Phys. Solids 4 (1956) 247-255.

15 Whitham, G.B.: Linear and nonlinear waves. Wiley, New York (1974).

16 Jeffrey, A. Quasilinear hyperbolic systems and waves, Pitman, London (1976).

17 Van Dyke, M.D.: Perturbation methods in fluid mechanics (annotated edition) Parabolic Press, Stanford, California (1975).

18 Cole, J.D.: Perturbation methods in applied mathematics, Blaisdell, Waltham, Mass. (1968).

19 Nayfeh, A.: Perturbation methods, Wiley, New York (1973)

20 Lin, C.C.: On a perturbation theory based on the method of characteristics. J. of Maths and Physics 33 (1954) 117-134.

21 Fox, P.A.: Perturbation theory of wave propagation based on the method of characteristics. J. of Maths & Physics 34 (1955) 133-151.

22 Richmond, O. and Morrison, H.L.: Application of a perturbation technique based on the method of characteristics to axisymmetric plasticity. J. Appl. Mech. 35 (1968) 117-122.

23 Hill, R.: The mathematical theory of plasticity, Clarendon Press, Oxford (1950).

24 Collins, I.F.: On the theory of rigid/perfectly plastic plates under uniformly distributed loads. Acta. Mech. 18 (1973) 233-254.

Allowing for a Variable Flow Stress in the Analysis of Metal Working Processes

P.L.B. OXLEY
School of Mechanical and Industrial Engineering,
University of New South Wales, Australia

Summary

Plane strain slip-line fields constructed from experimental flow fields are given for extrusion, machining and wedge indentation. In applying the stress equilibrium equations to the stress analysis of these fields it is shown that account must be taken of variations in flow stress caused by strain-hardening in order to satisfy stress and force equilibrium, the constant flow stress theory (Hencky equations) giving results which are seriously in error. An approximate machining theory is described in which variations in flow stress with strain-rate and temperature in addition to strain are considered. Comparisons are made between predicted and experimental results.

Introduction

The slip-line field proposed by Hill [1] for plane strain extrusion through a smooth wedge-shaped die of 30° half-angle with 50 per cent reduction in thickness is given in Fig.1. The slip-line field method which is possibly the most rigorous approach to analysing metal working processes so far devised has the advantages compared with other more approximate methods of predicting the form of the deformation and the associated stress distributions as part of the solution. The predicted deformation is in general in excellent agreement with that observed experimentally as is shown, for example, by comparing the deformation of an initially square grid calculated from Hill's slip-line field (Fig.1(c)) with the experimentally deformed grid for similiar extrusion conditions given in Fig.2(a). Disadvantages of the method are that so far it has only been found possible to apply it widely to plane strain, steady motion problems and in order to obtain theoretical solutions it has been found necessary to assume that the work material is an idealized rigid-plastic material which deforms at constant flow stress. The assumption of a constant flow stress means that little account can be taken of the influence of speed (strain-rate) or temperature on a process which severely restricts the practical value of the method. The main purpose of this paper is to consider how the slip-line field method might be extended to take account of a variable flow stress.

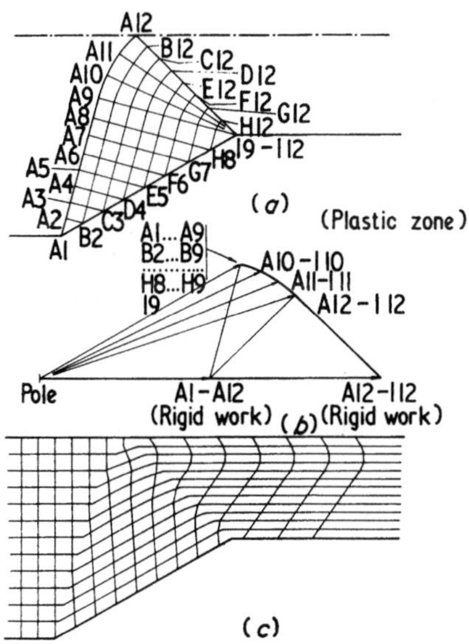

Fig.1 Theoretical slip-line field:
(a) slip-line field;
(b) hodograph;
(c) deformation of a square grid.

In calculating the distribution of mean compressive (hydrostatic) stress for the constant flow stress slip-line field in Fig.1 the well-known Hencky equations (stress equilibrium equations referred to slip-lines), that is

$$\begin{aligned} p+2k\psi &= \text{const} \quad \text{along a I-line} \\ p-2k\psi &= \text{const} \quad \text{along a II-line} \end{aligned} \qquad (1)$$

are used, where p is the hydrostatic stress which acts normal to the slip-lines, k is the shear flow stress (assumed constant) which acts parallel to the slip-lines and ψ is the rotation of the slip-lines measured from a fixed reference axis. From equations (1) it can be seen that the variation of p along a slip-line is directly related to the angle turned through by the slip-line. If k is not assumed constant then the stress equilibrium equations referred to slip-lines [2] [3] are

$$\frac{\partial p}{\partial s_1} + 2k\frac{\partial \psi}{\partial s_1} - \frac{\partial k}{\partial s_2} = 0 \quad \text{along a I-line}$$
$$\frac{\partial p}{\partial s_2} - 2k\frac{\partial \psi}{\partial s_2} - \frac{\partial k}{\partial s_1} = 0 \quad \text{along a II-line} \qquad (2)$$

where s_1 and s_2 are distances measured along the slip-lines. If k varies during deformation equations (2) show that the variation in p along a slip-line depends not only on the angle turned through by the slip-line but also on the rate of change of k with distance along slip-lines. It follows that the discontinuities in tangential velocity across the entry and exit slip-lines and the stress singularity at the exit die corner in Fig.1 are only valid if k is constant. It is well known that in real materials these idealizations are replaced by finite zones of intense shear. If it is assumed that in theory at least an equation of state exists (no history effects) and that the flow stress at a point can be uniquely related to the corresponding instantaneous values of strain, strain-rate and temperature then the variable flow stress terms in equations (2) can be expressed as

$$\frac{\partial k}{\partial s_{1,2}} = \frac{\partial k}{\partial \gamma}\frac{\partial \gamma}{\partial s_{1,2}} + \frac{\partial k}{\partial \dot{\gamma}}\frac{\partial \dot{\gamma}}{\partial s_{1,2}} + \frac{\partial k}{\partial T}\frac{\partial T}{\partial s_{1,2}} \qquad (3)$$

where γ, $\dot{\gamma}$ and T are the maximum shear strain, maximum shear strain-rate and temperature respectively. To date it seems that no theoretical or experimental (based on experimentally obtained flow fields) slip-line fields which take account of all the terms in equation (3) have been obtained. A small number of fields have however been constructed using slow speed experimental flow fields as a starting point in which case only the first term (strain-hardening term) need be considered. The first part of the paper considers this work and begins with a brief description of the techniques used in obtaining the experimental flow fields. In the second part an approximate machining theory which is based on the slip-line field work is given which takes account of variations in flow stress with strain, strain-rate and temperature. The work described is largely that of the author and his colleagues and because of space limitations no attempt is made to give a comprehensive review of the work of others. Such a review is however given in a previous paper [4].

Experimental Flow Fields

A widely used approach in obtaining flow fields for conditions approximating to plane strain is to divide the work-piece on what will be a typical plane of flow and to mark a grid on one of the faces. The two parts are then clamped together and the process is commenced. Once steady state conditions have been achieved the process is stopped as quickly as possible in order to 'freeze' the deformed grid so that it is representative of these conditions. The main difficulty with this method is to produce a grid which will remain attached to the deforming material and which is sufficiently accurate for the required measurements. The method that we have adopted is to use printed grids [5] [6]. With this method the central plane on which the grid is to be printed is fine machined and polished using standard techniques.

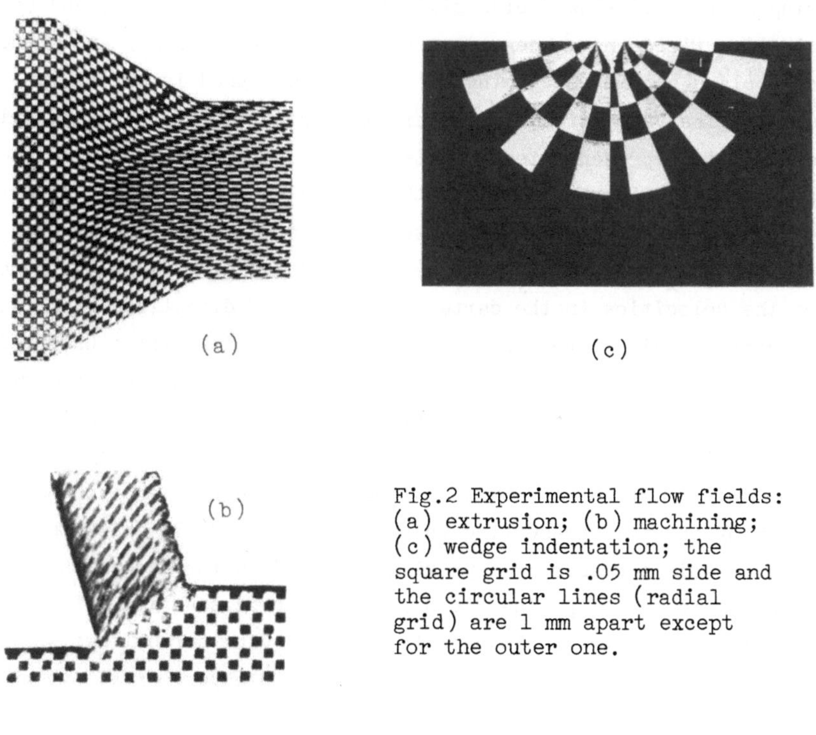

Fig.2 Experimental flow fields: (a) extrusion; (b) machining; (c) wedge indentation; the square grid is .05 mm side and the circular lines (radial grid) are 1 mm apart except for the outer one.

It is then thoroughly cleaned and spin coated with a layer of photosensitive plastic ('resist'). After drying the coated plane is clamped against a grid master consisting of alternate transparent and opaque regions in a special jig which locates the grid in the required position. The work-piece is then exposed to ultra-violet light thus polymerizing the resist in those regions where the grid master is transparent. After developing and washing the only resist that is left is that corresponding to these regions and in this way the required pattern is produced. Examples of deformed grids for extrusion, machining and wedge indentation are given in Fig.2 and it can be seen that the flow is clearly defined. In all cases the work material is an aluminium alloy (with normal strain-hardening characteristics) but the method has been applied successfully to other materials including steel. The speed for the extrusion and wedge indentation experiments was very low ($\approx$ .02 m/min) while that for machining was relatively high ($\approx$ 150 m/min).

Strain-Hardening Slip-Line Fields

Using slow speed flow fields similar to those in Figs.2(a) and (b) slip-line fields have been constructed for extrusion [7] and orthogonal (plane strain) machining [8]. The method used is briefly as follows. The velocities of flow are measured from the deformed grid (assumed to represent steady state flow with the originally horizontal lines which are parallel to the extrusion and machining directions taken as streamlines) and used to determine the strain-rate components $\dot{\varepsilon}_x = \partial u/\partial x$, $\dot{\varepsilon}_y = \partial v/\partial y$, $\dot{\gamma}_{xy} = \partial u/\partial y + \partial v/\partial x$ where u and v are the velocities in the cartesian coordinate directions x and y. The corresponding directions of maximum shear strain-rate (slip-line directions) are then found and used together with further information such as the estimated boundaries between rigid and plastic material (slip-lines) and frictionless surfaces and axes of symmetry which must be met at 45° by slip-lines, to construct an orthogonal system of curves representing the slip-line field. To check the accuracy of this necessarily approximate procedure the hodograph corresponding to the slip-line field is built up by starting with the rigid work velocity and is used to 'predict' the streamlines which can then be compared with the originally used experimental streamlines. Adjustments are then made to the slip-line field until agreement between predicted and experimental streamlines is acceptable.

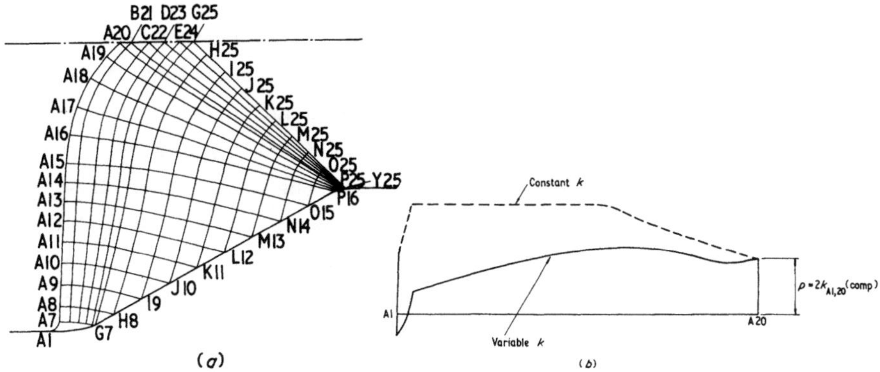

Fig.3. Extrusion: (a) slip-line field; (b) stress distributions.

Slip-line fields obtained in this way are given in Figs.3 and 4. It is interesting to note the similarity between that part of the extrusion field (Fig.3) contained by the slip-lines G7-G25 and G25-Y25 and the die face (which was approximately frictionless) and the ideal (constant flow stress) slip-line field given in Fig.1. It is

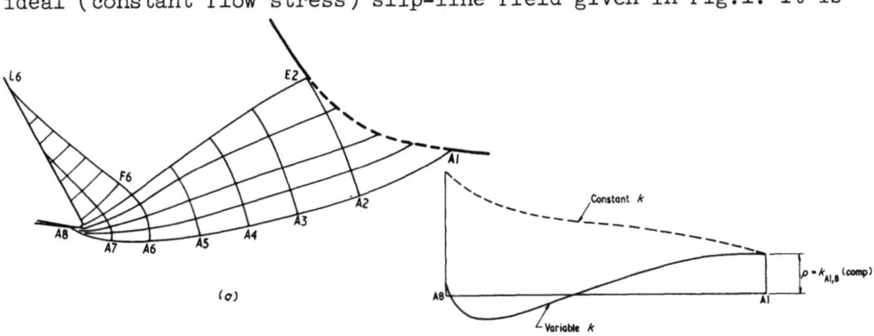

Fig.4. Machining: (a) slip-line field; (b) stress distributions.

clear however that the part of the field between the slip-lines A1-A20 and G7-G25 which replaces the velocity discontinuity of the ideal field is not consistent with equations (1) as it contains slip-lines of conflicting curvature. Before equations (2) can be applied to the stress analysis of a field it is necessary to know the distribution of shear flow stress k. This has been achieved for the fields in Figs.3 and 4 by first determining the distribution of maximum shear strain by integrating maximum shear strain-rate along streamlines with respect to time and then referring to a suitable stress/strain curve for the work material. If uniaxial stress/strain data are used then these are related to plane strain conditions in the usual way by working with an effective stress and strain. Using equations (2) it has been shown that the slip-line fields in Figs.3 and 4 are consistent for stress and for example the change in hydrostatic stress in working round the slip-line loop A19-G19-G16-A16-A19 (Fig.3) is very nearly zero while for the same loop equations (1) give a change of the order of k. The most striking difference between the stress distributions calculated from equations (1) and (2) is shown by the distributions of hydrostatic stress p on the entry slip-line to the extrusion and machining fields (Figs.3(b) and 4(b)). The distributions of p for a variable k give resultant forces which are in equilibrium with the resultants at the die and tool faces and which agree well with the experimentally measured resultants while the distributions of p for a constant k give results which are seriously in error.

Slip-line field theory has also been applied to certain non-steady motion problems and for example Hill, Lee and Tupper [9] have given a slip-line field for plane strain wedge indentation which is a class of problem in which the plastic region can be assumed to develop in such a way that the entire configuration remains geometrically similar. For such cases the stress and velocity are functions only of the single variable r/c where r is the position vector of an element measured from the origin of deformation and c is a characteristic length defining the stage of the deformation. Thus the same slip-line field applies throughout the deformation and if the characteristic length is made unity the figure representing the deformation is termed the unit diagram [9]. It can be shown that the path of an element in actual space is represented by a trajectory in the unit diagram and that all elements initially situated on the same radius through the origin of deformation follow the same trajectory in the unit diagram. Roth [10] has made use of this to check how accurately geometric similarity is maintained in actual wedge indentation experiments.

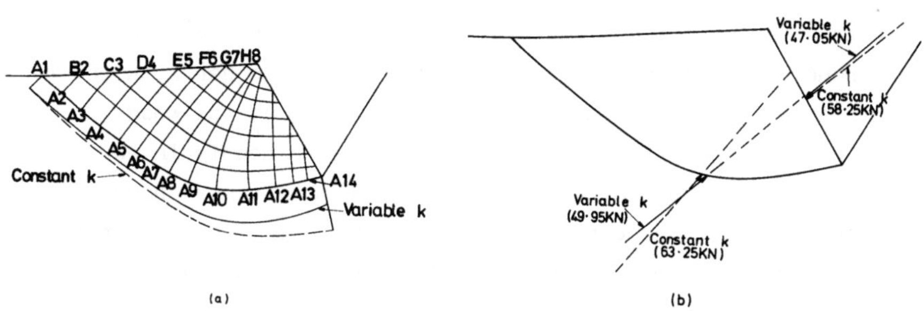

Fig.5. Wedge indentation: (a) slip-line field and stress distributions; (b) resultant forces.

Using a radial grid (Fig.2(c)) with its centre coinciding with the origin of deformation he has covered a range of indentation depths and has compared the deformed grids when scaled to the same indentation depth. His results show (for a range of wedge angles) that an initially radial line follows very nearly the same trajectory for all indentation depths thus indicating a close approximation to geometric similarity. A slip-line field (Fig.5) for one of the experimental conditions used by Roth has recently been completed. It is hoped to publish full details of this investigation shortly [11] but the following is a brief account of the method used and some of the main results. An incremental indentation is made (using a square grid) to determine the directions of maximum shear strain increment (slip-line directions) and the slip-line field is constructed using these together with the free surface condition etc. The hodograph is then built up using the slip-line field and the direction of the free surface trajectory at various points along it and small adjustments are made to the slip-line field until there is good agreement between the 'predicted' and experimental trajectories throughout the field. To determine the distribution of k for use in the stress analysis increments of maximum shear strain are summed along trajectories to give the distribution of maximum shear strain. The results of the stress analysis again show that as for extrusion and machining it is only by taking account of variations in k, with p determined from equations (2) that stress and force equilibrium can be satisfied. The distributions of p along the boundary slip-line calculated for both a variable and a constant k are given in Fig.5(a) while the corresponding resultants are given in Fig.5(b) together with the resultants calculated from the stress distributions at the wedge face. It can be seen that the resultants are much more nearly in equilibrium for a variable k than for a constant k. Also the former shows excellent agreement with the experimentally measured vertical component of force. Along the free surface p must be equal to k to satisfy equilibrium. Table 1 gives values of p at points along this surface calculated for a variable and a constant k by working from A1 along sliplines A1-A2-B2, A1-A3-C3 etc. together with the corresponding values of k. These results show that while the values of p for a variable k approximately satisfy the free surface condition there are large discrepancies for a constant k.

Table 1

Point on free surface	p MPa		k
	variable k	constant k	MPa
A1	189	189	189
B2	206	221	204
C3	214	241	217
D4	222	250	227
E5	248	280	238
F6	251	319	246
G7	263	335	252
H8	285	401	269

If carried out manually the above procedures are extremely time consuming and because of this computer-aided methods are being developed to speed up the work. A paper describing a computer-aided method for determining distributions of strain-rate and strain from an experimental flow field has already been published [12] and another describing recent developments including smoothing techniques is being prepared [13]. In essence the approach is as follows. A tracing is made of the originally horizontal and vertical lines with the former representing the streamlines of flow and with the intersection points of the two sets of lines giving points of equal time interval along the streamlines. Then using the digitized values of the grid intersection points together with the value of the equal time interval curve fitting techniques are applied to the streamlines and their normals and the corresponding velocity equations are derived from which by differentiation the components of strain-rate at the intersection points are determined. These are then used to calculate the directions of maximum shear strain-rate at the intersection points and by integrating the maximum shear strain-rate along streamlines with respect to time the distribution of maximum shear strain is obtained. Originally [12] second order polynomials were used in curve fitting but in the more recent work [13] these have been replaced by quintic splines in order to improve accuracy. Procedures have also been developed for smoothing irregularities in the flow patterns (introduced, for example, by individual grain distortion) by equalising the areas within the curvilinear quadrilaterals which make up the pattern, which should of course remain equal for plane strain and volume constancy. This in turn smooths the velocities along the streamlines and the consequent reduction in the irregularities of the calculated directions of maximum shear strain-rate means that a far better starting point for constructing the slip-line field exists. Routines are now being developed to construct the slip-line field and corresponding hodograph and to carry out the stress analysis. Although this work is initially aimed at the analysis of experimental flow fields it is hoped that it might eventually provide theoretical solutions by starting with a 'guessed' flow field and then carrying out iterations until a solution which is acceptable for stress and velocity is obtained. It is also hoped to apply the method to conditions where strain-rate and temperature (and possibly history effects) as well as strain influence flow stress and with this in mind a finite element method for calculating temperature distributions has been developed [14].

Approximate Machining Theory

An approximate machining theory which incorporates some of the features of the slip-line field analysis for machining, but in which account is taken of variations in flow stress with strain-rate and temperature in addition to strain, has been developed with a view to predicting cutting forces etc. for practical cutting conditions. The model of chip formation used in this analysis is given in Fig.6. In this the plane AB is found from the same construction as used in the

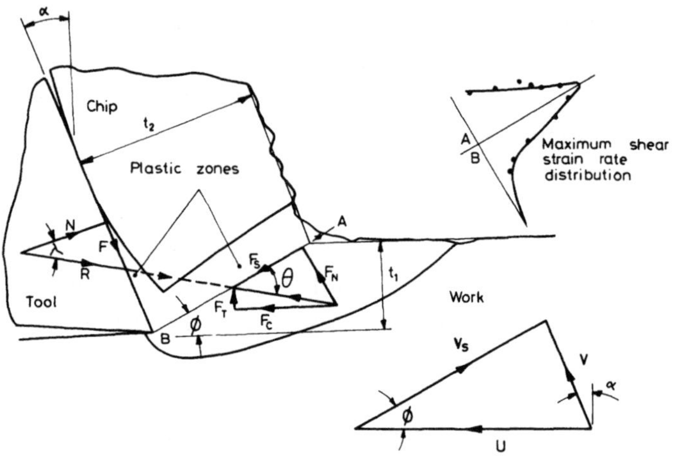

Fig.6. Model of chip formation used in approximate machining theory.

well-known Merchant [15] shear plane model in which the rigid work velocity U is changed to the rigid chip velocity V instanteously at AB which requires a discontinuity in tangential velocity across AB (V_S in Fig.6). With the present model AB is again assumed to be a direction of maximum shear stress and maximum shear strain-rate but in this case is considered to be a straight slip-line near the centre of a finite plastic zone in which the change in velocity from U to V occurs along smooth streamlines. The tool-chip interface is also assumed to be a direction of maximum shear stress and maximum shear strain-rate with plastic flow occurring in the bulk chip material over the entire contact length. Briefly, the method of solution is to determine the shear angle ϕ (Fig.6), which defines the geometry of the process, by analysing the stresses along AB and the tool-chip interface and by selecting ϕ so that the resultant forces corresponding to these distributions are in equilibrium. Experiments have shown that the strain-rate distribution is nearly symmetrical about AB (Fig.6) and that the average value along AB can be estimated from the equation

$$\dot{\gamma}_{AB} = CV_S/\ell \tag{4}$$

where V_S (Fig.6) is the shear velocity, ℓ is the length of AB and C is taken to be a material constant ($\approx$ 5.9 for plain carbon steels). The average strain-rate along the tool-chip interface is estimated from the equation

$$\dot{\gamma}_{int} = V/\delta t_2 \tag{5}$$

where V is the chip velocity and δt_2 (t_2 is the chip thickness) is the width of the rectangular plastic zone which is used to represent the flow at the interface. Substituting typical values for velocity etc. in equations (4) and (5) shows that the strain-rates in machining are of the order 10^4 to 10^6/s which is well above the values encountered in normal materials testing. The equations used for calculating temperature and strain (for a full account of the machining theory including these equations readers are referred to the paper by Oxley and Hastings [16]) indicate that the temperatures at AB and the tool-chip interface are of the order 250°C and 1200°C respectively while the corresponding strains vary from unity to exceedingly high values. One of the main problems in making predictions for the machining process is the lack of data on flow stress etc. for these extreme conditions.

In the main predictions have so far been restricted to plain carbon steels for which Oyane et al [17] have provided flow stress results from high speed compression tests for steels with carbon content in the range 0.16 to 0.55 per cent. In these tests the strain-rate ($\approx$ 450/s)was lower than in machining but the testing temperatures covered a wide range (0 to 1100°C) which makes it possible to extrapolate the compression test results into the machining range by using the velocity modified temperature concept of MacGregor and Fisher [18] which can be expressed in the form

$$T_{mod} = T(1-\nu \lg \dot{\varepsilon}/\dot{\varepsilon}_0) \tag{6}$$

where T(K) is the temperature and $\dot{\varepsilon}$ the strain-rate with ν and $\dot{\varepsilon}_0$ constants. This parameter has been used many times before to represent the opposing effects of strain-rate and temperature but normally in situations where temperature is the main variable and there are relatively small variations with strain-rate, with the strain-rates not very high. The results of Oyane et al are not suitable for testing the validity of equation (6) for the more extreme conditions of machining as the strain-rate is constant. However, some results of Campbell and Ferguson [19] which show the effects of wide ranges of strain-rate (10^{-4} to 10^5/s) and temperature (195 to 713K) on the lower yield point of a 0.12 per cent carbon steel are suitable for this purpose. These are shown plotted against T_{mod} with $\nu = 0.09$ and $\dot{\varepsilon}_0 =$ 1/s in Fig.7 and it can be seen that they fit a single curve exceptionally well thus giving some support to the use of equation (6) for conditions appropriate to machining. The results of Oyane et al show that for a given temperature (and strain-rate) the experimental stress/strain points can be fitted by the frequently used empirical relation

$$\sigma = \sigma_1 \varepsilon^n \tag{7}$$

where σ and ε are the uniaxial flow stress and strain and σ_1 and n are material 'constants' which define the particular stress/strain curve. The values of σ_1 and n for the 0.16 per cent carbon steel are plotted against T_{mod} in Fig.7 and can be seen to show a clear dynamic strain-ageing region, where flow stress increases with increase in temperature, which is typical for this kind of steel. Similar results for the other carbon steels have been obtained and all of the curves have been represented mathematically so that for given values of strain, strain-rate and temperature the flow stress can be determined for the range of steels considered. In the calculations allowance has also been made for variations in specific heat and thermal conductivity with temperature and of thermal conductivity with chemical composition.

In calculating ϕ and hence the cutting forces etc. for given cutting conditions the outline of the method used is as follows. The magnitude and direction of the resultant cutting force are determined from the following equations (derived by analysing the stresses along AB i.e. by applying the relevant equation of equations (2)) for an assumed value of ϕ

$$\tan \theta = 1 + 2\left(\tfrac{1}{4}\pi - \phi\right) - Cn \tag{8}$$

$$R = \frac{k_{AB} t_1 w}{\sin \phi \cos \theta} \tag{9}$$

where θ (Fig.6) is the angle made by R with AB, C is the strain-rate constant in equation (4), n is the strain hardening index in equation (7), k_{AB} is the shear flow stress corresponding to the strain, strain-rate and temperature along AB, t_1 is the undeformed chip thickness, and w is the width of cut. The resultant R is then resolved to give the frictional component F at the tool-chip interface which is divided by the contact area to give the resolved shear stress. The corresponding strain-rate and temperature at the tool-chip interface are then calculated to find the shear flow stress at the interface and ϕ is adjusted until the resolved shear stress and shear flow stress are equal thus giving the equilibrium value of ϕ. Fig.8 gives a comparison between predicted and experimental values of shear angle and cutting forces for a 0.19 per cent plain carbon steel and it can be seen that the agreement is good with the influence of cutting speed accurately predicted. The broken lines indicate when the calculated conditions at the interface are in the dynamic strain-ageing range.

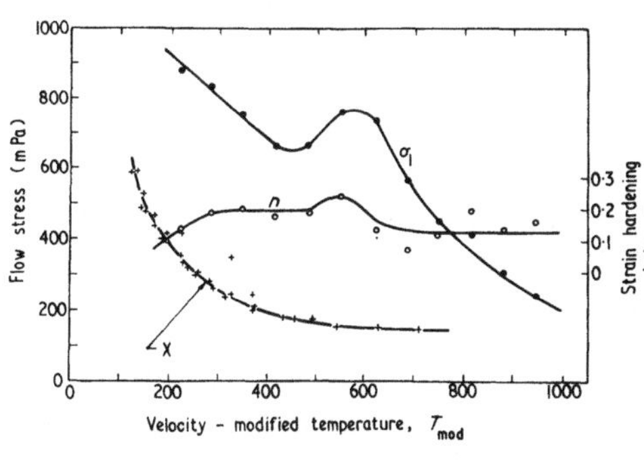

Fig.7. Flow stress results used in approximate machining analysis; X lower yield stress results.

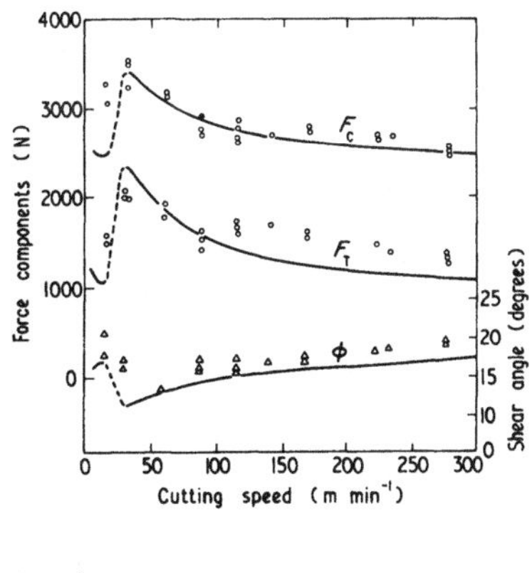

Fig.8. Predicted and experimental cutting force and shear angle results.

It has been suggested by a number of workers that this might lead to a built-up edge being formed on the cutting tool and some evidence of this was found in the experiments. Similar good agreement between predicted and experimental results (including temperature results) has been found for a range of steels and cutting conditions. In the earlier work C and δ have been treated as empirical constants with $C \approx 5.9$ and $\delta \approx 0.05$ but recently [16][20] a theoretical basis for their determination has been established.

Concluding Remarks

In addition to showing how a more realistic account can be taken of material properties in the analysis of metal working processes it is hoped that the work described might also help to encourage closer co-operation between workers in continuum and crystal mechanics. This is clearly desirable once the flow stress is no longer assumed constant and it is necessary to know, for example, how strain-rate and possibly strain-rate history affect flow stress.

References

1 Hill, R.: A theoretical analysis of the stresses and strains in extrusion and piercing. J. Iron Steel Inst. 158 (1948) 177.

2 Christopherson, D.G.; Oxley, P.L.B.; Palmer, W.B.: Orthogonal cutting of a work-hardening material. Engineering 186 (1958) 113-115.

3 Kuznetsov, A.I.: Arch. Mech. Stos. 10 (1958) 447.

4 Oxley, P.L.B.: Allowing for strain-rate effects in the analysis of metal working processes. Inst. Phys. Conf. Ser. No.21 (1974) 359-381.

5 Davies, R.W.; Hastings, W.F.: An orthogonal grid technique for the determination of strain-rates in metal deformation processes. Harold Armstrong Conf. on Production Science in Industry. Inst. Engrs. Aust. (1971) 310-321.

6 Farmer, L.E.; Fowle, R.: Plane strain extrusion of aluminium - description of photographic method for studying plastic flow. Report No.IE/I University of New South Wales (1974).

7 Farmer, L.E.; Oxley, P.L.B.: A slip-line field for plane-strain extrusion of a strain-hardening material. J. Mech. Phys. Solids. 19 (1971) 369-388.

8 Roth, R.N.; Oxley, P.L.B.: Slip-line field analysis for orthogonal machining based on experimental flow fields. J. Mech. Eng. Sci. 14 (1972) 85-97.

9 Hill, R.; Lee, E.H.; Tupper, S.J.: The theory of wedge indentation of ductile materials. Proc. Roy. Soc. A188 (1947) 273-289.

10 Roth, R.N.: The experimental determination of trajectories in geometrically similar plastic flows. J. Strain An. 13 (1978) 29-34.

11 Bhasin, Y.P.; Oxley, P.L.B.; Roth, R.N.: A slip-line field for plane-strain wedge indentation of a strain-hardening material. To be published.

12 Farmer, L.E.; Oxley, P.L.B.: A computer-aided method for calculating the distributions of strain-rate and strain from an experimental flow field. J. Strain An. 11 (1976) 26-31.

13 Farmer, L.E.; Conning, S.: Numerical smoothing of flow patterns. To be published.

14 Tay, A.O.; Oxley, P.L.B.: Determination of the temperature distribution generated during extrusion using the finite element method. Conf. on finite element methods. Inst. Engrs. Aust.(1976).

15 Merchant, M.E.: Mechanics of metal cutting process. J. Appl. Phys. 16 (1945) 267.

16 Oxley, P.L.B.; Hastings, W.F.: Minimum work as a possible criterion for determining the frictional conditions at the tool/chip interface in machining. Phil. Trans. Roy. Soc. A282 (1976) 565-584.

17 Oyane, M.; Takashima, F.; Osakada, K.; Tanaka, H.: The behaviour of some steels under dynamic compression. 10th Japan Congress on Testing Materials (1967) 72.

18 MacGregor, C.W.; Fisher, J.C.: A velocity-modified temperature for the plastic flow of metals. J. Appl. Mech. Trans. A.S.M.E. 13 (1946) A11-A16.

19 Campbell, J.D.; Ferguson, W.G.: The temperature and strain-rate dependence of the shear strength of mild steel. Phil. Mag. 21 (1970) 63.

20 Oxley, P.L.B.; Hastings, W.F.: Predicting the strain-rate in the zone of intense shear in which the chip is formed in machining from the dynamic flow stress properties of the work material and the cutting conditions. Proc. Roy. Soc. A356 (1977) 395-410.

Plastic Deformations with Free Boundaries – A Finite Element Approach

W. KLIE

Institut für Mechanik der
Universität Hannover, BRD

Summary

The problem of free surfaces in plastic flow has been treated by a new finite element model. A variational statement which is close to the upper bound theorem of plasticity has been generalized by introducing the mean stress and the surface tractions as Lagrangean multipliers. The main part of the contribution is the discussion of the model. Also, some results for plane upsetting are presented.

Introduction

A new field of application for variational principles in plasticity has been opened by the finite element concept. LUNG [1] presented a finite element procedure for large deformations of a rigid-plastic material which was subsequently applied to plane metal forming processes under steady-state conditions [2]. This method was extended to axisymmetric deformation [3].

The method is based on a variational principle which is close to the upper bound theorem of plasticity. A fundamental discussion of the functional was presented by HILL [4].

In this paper a generalized principle is used. All subsidiary conditions for the velocity field equations have been added to the variational statement by means of the Lagrangean multiplier method [5].

Boundary-value problem for a rigid-plastic body

In the flow theory of plasticity the velocity vector v_i constitutes a boundary value problem in the plastic region. The velocity field has to satisfy the relations (1) between the strain rates $\dot{\varepsilon}_{ij}$ and the velocity gradients, the condition of incompressibility (2) and the velocity boundary conditions (3)

$$\dot{\varepsilon}_{ij} = \frac{1}{2}(v_{i,j} + v_{j,i}) \tag{1}$$

$$\dot{\varepsilon}_{ii} = 0 \tag{2}$$

$$u_i = v_i \tag{3}$$

The deviatoric stresses σ'_{ij} are associated with the strain rates $\dot{\varepsilon}_{ij}$ by the non-negative scalar material function ϕ. The stresses simultaneously fulfil the yield criterion which is given in (5).

$$\sigma'_{ij} = \Phi \dot{\varepsilon}_{ij} \tag{4}$$

$$\frac{3}{2}\sigma'_{ij}\sigma'_{ij} = Y^2 \tag{5}$$

The material function ϕ depends on the local yield stress Y and on the second invariant of the strain rate tensor. So, it substantially reflects the material nonlinearity of the plastic flow problem.

The stress field has to be in equilibrium

$$\left.\begin{aligned} \sigma_{ij,j} &= 0 \\ \sigma_{ij} &= \sigma_{ji} \end{aligned}\right\}, \tag{6}$$

and has to satisfy the stress boundary conditions

$$f_i = \sigma_{ij} n_j \tag{7}$$

n_j is the outer normal of the body surface.

Variational principles related to the upper bound theorem of plasticity and nonlinear set of algebraic equations

The functional for the variational statement can be obtained from the upper bound theorem of plasticity. It was given by MARKOV [6] in 1947 and states, that among admissible velocity fields the true velocity field renders the functional (8) an absolute minimum.

$$\sqrt{\tfrac{2}{3}}\, Y \int_V \sqrt{\dot{\varepsilon}_{ij}\,\dot{\varepsilon}_{ij}}\, dV - \int_{S^\sigma} f_i v_i \, dS \Rightarrow \mathrm{MIN} \tag{8}$$

Admissible velocity fields must have piecewise first derivatives with respect to the space coordinates x_i and they must satisfy the strain rate - velocity relations (1), the incompressibility condition (2) and the velocity boundary conditions (3).

There are different procedures to construct admissible velocity fields. Generally speaking, the aim is to find a discrete analogue to the continous relations set up by the boundary value problem. If we adopt the MARKOV-principle, the discrete model is obtained by replacing the velocities by interpolation functions which have to satisfy the so called conformity relations, given by the principal restrictions (1) - (3) for the velocity field.

The relations (1) are satisfied identically by replacing the $\dot{\varepsilon}_{ij}$ by the velocity gradients, and for polynomials having piecewise first derivatives. Unfortunately, however, it is in general not possible to find interpolation functions which fulfil the incompressibility condition. A way out of this dilemma is to add the condition of incompressibility to the original functional by means of Lagrangean multipliers. This was done in 1971 by LUNG [1], who used the functional in connection with the finite element method. The multiplier means physically the negative hydrostatic pressure.

The velocity boundary conditions (3) can be handled in a different manner. In the finite element displacement approach usually the coefficients in the velocity interpolation functions stand for the displacement velocities at discrete points on the boundaries of the elements. By the well known procedure of reassembling the continuity of velocites at the nodes is obtained.

A more general approach will be obtained, if the velocity boundary conditions are added to the original functional. Again, the Lagrangean multiplier method is used.

$$\sqrt{\tfrac{2}{3}}\,Y\int_V\sqrt{\dot\varepsilon_{ij}\dot\varepsilon_{ij}}\,dV-\int_{S^\sigma} f_i v_i\,dV+\underbrace{\int_V\dot\varepsilon_{ii}\sigma^m dV}_{\text{incompressibility}}\underbrace{-\int_{S^v}\sigma_{ij}n_j(v_i-u_i)dS}_{\text{continuity}}\Rightarrow \text{STAT} \qquad (9)$$

The multipliers are now the surface tractions $\sigma_{ij}\,n_j$. So, we obtain the generalized functional (9); if the strain rates are replaced by the velocity gradients, there are no more restrictive conditions left for the velocity field equations.

To obtain algebraic equations from the generalized functional, interpolation functions (10) for the velocities in the field, the surface tractions and the mean stress are set up.

$$\left.\begin{aligned} v_i &= n^v_{il}\,a^v_l \;; \quad l = 1,2,\cdots,N^v \\ f_i &= n^f_{il}\,a^f_l \;; \quad l = 1,2,\cdots,N^f \\ \sigma^m &= n^{\sigma m}_l\,a^{\sigma m}_l \;; \quad l = 1,2,\cdots,N^{\sigma m} \end{aligned}\right\} \qquad (10)$$

The polynomials generally are composed of coordinate functions n_ℓ and algebraic numbers a_ℓ . The velocity and the mean stress approximation functions are set up in terms of the cartesian space coordinates x_i. In (11), the coordinate functions for the polynomial representation of the velocity field are given. NG^v is the degree of the approximation function. Likewise, the hydrostatic pressure was approximated. On the boundaries Legendre polynomials were used for coordinate functions (12), set up in terms of the tangential surface coordinate ξ. ξ is a normalized coordinate defined in the interval from -1 to +1. NG^f is the degree of the surface traction approximation functions.

$$n^v_{il} = (x_1)^m \; (x_2)^n \quad , \qquad m,n = 0,1,\dots,NG^v \quad m+n \leqq NG^v \tag{11}$$

$$n^f_{il} = \frac{1}{2^\nu \nu!} \frac{d^\nu \left[(\xi^2-1)^\nu\right]}{d\xi^\nu} \quad , \qquad \text{for } \{\xi \,/ -1 \leqq \xi \leqq 1\} \quad \nu = 0,1,\dots,NG^f \tag{12}$$

By independent variation of the functional with respect to the coefficients a^v, a^f and $a^{\sigma m}$ the nonlinear set of algebraic equations (13) is obtained. In (13 a) this set of equations is given in matrix notation.

$$\begin{bmatrix} \int_V \Phi \, n^v_{i,jl_2} n^v_{i,jl_1} dV & \int_V n^v_{i,il_2} n^{\sigma m}_n dV & -\int_{S^v} n^v_{il_2} n^f_{im} dS \\ \int_V n^{\sigma m}_n n^v_{i,il_2} dV & 0 & 0 \\ -\int_{S^v} n^f_{im} n^v_{il_2} dS & 0 & 0 \end{bmatrix} \begin{bmatrix} a^v_{l_2} \\ a^{\sigma m}_n \\ a^f_m \end{bmatrix} = \begin{bmatrix} \int_{S^\sigma} f_i n^v_{il_2} dS \\ 0 \\ -\int_{S^v} n^f_{im} u_i dS \end{bmatrix} \tag{13}$$

$$\text{with} \quad \Phi = \tfrac{2}{3} Y \left(\tfrac{2}{3} n^v_{i,jl_1} a^v_{l_1} n^v_{i,jl_2} a^v_{l_2}\right)^{-\frac{1}{2}}$$

$$\begin{bmatrix} \underline{H}^{vv} & \underline{H}^{v\sigma m} & \underline{H}^{vf} \\ \underline{H}^{v\sigma mT} & \underline{0} & \underline{0} \\ \underline{H}^{vfT} & \underline{0} & \underline{0} \end{bmatrix} \begin{bmatrix} \underline{a}^v \\ \underline{a}^{\sigma m} \\ \underline{a}^f \end{bmatrix} = \begin{bmatrix} \underline{h}^\sigma \\ \underline{0} \\ \underline{h}^v \end{bmatrix} \tag{13 a}$$

Plane upsetting with sticking friction - An example

So far, we have developped a most general tool for solving plastic flow problems. For practical use, there are some more arrangements required. Considering the finite element displacement method, convergence is obtained by mesh refinement; if any characteristic element dimension tends to zero, the discrete model should solve the continuous problem exactly within the limits of its assumptions. If we employ the generalized principle, we can use any approximation function without subsidiary conditions. The convergence in the sense of the Ritz procedure can be obtained independently from the mesh refinement by systematic increase of the degree of the interpolation functions.

For the problem of plane upsetting global approximation functions were used without any subdivision of the domain. For simplicity, only the first deformation path is considered here. So the boundary of the domain is given by four straight lines (Fig. 1). There are mixed boundary conditions at the boundaries 1 and 2. Because of symmetry, for both coordinate directions either the velocity or the surface traction component vanishes. The boundary 3 is stress

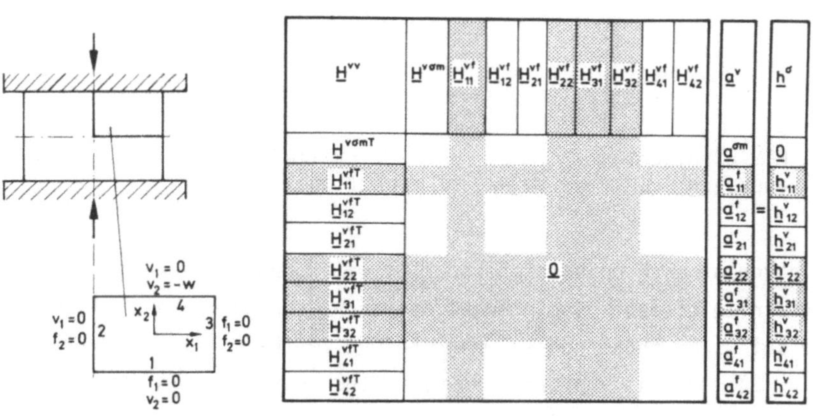

Fig. 1 Solution of the hyper matrix equation
Plane upsetting, sticking friction

free, and we have pure velocity conditions at the boundary 4.

The matrix equation for the upsetting problem is shown in Fig. 1. The shaded rows and columns are eliminated from the equations by prescribing stress boundary conditions. This leads to a considerable reduction of the number of unknowns, particularly if free surface problems are considered. The reduced set of matrix equations will be solved by inverting the $\underline{H}$-matrix.

Continuity of velocities at boundaries

The most questionable point in this new approach was, whether it was possible or not to force the velocity functions to satisfy the boundary conditions in a proper way. The continuity restriction (3)

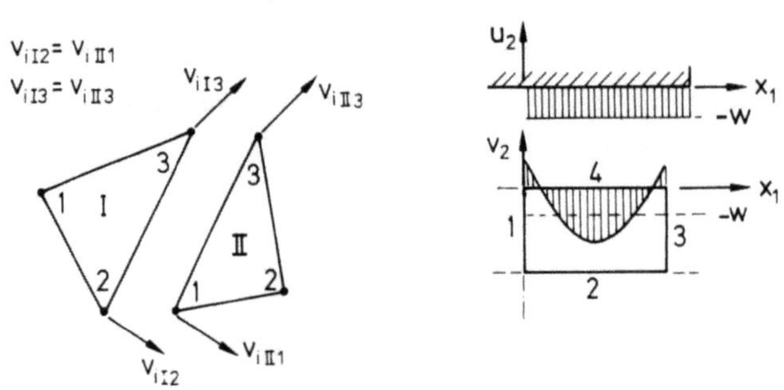

FEM-displacement approach

Generalized functional

Fig. 2 Continuity of velocities at boundaries

is given at each point at boundaries with prescribed velocities. In the finite element displacement approach, the boundary conditions are set up at those nodes, which belong to the considered boundary (Fig. 2). In this approach, however, the continuity is enforced by an integral relation (14). If we have a constant

$$-\int_{S^v} n^f_{im} n^v_{il} a^v_l \, dS = -\int_{S^v} n^f_{im} u_i \, dS \tag{14}$$

velocity, say $-w$ prescribed, it could happen that a parabolic velocity profile is obtained as result (Fig. 2). This, in fact, satisfies the integral relation (14), but not the local velocity condition. Obviously, it depends on the degrees of the approximation functions, whether or not the velocity boundary conditions are fulfilled satisfactorily.

In Fig. 3 the velocity profiles at the three boundaries, where they are prescribed, are shown. The left diagram shows the velocity profiles for approximation functions of the degree $NG^v = 4$ for the velocities and $NG^{\sigma m} = 3$ for the mean stress. The degree of the surface traction approximation was variated from $NG^f = 1$ to

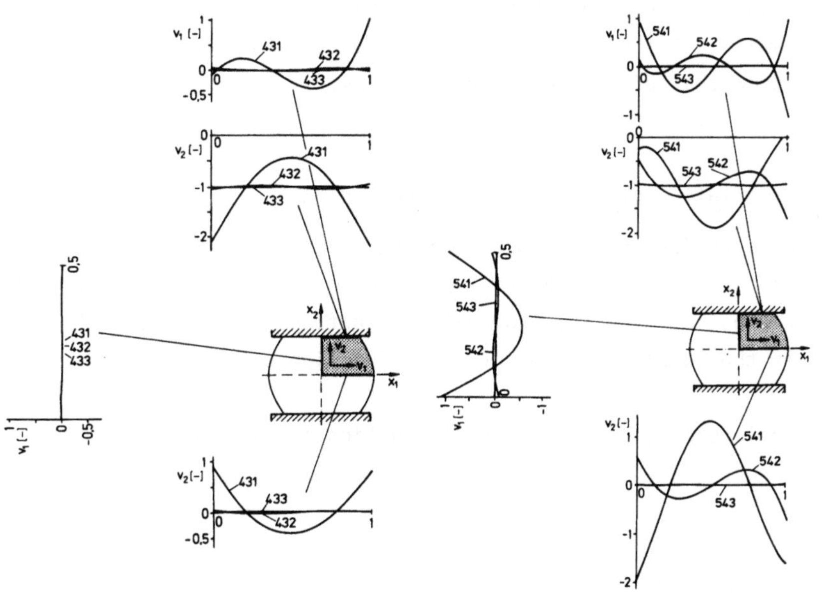

Fig. 3 Continuity of velocities at boundaries
Plane upsetting, sticking friction

Left: $NG^v=4$, $NG^{\sigma m}=3$, $NG^f=1,2,3$; Right: $NG^v=5$, $NG^{\sigma m}=4$, $NG^f=1,2,3$

$NG^f = 3$. In this case the material function ϕ was set to a constant mean value all over the domain. It is seen, that only the first order approximation, $NG^f = 1$, leads to a considerable error of the velocity curve. The right diagram shows the profiles for a fifth order approximation for the velocities and a fourth order approximation for the mean stress. The deviations of the obtained velocity profiles from the prescribed constant values are greater for the first and the second order approximation for the surface tractions. The degree $NG^f = 3$ again gives a very smooth profile. So, generally speaking, there must be always a certain relation between the degrees of the velocity and the surface traction approximation to fulfil the boundary conditions satisfactorily.

A very close relation is set up between the polynomials for the velocities and the mean stress. As in both cases the same power series were used, the volume constancy is obtained exactly in each point of the domain, if the degree of the mean stress function is just one less than that of the velocity functions.

Local variation of the material function ϕ

The results for the examination of the velocity continuity were obtained for a constant material function ϕ in the domain. Evidently, the use of global approximation functions will be succesful only, if the local variation of the ϕ-function can be managed.

The nonlinearity of the material function is handled by an iteration procedure.

$$\Phi^{(n)}(x_i) = \frac{2}{3} \frac{Y\,(v_i^{(n-1)})}{\bar{\dot{\varepsilon}}\,(v_i^{(n-1)})} \tag{15}$$

The ϕ-function is calculated with (15) from the solution of the previous, (n-1)-th iteration path. The process will be continued until a stationary value has been found.

To describe the local variation of ϕ , an interpolation polynomial (16) was set up

$$\Phi(x_i) = n_\mu^\Phi a_\mu^\Phi \;, \quad \mu = 1,2,\cdots,N^\Phi \qquad , \tag{16}$$

where N^ϕ is the number of coefficients in the approximation function. The ϕ-function was calculated at N^S points in the domain

$$\Phi_\alpha = n_{\alpha\mu}^\Phi a_\mu^\Phi \;, \alpha = 1,2,\cdots,N^S > N^\Phi \quad , \tag{17}$$

and for $N^S > N^\phi$ the coefficients a_μ^ϕ were determined by the least square method.

$$\left.\begin{aligned} &\Delta_\alpha \Delta_\alpha \Rightarrow \mathrm{Min} \\ &\quad \Delta_\alpha = n_{\alpha\mu}^\Phi a_\mu^\Phi - \Phi_\alpha \\ &\quad a_\mu^\Phi = (n_{\mu\alpha}^\Phi n_{\alpha\mu}^\Phi)^{-1} n_{\mu\alpha} \Phi_\alpha \end{aligned}\right\} \tag{18}$$

So, after replacing of the function ϕ by the polynomial (16) and exchanging the integration and the summarizing, the volume integral over a locally variable function ϕ is led back to the integration of coordinate functions with constant coefficients in (19).

$$\int_V \Phi(x_i) n_{i,jl_2}^v a_{l_2}^v n_{i,jl_1}^v dV = \sum_{\mu=1}^{N^\Phi} a_\mu^\Phi \int_V n_\mu^\Phi n_{i,jl_2}^v a_{l_2}^v n_{i,jl_1}^v dV \tag{19}$$

In Fig. 4 the ϕ-function is shown together with the polynomial representation. For coordinate functions n_μ^ϕ the complete power series (12) were taken. The results are drawn up over the (x_1, x_2)-plane with the boundaries of the domain given by the numbers 1 up to 4. The approximation degrees were $NG^v = 4$ for the velocities, $NG^{\sigma m} = 3$ for the mean stress and $NG^f = 2$ (left diagram) and $NG^f = 3$ (right diagram) for the surface traction. The degree of the ϕ-polynomial was $NG^\phi = 4$ in both cases.

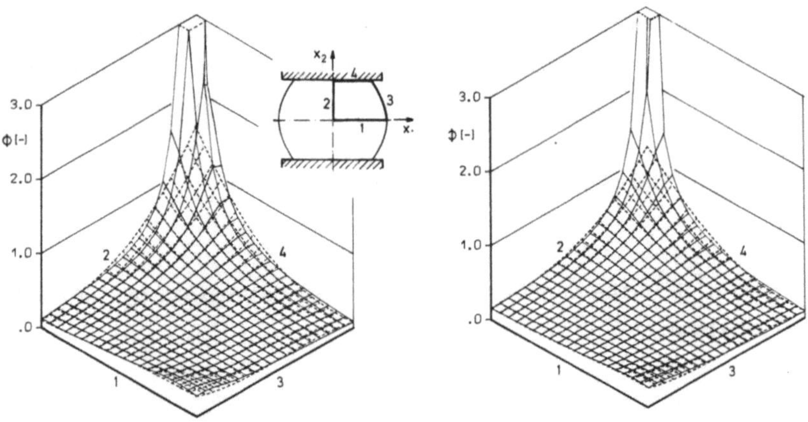

Fig. 4 Local variation of the material function ϕ
Plane upsetting, sticking friction

Left: $NG^v=4$, $NG^{\sigma m}=3$, $NG^f=2$, $NG^\phi=4$
Right: $NG^v=4$, $NG^{\sigma m}=3$, $NG^f=3$, $NG^\phi=4$

The dotted lines show the ϕ-approximation which is still poor in the zone where the ϕ-function itself tends to infinite. A better approximation could be obtained by the use of other coordinate functions which describe singularities in a better way; a more safe approach, however, is to subdivide the domain just for the calculation of the volume integral. In each subdomain the ϕ-function will be approximated by linear functions.

The following two figures show the equivalent strain rates and the mean stress again drawn up over the (x_1, x_2)-plane. The degrees of the polynomials for the velocities, the mean stress and the surface tractions are $NG^v = 4$, $NG^{\sigma m} = 3$ and $NG^f = 2$. The left diagram of Fig. 5 shows the equivalent strain rates for a linear ϕ-approximation and the right one that of fourth degree. Obviously, the linear approximation equalizes the strain rates in the domain. High values are lowered and low values are computed too high. If the fourth order approximation for the ϕ-function is used, the expected maximum values are obtained along the diagonal of the domain, where the velocity gradients are known to be very high.

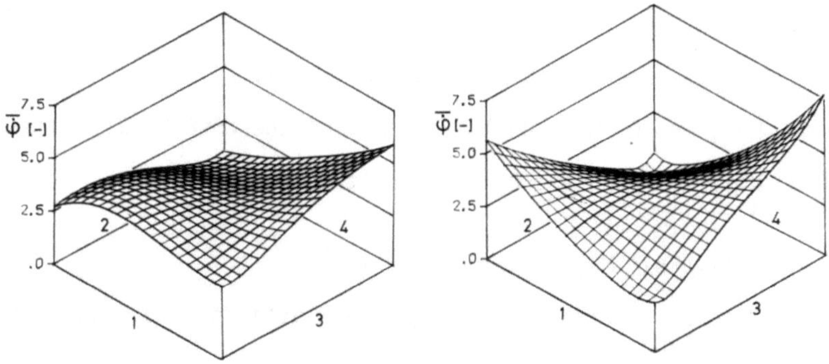

Fig. 5 Equivalent strain rate $\dot{\bar{\varepsilon}}$
Plane upsetting, sticking friction
$NG^v=4$, $NG^{\sigma m}=3$, $NG^f=2$, Left: $NG^\phi=1$, Right: $NG^\phi=4$

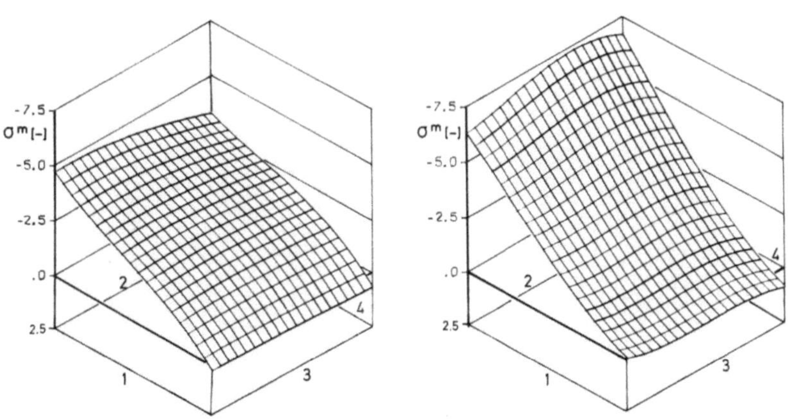

Fig. 6 Mean stress σ^m
Plane upsetting, sticking friction
$NG^v=4$, $NG^{\sigma m}=3$, $NG^f=2$, Left: $NG^\phi=1$, Right: $NG^\phi=4$

The mean stress in the domain (Fig. 6) shows the typical behaviour. The values increase almost linearly from zero value at the free surface to the maximum value at the symmetry line. The linear approximation of the ϕ-function again equalizes the curve. At the free surface the mean stress is positive in the zone near the die wall; there is no physical explanation of this fact. It is supposed, that the ϕ-approximation which is still unsatisfactory, is responsible.

Literature

1 M. Lung, Dissertation, TU Hannover, 1971.

2 M. Lung and O. Mahrenholtz, Trans. of the CSME 2, 31, 1973-74.

3 W. Klie, M. Lung and O. Mahrenholtz, Mech.Res.Com. 1, 315, 1974.

4 R. Hill, Phil.Mag. 42, 868, 1951.

5 K. Washizu, Variational Methods in Elasticity and Plasticity, Oxford, 1968.

6 A.A. Markov, Prikl.Mat.i.Mekh. 11, 338, 1947.

Finite-Element Formulation for the Analysis of Plastic Deformation of Rate-Sensitive Materials in Metal Forming

S. I. OH, N. REBELO, and S. KOBAYASHI

Department of Mechanical Engineering
University of California
Berkeley, California 94720 U.S.A.

Summary

The objective of the present research is to establish the finite-element formulation for the analysis of large plastic deformation of rate-sensitive materials. The behavior of the material is characterized as rigid-viscoplastic. First, an approach to the construction of constitutive equation for rigid viscoplastic materials is described. Then, the extremum principle and uniqueness of rigid-deforming boundary are presented as the basis for the finite-element formulation of flow behavior analysis. The finite-element method code is written and applied to the analysis of simple compression and ring compression. The effect of rate-sensitivity of materials on flow behavior is revealed and computed results are compared with experiments found in the literature.

1. Introduction

The mathematical theory of plasticity adequately describes the time-independent aspect of the behavior of materials, but is inadequate for analysis of time-dependent behavior. An approach to achieving a satisfactory formulation for time-dependent behavior has been to generalize plasticity to cases within the strain-rate sensitive range. One such generalization has been provided by the theory of viscoplasticity. The history of this theory dates back to as early as 1922 [1]. Since then, generalizations of early versions have been achieved and various forms of the theory of viscoplasticity have been provided (summarized, for example, by Perzyna [2] and by Cristescu [3]).

Considerable development of the finite-element method in recent years has removed many barriers in applying theories to various problems. A finite-element formulation of viscous materials, such as Bingham plastic or power law pseudoplastic, in isothermal processes, was given by Lyness et al. [4]. A finite-element formulation of elastic-viscoplastic materials was proposed by Zienkiewicz et al. [5] for the infinitesimal deformation and by Nagarajan and Popov [6], based on the finite strain theory. In the area of metal-working analysis, however, the treatment of time-dependent behavior of materials within the framework of the theory of viscoplasticity is still in its beginning [7]-[16]. Cristescu [7], [8] applied the theory to the upper-bound approach to drawing. Zienkiewicz and Cormeau [9], [10] have solved several problems in structural engineering with an elasto-viscoplastic formulation. Zienkiewicz and Godbole [11], [12] and Zienkiewicz, Jain, and Oñate [13] have shown the feasibility of the finite-element approach to rigid viscoplastic materials by treating them as non-Newtonian viscous fluids. Price and Alexander [14], [15] have applied this formulation to creep forming.

The present research is aimed at establishing the finite-element formulation for the analysis of large plastic-deformation of rate-sensitive materials. Although the residual stress calculation is an important aspect in metal-working, equally important is the flow analysis during deformation. An accurate determination of flow characteristics under various process parameters is a prime requirement for the proper design and control of any metal-working process. Therefore, an analysis method in this field of application can only be justified by its solution accuracy and computation efficiency. With this view, the behavior of materials is characterized as rigid-viscoplastic.

In this paper, we first describe an approach to the construction of a constitutive equation for rigid viscoplastic materials. Then, the extremum principle and uniqueness of the rigid-deforming boundary are presented as the basis for the finite-element formulation of flow behavior analysis. The finite-element method code is written and applied to the analysis of simple compression and ring compression. Including the effect of temperature on deformation characteristics will be the next step of this investigation.

2. Theoretical foundations

2.1 Constitutive equations What follows is based on work by Perzyna [17]. We consider a rigid viscoplastic material and apply the infinitesimal theory for each incremental deformation by successive updating of the coordinates.

A static yield function, assumed regular and convex,

$$F(\sigma_{ij},\varepsilon^{p}_{k\ell}) = \frac{f(\sigma_{ij},\varepsilon^{p}_{k\ell})}{K} - 1 \tag{1}$$

is introduced, where $f(\sigma_{ij},\varepsilon^{p}_{k\ell})$ is a function of the state of stress, σ_{ij}, and the state of plastic strain, $\varepsilon^{p}_{k\ell}$. The strain-hardening parameter, K, is given by

$$K = K(w_p) = K\left(\int_0^{\varepsilon^p_k} \sigma_{ij}\, d\varepsilon^p_{ij}\right) \tag{2}$$

The function F is such that for $F < 0$ the material is rigid, for $F = 0$, on the verge of yielding with practically zero strain-rate, and for $F > 0$, the material has yielded (with a certain strain-rate).

The constitutive equation, using the yield function and associated flow rule, is expressed as

$$\dot{\varepsilon}_{ij} = \gamma''\langle \phi(F) \rangle \frac{\partial F}{\partial \sigma_{ij}} \tag{3}$$

or

$$\dot{\varepsilon}_{ij} = \gamma'\langle \phi(F) \rangle \frac{\partial f}{\partial \sigma_{ij}} \tag{4}$$

where

$$\langle\phi(F)\rangle = \begin{cases} 0 & \text{for } F \leq 0 \\ \phi(F) & \text{for } F > 0 \end{cases} \tag{5}$$

and γ is a viscosity constant of the material. The specific form of the function $\phi(F)$ is to be chosen according to the test results of the material behavior. This constitutive relation simply states that strain-rate is a function of the excess of stress over the static yield stress.

Assuming isotropy (and isotropic work-hardening), incompressibility, and the Huber-Mises yield criterion, we have that

$$f = (J_2)^{1/2} \tag{6}$$

where $J_2 = \frac{1}{2}\sigma'_{ij}\sigma'_{ij}$ is the second invariant of the deviatoric stress tensor. Then, Eq. (4) becomes

$$\dot{\varepsilon}_{ij} = \gamma'\left\langle\phi\left(\frac{\sqrt{J_2}}{k} - 1\right)\right\rangle \frac{\sigma'_{ij}}{2\sqrt{J_2}} \tag{7}$$

Squaring both sides,

$$\varepsilon'_{ij}\cdot\varepsilon'_{ij} = \gamma'^2 \left\langle\phi\left(\frac{\sqrt{J_2}}{k} - 1\right)\right\rangle^2 \frac{\sigma'_{ij}\cdot\sigma'_{ij}}{4J_2},$$

we obtain

$$\sqrt{I_2} = \frac{\gamma'}{2}\left\langle\phi\left(\frac{\sqrt{J_2}}{k} - 1\right)\right\rangle \tag{8}$$

where $I_2 = \frac{1}{2}\dot{\varepsilon}_{ij}\dot{\varepsilon}_{ij}$ is the second invariant of the strain-rate tensor. Substituting Eq. (8) into Eq. (7), we get

$$\dot{\varepsilon}_{ij} = \frac{\sqrt{I_2}}{\sqrt{J_2}}\sigma'_{ij}$$

or, using effective stress and strain instead of $\sqrt{J_2}$ and $\sqrt{I_2}$,

$$\dot{\varepsilon}_{ij} = \frac{3}{2} \frac{\dot{\bar{\varepsilon}}}{\bar{\sigma}} \sigma'_{ij} \tag{9}$$

For metal-forming processes (excluding possibly explosive forming), experimental data found in the literature, of which [19]-[20] are just examples, can be realistically represented in the form (which we adopted in the present studies),

$$\sigma = Y[1 + (\frac{\dot{\varepsilon}}{\gamma})^n] \tag{10}$$

$Y = Y(\varepsilon)$ being the static yield stress. Then, from Eq. (9),

$$\sigma'_{ij} = \frac{2}{3} Y[1 + (\frac{\dot{\bar{\varepsilon}}}{\gamma})^n] \frac{\dot{\varepsilon}_{ij}}{\dot{\bar{\varepsilon}}} \tag{11}$$

In this particular case, by applying (8) to uniaxial stress, the form of ϕ can be shown to be

$$\phi(x) = x^{1/n} \tag{12}$$

and

$$\sqrt{3}k = Y$$

2.2 Extremum principle. The deformation process of the rigid-viscoplastic materials is associated with the following boundary value problems. At a generic stage in the process of quasistatic distortion, the shape of the body, the internal distribution of temperature, the state of inhomogeneity, and the current values of material parameters are supposed to be given or to have been determined already. The velocity vector, $\underset{\sim}{v}$, is prescribed on a part of the surface, S_V, together with traction, $\underset{\sim}{F}$, on the remainder of the surface, S_F. Equilibrium equations and compatibility equations are satisfied by the stress and velocity field solutions. To solve this boundary value problem, we consider the extremum principle derived by Hill [18].

Among all the possible constitutive equations, we will restrict our attention to the case where there exists the work function $E(\dot{\varepsilon}_{ij})$, such that

$$\sigma'_{ij} = \frac{\partial E}{\partial \dot{\varepsilon}_{ij}} \tag{13}$$

and where E is convex.

The existence of work function $E(\dot{\varepsilon}_{ij})$ is ensured, if $\underset{\sim}{\sigma}'$ is a single-valued function of $\dot{\underset{\sim}{\varepsilon}}$, satisfying $\partial\sigma'_{ij}/\partial\dot{\varepsilon}_{k\ell} = \partial\sigma'_{k\ell}/\partial\dot{\varepsilon}_{ij}$. Also, E is a convex function, if

$$E(\dot{\underset{\sim}{\varepsilon}}^*) - E(\dot{\underset{\sim}{\varepsilon}}) \geq (\dot{\varepsilon}^*_{ij} - \dot{\varepsilon}_{ij})\frac{\partial E}{\partial \dot{\varepsilon}_{ij}} \tag{14}$$

With this restriction,† we have the following relation, neglecting the body force,

$$\int E(\dot{\underset{\sim}{\varepsilon}}^*)\, dV - \int \underset{\sim}{F}\cdot\underset{\sim}{v}^*\, dS_F \geq \int E(\dot{\underset{\sim}{\varepsilon}})\, dV - \int \underset{\sim}{F}\cdot\underset{\sim}{v}\, dS_F \tag{15}$$

where $\dot{\underset{\sim}{\varepsilon}}$, $\underset{\sim}{v}$ are actual quantities and the starred quantities are kinematically admissible ones. It can be shown that the solution is uniquely determined at points of the body where E is strictly convex but not necessarily in all respects elsewhere. It can also be proved that the relation (15) holds even for the deformation process in which a rigid zone is involved, although Eq. (13) is not defined at the origin of the strain-rate space.

3. Finite-element formulation

The rigid-viscoplastic finite-element formulation is an extension of the rigid-plastic counterpart derived by Lee and Kobayashi [21]. According to the extremum principle, the minimum of the functional

$$\pi = \int_V E(\dot{\underset{\sim}{\varepsilon}})\, dV - \int_F \underset{\sim}{F}\cdot\underset{\sim}{v}\, dS_F \tag{16}$$

with respect to the admissible velocity field, $\underset{\sim}{v}$, with incompres-

†The assumption of convexity may not be a too severe restriction, in view of the experimental observations [19], [20].

sibility constraint, is searched. By introducing a Lagrangean multiplier, this problem is changed to the stationary value problem without constraints of

$$\pi = \int_V E(\dot{\underset{\sim}{\varepsilon}})\, dV - \int_{S_F} \underset{\sim}{F}\cdot\underset{\sim}{v}\, dS_F + \int_V \lambda\dot{\varepsilon}_{ii}\, dV \tag{17}$$

The Lagrangean multiplier λ can be shown to be equal to the hydrostatic stress component. For interface friction the velocity-dependent frictional stresses developed by Chen and Kobayashi [22] were used:

$$f = -mk \left\{ \left(\frac{2}{\pi}\right) \tan^{-1}\left(\frac{|\underset{\sim}{V}_R|}{a|\underset{\sim}{V}_D|}\right) \right\} \underset{\sim}{t} \tag{18}$$

where m is the friction factor ($0 \leq m \leq 1$);

k is the shear yield stress;

$|\underset{\sim}{V}_R|$ is the magnitude of relative velocity between die and workpiece;

a is a constant several orders of magnitude less than the die velocity (was taken as 10^{-5});

$|\underset{\sim}{V}_D|$ is the magnitude of die velocity;

and the functional (17) was modified to

$$\pi = \int_V E(\dot{\underset{\sim}{\varepsilon}})\, dV + \int_V \lambda\dot{\varepsilon}_{ii}\, dV - \int_{S_F} \underset{\sim}{F}\cdot\underset{\sim}{v}\, dS_F - \int_{S_c} \left(\int_0^{|\underset{\sim}{V}_R|} \underset{\sim}{f}\cdot d\underset{\sim}{v}_R \right) dS_c \tag{19}$$

S_c being the part of the boundary where frictional forces occur. The proof is identical to the one presented in [22].

The finite-element implementation was made for axisymmetric problems using 4-node quadrilateral isoparametric elements; the field variable is the velocity $\underset{\sim}{v}$, and the Lagrangean multiplier was taken as a constant within each element. Accordingly, there were 8 nodal point velocity components u_i plus λ as unknowns in each element.

Applying the stationary condition, a set of algebraic nonlinear equations in $\underset{\sim}{u}$ and λ resulting from the assemblage for elements N of

$$\frac{\partial \pi^N}{\partial \underset{\sim}{u}} = 0$$
$$\frac{\partial \pi^N}{\partial \lambda} = 0 \tag{20}$$

was obtained and solved in the following way: assume a solution $\underset{\sim}{u}_0$ close to the actual solution. Then, using the first two terms of a Taylor expansion

$$\left.\frac{\partial \pi^N}{\partial u_i}\right|_{\underset{\sim}{u}=\underset{\sim}{u}_0} + \left.\frac{\partial^2 \pi^N}{\partial u_i \partial u_j}\right|_{\underset{\sim}{u}=\underset{\sim}{u}_0} \Delta u_j = 0$$
$$\left.\frac{\partial \pi^N}{\partial \lambda}\right|_{\underset{\sim}{u}=\underset{\sim}{u}_0} + \left.\frac{\partial^2 \pi^N}{\partial \lambda \partial u_j}\right|_{\underset{\sim}{u}=\underset{\sim}{u}_0} \Delta u_j = 0 \tag{21}$$

$\Delta\underset{\sim}{u}$ and λ were obtained. The new approximation to the solution becomes $\underset{\sim}{u}_0 + \Delta\underset{\sim}{u}$, and this process is repeated until the desired accuracy is reached.

The problem of non-uniqueness of stress in the rigid zones, and the difficulty in determining the minimum of Eq. (15) under such conditions was numerically handled by considering an effective strain-rate offset, several orders of magnitude smaller than the average strain rates. Below that strain rate, the stress was considered to vary linearly down to zero, as shown schematically in Figs. 1 and 2.

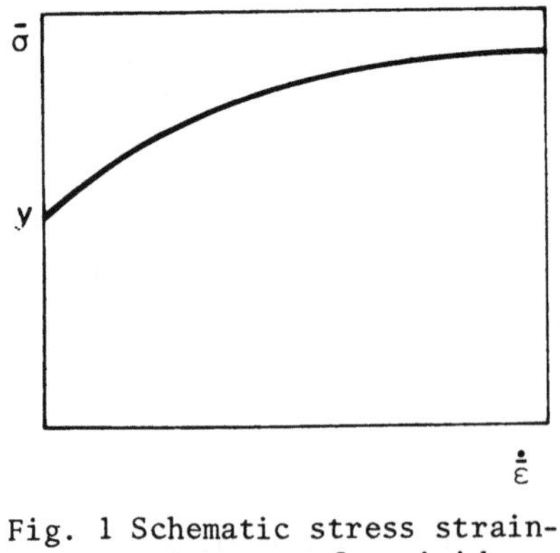

Fig. 1 Schematic stress strain-rate curve for rigid-viscoplastic materials.

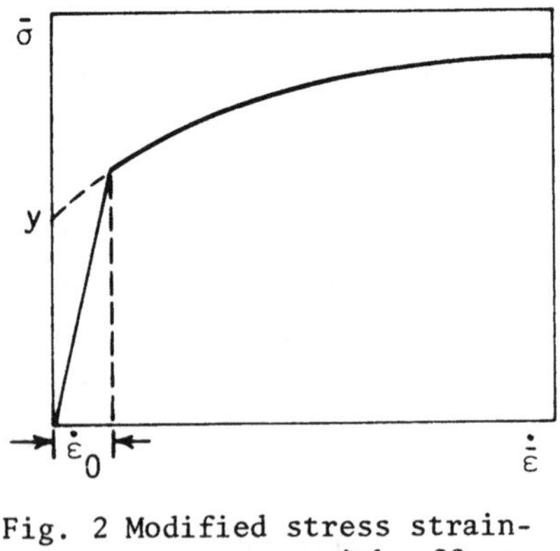

Fig. 2 Modified stress strain-rate curve with offset strain rate.

Whenever, $\dot{\bar{\varepsilon}} < \dot{\varepsilon}_0$ in one element, Eq. (11) becomes

$$\sigma'_{ij} = \frac{2}{3} Y\left(1 + \frac{\dot{\varepsilon}_0^n}{\gamma^n}\right) \frac{\dot{\varepsilon}_{ij}}{\dot{\varepsilon}_0} \tag{22}$$

and Eqs. (21) are accordingly modified. For all practical purposes, an element may be considered to be rigid if $\dot{\bar{\varepsilon}} < \dot{\varepsilon}_0$.

4. Applications

There are few available references in the literature where strain-rate effects in non-uniform deformation are explicitly put in evidence. Among them, the work by Klemz and Hashmi [23] on simple upsetting of lead cylinders at room temperature provides good experimental results for comparison with the finite-element solution, since lead is quite strain-rate sensitive at room temperatures. Another experimental result was presented by Nagpal, Lahoti, and Altan [24] with the upper-bound and finite-difference analysis of flow and temperature in ring compression. Although the present finite-element method does not include the temperature calculation, simulation of ring compression is performed because it involves unique features of the boundary-value problems in metal-working.

4.1 Simulation of compression of a lead solid cylinder by a drop hammer A cylinder having a diameter of 1.00 (in.) and a height of 0.95 (in.) was considered. The stress strain strain-rate data for pure lead at room temperature were taken from [23] and from experimental results by Loizou and Sims [25]. The static stress-strain curve was approximated according to the expressions given in Table I. Then, for several strains, curves of the form $\sigma = Y[1 + (\dot{\varepsilon}/\gamma)^n]$ were fitted to experimental data, and the obtained values of n and γ^n were interpolated in the program for intermediate strain values (Table I). The considered offset value $\dot{\varepsilon}_0$ of Eq. (22) was 1. The friction factor m of Eq. (18) was taken as 0.06 (m = a fraction of the static yield shear strength), in agreement with the assumed value in [23].

Table I Stress strain strain-rate properties of lead at room temperature, used in calculations.

ε	.105		.225	.6
Y(psi)	$1000+$ $+13212.4\varepsilon$	$2387.3+6878.3(\varepsilon-.105)-28659.7(\varepsilon-.105)^2$	$2650+$ $+150\cos[\pi(\varepsilon-.225)/.375]$	2500
n	$.015+.1525\varepsilon$		$-.06+1.727-6.117\varepsilon^2+$ $+6.104\varepsilon^3$	.093
γ^n	$6.192-15.355\varepsilon$		$6.67-24.54\varepsilon+34.96\varepsilon^2-$ $-15.86\varepsilon^3$	1.105

The calculations simulate the deformation of the cylinder by a drop-hammer, whose tup mass is 35.5 lb, and hits the specimen at a speed of 30 ft s^{-1}. The assumption of a quasi-static process is justified for these velocities [26], [27]. Accordingly, the calculated work done by the contact forces between tup and workpiece in each step of deformation was subtracted to the tup kinetic energy, and the process ended when the tup came to a stop.

For comparison, the calculations were repeated without introducing strain-rate effects (making $\gamma = \infty$ in the formulation).

This non-steady-state deformation problem was analyzed in a step-by-step manner by treating it quasi-linearly during each incremental deformation. The reduction in height at each step is 1%. The solution of velocity field for uniform compression was used as an initial guess. The measure of the convergence is represented by the quantity $\|\Delta\underset{\sim}{u}\|/\|\underset{\sim}{u}\|$, where $\Delta\underset{\sim}{u}$ is the error vector found during iterations and the Euclidean vector norm is defined by $\|u\| = \sqrt{\sum_{i=1}^{N} |\underset{\sim}{u}_i|^2}$, where N = total number of nodal points. The solution obtained from the previous step was then used as an initial guess for the subsequent step. Normally, 7 iterations were required for the first step to reach the accuracy of $\|\Delta\underset{\sim}{u}\|/\|\underset{\sim}{u}\| \leq 0.00005$. For subsequent steps, only 2 to 4 iterations were necessary to reach the same accuracy.

The calculated overall quantities are compared with experimental data and with the numerical dynamic analysis of Klemz and Hashmi

[23] in Figs. 3 and 4. Some observations can be made from the figures. The result for non-strain-rate-sensitive analysis (Fig. 3) tends to follow those by the analysis of Klemz and Hashmi, which did not take into account strain-rate effects, but was based on a dynamic analysis. The agreement of the two indicates that the dynamic effect is negligible. However, both solutions show poor agreement with experiments. The analysis which includes strain-rate sensitivity shows good agreement with experiments (Fig. 4). No "rigid" zones were formed at such a low value of friction in either rate-sensitive or rate-insensitive cases.

4.2 Simulation of compression of an aluminum ring at elevated temperature by a mechanical press A standard ring having an outside diameter of 54 mm and proportions o.d.:i.d.:height, 6:3:2, was considered. Using experimental data obtained by U. S. Lindholm [28] on Aluminum 1100 at 400°C (~672°K) and the curves given in [29] for Aluminum 1100, an approximate surface representing stress vs. strain and strain-rate (at 400°C) was obtained. This surface is characterized by the stress-strain curve at zero strain-rate and by the values of n and γ^n given in Table II for several values of strain. The computer program interpolates for intermediate values of strain.

Table II Stress-strain strain-rate properties of Al 1100 at 400°C used in calculations.

ε	.05	.10	.15	.25	.50	.70
n	.346	.226	.200	.419	.507	.502
γ^n	7.908	3.936	4.162	4.228	5.060	4.509
Y	$2 + 10.86\varepsilon - 29.84\varepsilon^2 + 12.7\varepsilon^3$			$3 + 2.765(\varepsilon - .15) - 5.98(\varepsilon - .15)^2 + 13.866(\varepsilon - .15)^3$		3.4

The offset value ε_0 of Eq. (22) was taken as one-tenth of the ratio of the present die velocity to the starting velocity. Three conditions of friction were considered: (i) m = 1, with k being the static shear yield stress (see Eq. (18)); (ii) m = 1, with k being

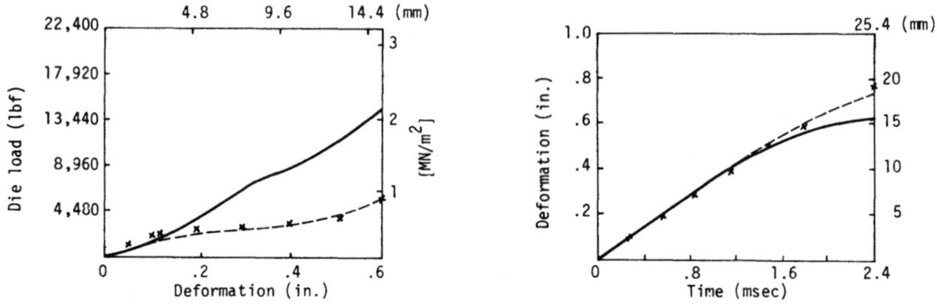

Fig. 3 Comparison of the finite-element rigid-plastic analysis (xxxx) with the dynamic analysis (– – –) [23] and experiment (——) [23].

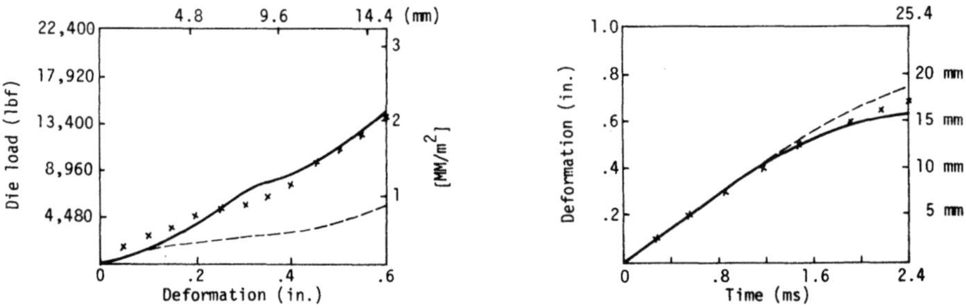

Fig. 4 Comparison of the finite-element rigid-viscoplastic analysis (xxxx) with the dynamic analysis (– – –) [23] and experiment (——) [23].

the rate-sensitive shear yield stress; and (iii) sticking condition. As it can be easily seen, case (ii) would make the $\partial^2\pi^N/\partial u_i \partial u_j$ matrices nonsymmetric, and hence the solution of the equations more expensive. Therefore, an approximation was made, by considering k as the shear yield stress corresponding to the effective strain-rate value in the previous step of deformation. The value m = 1 was the one used by Nagpal et al. [24] in their calculations, and the experiment was performed without lubricant between dies and workpiece. It should be mentioned that the experimental data with which calculations are confronted correspond to a compression of the ring up to 50% deformation with a mechanical press, the specimen being initially at 427°C and the dies at room temperature. Under these circumstances, agreement between theory and experiment is not expected because calculations are performed at a constant temperature. However, a general trend of deformation and the role of friction in ring compression may be revealed.

The calculations simulate a mechanical press, by changing the die velocity in each step of deformation, according to an empirical expression

$$v = 0.1 \times N \times \sqrt{S \times W - W^2} \tag{23}$$

where N is the number of strokes per minute (= 90);
S is the die traveling distance (= 254 mm);
W is the distance from the lower position (= 9 mm, initial value)

The reduction in height at each step was again 1%. The initial guess for the velocity field used in the cases of m = 1 was obtained from other solutions for lower m values. In the sticking case, zero radial velocity and uniform compression in the z velocity were used. Between 10 and 30 iterations were required for the first step, and between 4 and 17 for subsequent steps, in order to reach an accuracy of $\|\Delta \underset{\sim}{u}\| / \|\underset{\sim}{u}\| \leq .00005$. Some additional iterations were required during steps where inside or outside folding was formed.

The changes of minimum internal diameter as functions of reduction in height are compared in Fig. 5 with the upper-bound solutions of Nagpal et al. and the point produced by their experiment. It is seen that the case of m = 1 with static shear yield stress produced clearly unacceptable results, and would be equivalent to a much lower real m value. The case for m = 1 with strain-rate-sensitive shear stress gives more reasonable solutions when compared with experiments. The apparent good agreement between the sticking condition and experiment is fortuitous. According to Male and dePierre's experiments [30], on aluminum rings, sticking is only observed at temperatures of 600°C. The experiment used for comparison was done with cold dies; this situation produces much more bulging than the present calculations, since the material near the dies, upon cooling, will become more resistant to deformation.

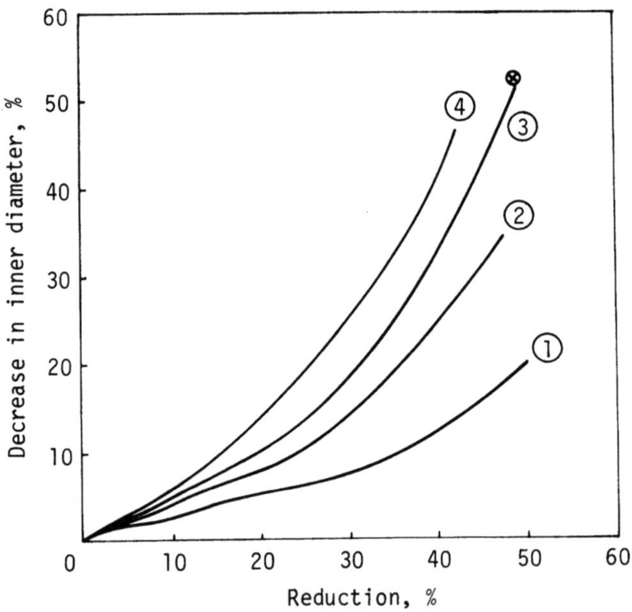

Fig. 5 The changes of inner diameter as functions of reduction in height: ① m = 1, static shear yield stress; ② m = 1, rate-dependent shear yield stress; ③ sticking; ④ upper-bound solution; and experiment (⊗).

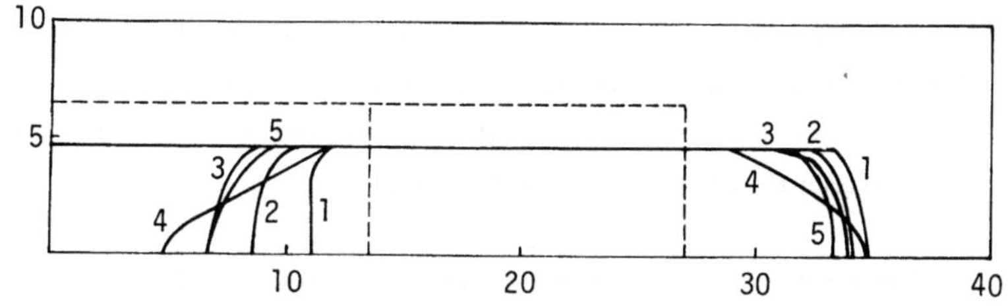

Fig. 6 Bulge profiles at 48.5% reduction in height: 1 m = 1, static shear yield stress; 2 m = 1, rate-dependent shear yield stress; 3 sticking; 4 upper-bound solution; and 5 experiment.

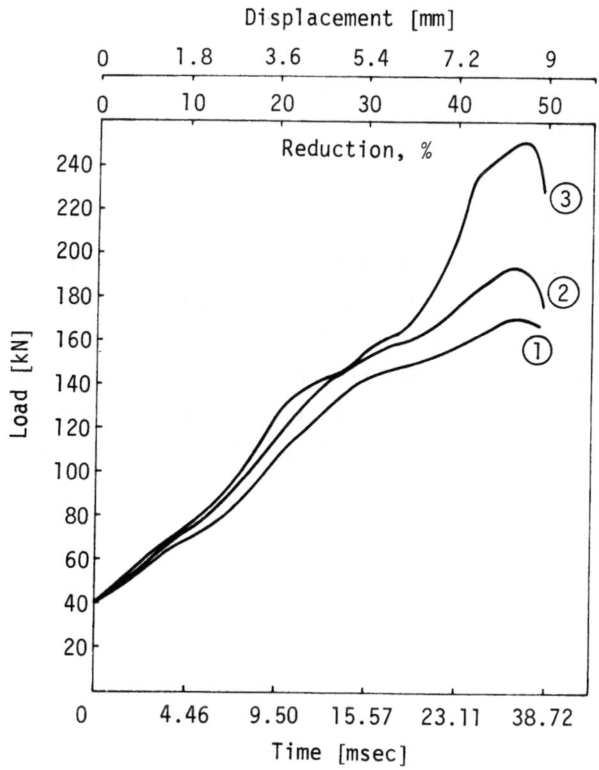

Fig. 7 Load-displacement curves for three interface conditions. ① m = 1, static shear yield stress; ② m = 1, rate-dependent shear yield stress; and ③ sticking.

The observations above apply to Fig. 6, where the profiles at 48.5% deformation of the several solutions are compared to the measured one. It is seen, too, that the qualitative agreement of the finite-element solution is fairly good.

Load displacement (or time) curves are presented in Fig. 7. The only comment that can be made is that the order of magnitude is as expected. The drop in load towards the end is a result of the drastic decrease in velocity of the die according to Eq. (23), which produces a decrease in yield stress of the rate-sensitive material.

Detailed information about the deformation process is obtained with the finite-element method. As an example, the grid distortion at 50% deformation is shown in Fig. 8.

5. Concluding remarks

It is concluded that the rigid-viscoplastic finite-element formulation is adequate for flow analysis of rate-sensitive materials. The use of velocity-dependent frictional stress for the interface condition has made it possible to analyze the ring compression problem, where the direction of frictional stress is not known a priori. The next step is to include the temperature calculation in the finite-element scheme.

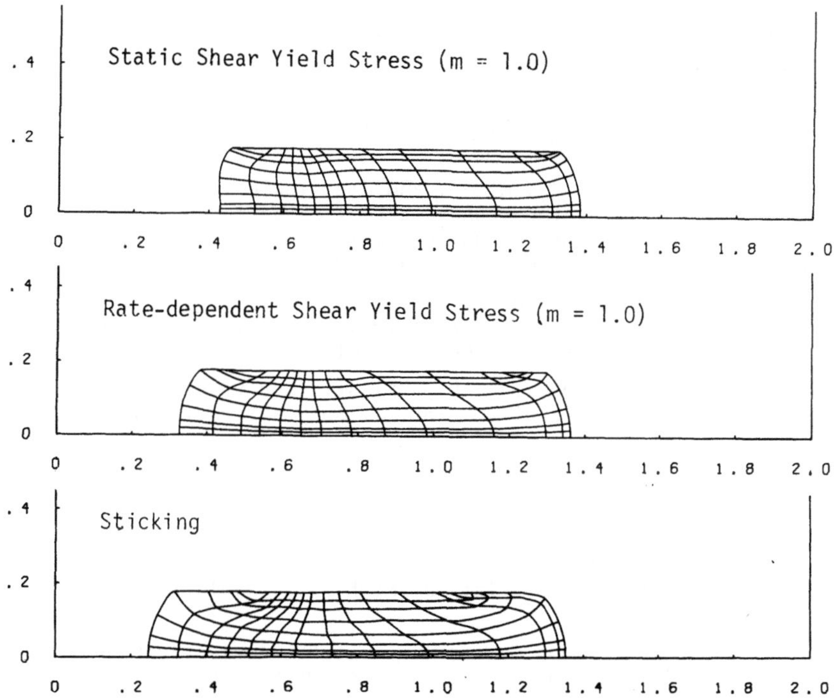

Fig. 8 Grid distortions at 50% reduction for three friction conditions.

Acknowledgements

The authors wish to thank the U.S. Air Force Materials Laboratory, Wright-Patterson Air Force Base, Ohio, for its contract F33615-77-C-5111 under which the present investigation was possible. The senior author (Shiro Kobayashi) wishes to thank the Miller Institute, University of California, Berkeley, for its support through a research professorship. The authors are also grateful to Ikuko Workman for typing the manuscript.

References

1. Bingham, E. C., Fluidity and Plasticity, New York, p. 125, 1922.
2. Perzyna, P., "The study of the dynamic behavior of rate-sensitive plastic materials," Arch. Mech. Stos., vol. 15, p. 113, 1963.
3. Cristescu, N., Dynamic Plasticity, North Holland Press, Amsterdam, 1967.
4. Lyness, J. F., Owen, D. R. J., and Zienkiewicz, O. C., "Finite-element analysis of the steady-state flow of non-Newtonian fluids through parallel-sided conduits," Int. Sym. on Finite-element Method in Fluid Problems, Swansea, 1974.
5. Zienkiewicz, O. C., Owen, D. R. J., and Cormeau, I. C., "Analysis of viscoplastic effects in pressure vessels by the finite-element method," Nuclear Engr. and Design, vol. 28, p. 278, 1974.
6. Nagarajan, S. and Popov, E. P., "Plastic and viscoplastic analysis of axisymmetric shells," Int. J. Solids Structures, vol. 11, p. 1, 1975.
7. Cristescu, N., "Plastic flow through conical converging dies, using a viscoplastic constitutive equation," Int. J. Mech. Sci., vol. 17, pp. 425-433, 1975.
8. Cristescu, N., "Drawing through conical dies--an analysis compared with experiments," Int. J. Mech. Sci., vol. 18, pp. 45-49, 1976.
9. Zienkiewicz, O. C. and Cormeau, I. C., "Viscoplasticity solution by finite-element process," Arch. Mech., vol. 24, p. 873, 1972.
10. Zienkiewicz, O. C. and Cormeau, I. C., "Viscoplasticity, plasticity, and creep in elastic solids--a unified numerical solution approach," Int. J. Numerical Methods in Engr., vol. 8, p. 821, 1974.
11. Zienkiewicz, O. C. and Godbole, P. N., "Viscous, incompressible flow with special reference to non-Newtonian (plastic) fluids," Chapter 2, Finite Elements in Fluids, ed. by R. H. Gallagher et al., Wiley, 1975.
12. Zienkiewicz, O. C. and Godbole, P. N., "Flow of plastic and viscoplastic solids with special reference to extrusion and forming processes," Int. J. Num. Meth. Engr., vol. 8, p. 3, 1974.
13. Zienkiewicz, O. C., Jain, P. C., and Oñate, E., "Flow of solids during forming and extrusion: some aspects of numerical solutions," Int. J. Solids and Structures, vol. 14, p. 15, 1978.
14. Price, J. W. H. and Alexander, J. M., "A study of the isothermal forming or creep forming of a titanium alloy," Proceedings of the 4th N.A.M.R.C., p. 46, 1976.

15. Price, J. W. H. and Alexander, J. M., "The finite-element analysis of two high-temperature metal deformation processes," 2nd Int'l Symposium on F.E.M. in Flow Problems, p. 717, 1976.

16. Wang, N. M. and Wenner, M. L., "Elastic-viscoplastic analysis of simple stretch-forming problems," Proc. General Motors Research Symposium on Mechanics of Sheet-Metal Forming, Oct. 1977.

17. Perzyna, P., "Fundamental problems in viscoplasticity," Adv. in Applied Mechanics, vol. 9, p. 243, 1966.

18. Hill, R., "New horizons in the mechanics of solids," J. Mech. Phys. of Solids, vol. 5, p. 66, 1956.

19. Hauser, F. E., Simmons, J. A., and Dorn, J. E., "Strain-rate effects in plastic wave propagation," Technical Report No. 3, University of California, June 1960.

20. Lindholm, U. S. and Yeakley, L. M., "Dynamic deformation of single and polycrystalline aluminum," J. Mech. Phys. Solids, vol. 13, p. 41, 1965.

21. Lee, C. H. and Kobayashi, S., "New solutions to rigid-plastic deformation problems using a matrix method," Trans. ASME, J. Engr. Ind., vol. 95, p. 865, 1973.

22. Chen, C. C. and Kobayashi, S., "Rigid-plastic finite-element analysis of ring compression," to be presented at the 1978 ASME Winter Annual Meeting.

23. Klemz, F. B. and Hashmi, S. J., "Simple upsetting of cylindrical billets: experimental investigation and theoretical prediction," 18th MTDR Conference, London, 1977.

24. Nagpal, V., Lahoti, A. D., and Altan, T., "A numerical method for simultaneous prediction of metal flow and temperatures in upset forging of rings," ASME paper no. 77-WA/PROD-35.

25. Loizou, N. and Sims, R. B., "The yield stress of pure lead in compression," J. of Mech. Phys. of Solids, vol. 1, p. 234, 1953.

26. Lippman, H., Proc. 7th Int'l MTDR Conference, Birmingham, London, p. 53, 1966.

27. Samanta, S. K., "Resistance to dynamic compression of low-carbon steel and alloy steels at elevated temperatures and at high strain rates," Ind. J. Mech. Sci., vol. 10, p. 613, 1968.

28. Lindholm, U. S., "Experiments in dynamic plasticity," Symposium on Mechanical Behavior of Materials Under Dynamic Loads, 1968.

29. Altan, T. and Boulger, F. W., "Flow stress of metals and its application in metal-forming analysis," J. of Engr. for Ind., vol. 95, p. 1009, 1973.

30. Male, A. T. and dePierre, V., "The validity of mathematical solutions for determining friction from the ring compression test," J. of Lubr. Tech., p. 389, 1970.

Theoretical Solutions of Some Plastic Working Processes in the Light of Experimental Evidence

L. DIETRICH and H. PETRYK
Institute of Fundamental Technological Research, Warsaw, POLAND

Summary

The normal stresses acting on the tool-material interface during compression of truncated cones, cylinders, rings and blocks were measured. The results of experiments have been compared with the theoretical stress distributions calculated for the rigid-perfectly plastic model of the material.

Introduction

Solutions of many practical plastic working processes can be obtained by means of the theory of plasticity. However, the solutions must be treated with reasonable caution since they are based on the drastic assumptions concerning behaviour of metals. The usefulness of the theory in solving practical problems should be verified experimentally. Usually experimental verification of the theoretical solutions is limited to the comparison of the values of total forces. Much more information can of course provide an experiment which enables us to verify local quantities.

In the paper the solutions based on the plasticity theory of some boundary value problems for both axially-symmetric and plane strain conditions are verified by comparison of the normal stress distribution over the tool-material interface. The method made use of pressure sensitive pins placed in the steel platens. The measurements of the contact stress distribution by using a similar technique have been performed before in a few papers, mainly for the upsetting of cylinders or blocks. The papers by Unksov [1] or van Rooyen and Backofen [2] can be mentioned, where general agreement was found between the plane strain theory and experimental results for short specimens. In the paper [2] the plane strain solution was also found to describe the pressure distribution over the interface in axial-symmetric compression. An analysis of the pressure sensitive pins technique applied for measurements of the contact stress distribution in different papers has been performed by Löwen [3].

Experimental Procedure and Specimens

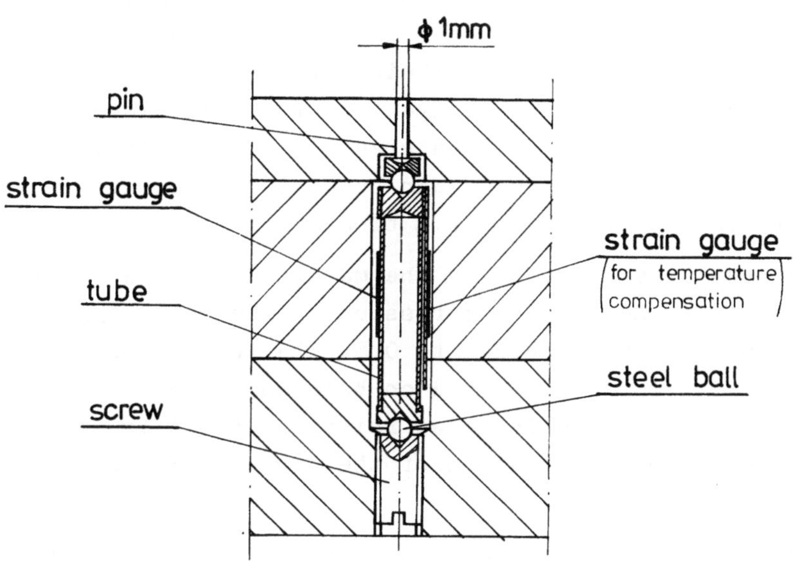

Fig.1. Schematic diagram of a pressure sensitive pin

The experimental device for measurement of the normal stresses acting on the interface was similar to that one used in paper [4]. The main element of the pressure sensitive pin transducer which is shown in Fig.1 has the form of a small tube /3 mm I.D., 3.3 mm O.D./ on which the active semiconductor strain gauge has been cemented. Electric signals from all transducers built up in the steel plate at different distances from the centre were recorded in sequence by using the automatic chanel selector, digital voltmeter and printer. The current diameter of specimen at the interface was also measured during compression by three displacement transducers.

The specimens made of lead were carefully machined from extruded rods. The solid cylindrical specimens were of initial diameter d_o=24 mm with different d_o/h_o ratios over the range of 0.5 to 5, h_o being the initial height of a specimen. The hollow cylindrical specimens had dimensions of 11 mm I.D., 32 mm O.D. /D_o/d_o= 2.9/ and D_o/h_o ratio in the range of 1 to 5. In the case of blocks the initial width-to-height ratio, b_o/h_o, has been changed also in the range of 1 to 5. The length l_o of the blocks was five times greater than its width b_o= 24 mm, so the plane strain conditions were satisfied in the central part of a block.

The stress-strain curve for the material used have been determined by means of compression of the solid cylinder with d_o/h_o ratio equal to 0.5. The teflon small squares /3x3 mm/ were placed on both end-faces of the specimen in order to decrease friction. As it is seen from Fig.2 the distribution of the stresses was almost uniform.

The cyclic waves on the stress-strain curve caused by dynamic recrystallization have been observed at some strain rates. However, it was found that the character of the measured stress distribution has not been affected by this phenomenon.

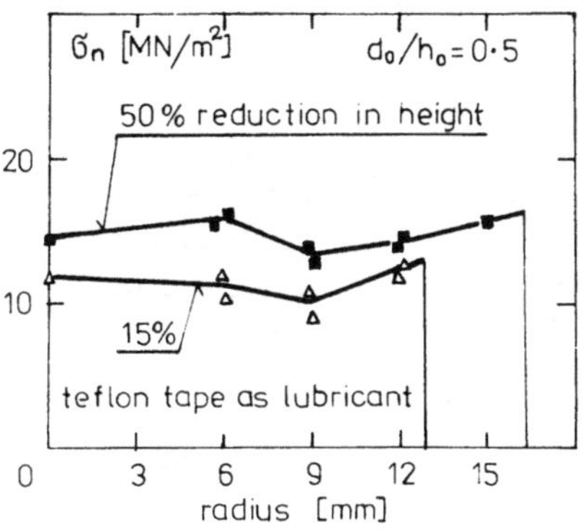

Fig.2. Distributions of normal stresses σ_n on the interface during compression of cylinder

All other specimens were compressed under dry friction conditions. In order to check the correctness of the stress measurements the force calculated by integration of the stress distribution for the cylinders was compared with the measured total force. In all cases the agreement was within 10 %.

Experimental Results and Comparison with Theory

Compression of a truncated cone

The first of the problems considered was the compression of a truncated cone by a flat die. The theoretical solution of this problem can be obtained by using the slip-line theory under the assumptions of Tresca yield criterion and Haar-Kármán hypothesis [5].

The slip-line field solution shown in Fig.3 has been constructed in [6] for Coulomb friction on the interface OA with the coefficient $\mu = 0.1$. When $\mu = 0$ then the rigid region OEF disappears and the solution is of Shield's type [5]. On the other hand, when μ is sufficiently large then the rigid region OEF covers the whole interface OA [7]. Since the stresses in rigid regions

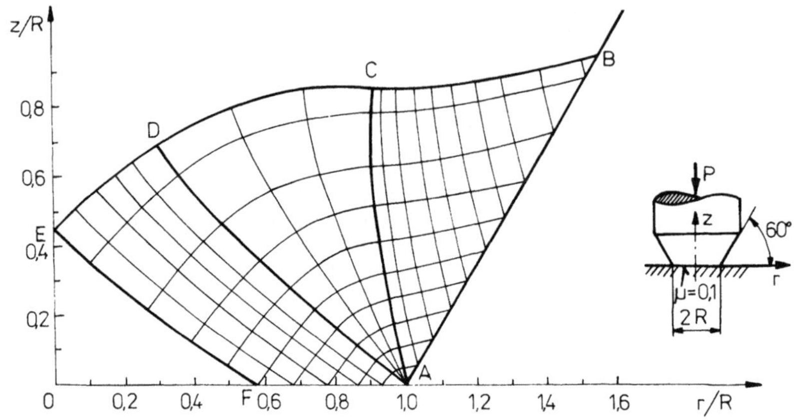

Fig.3. Slip-line solution for compression of a truncated cone [6]

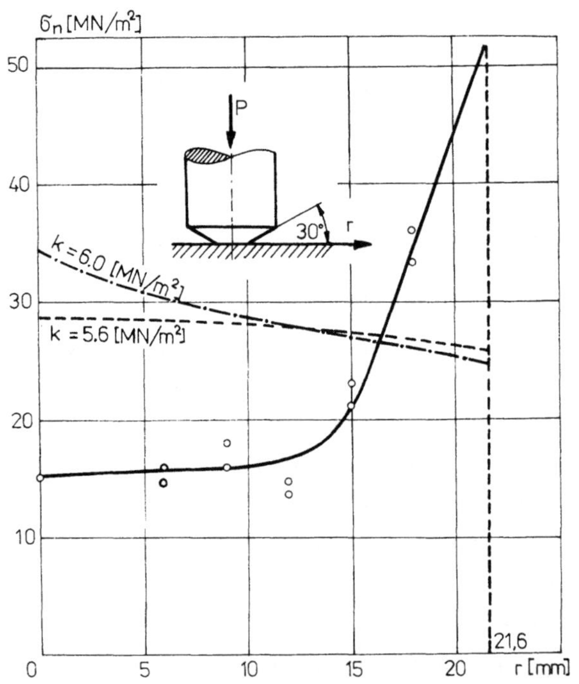

Fig.4. Theoretical and experimental normal stress distribution over the interface of the truncated cone [8]

are not prescribed, the normal stresses acting on the interface can be calculated directly only there, where the plastic region touches the die. However, assuming that the rigid region OEF is also in the plastic state, we can extend there the slip-line field and determine the normal stresses on the remaining part of interface OA. The distribution of the normal stresses calculated in this manner is found to be weakly dependent on the coefficient of friction.

In Fig.4 the dotted and broken lines show the theoretical normal stress distribution over the interface for two limiting cases $\mu = 0$ and $\mu \geqslant 0.152$ respectively when the cone semiangle is equal to 60° [4]. The solid line has been drawn by the experimental points obtained in the case of a well lubricated contact surface. In the calculations the value of the yield shear stress k has been so chosen that the total force for both the theoretical and experimental distributions is equal. It is seen that on the contrary to the theoretical prediction the maximum value of the normal stress occurs at the external radius of the interface. A similar disagreement has been observed also for the other specimens [4]. It is worthy to underline that the theory predicts well the values of total forces which cause yielding of a cone [4].

The question arises whether the disagreement between the stress distribution calculated on the basis of the rigid-perfectly plastic model of the material, and that one determined experimentally will also be observed for other plasticity problems. In order to answer the question some other series of experiments as described above have been performed.

Compression of an axi-symmetrical cylinder

Some of the results obtained for cylindrical specimens

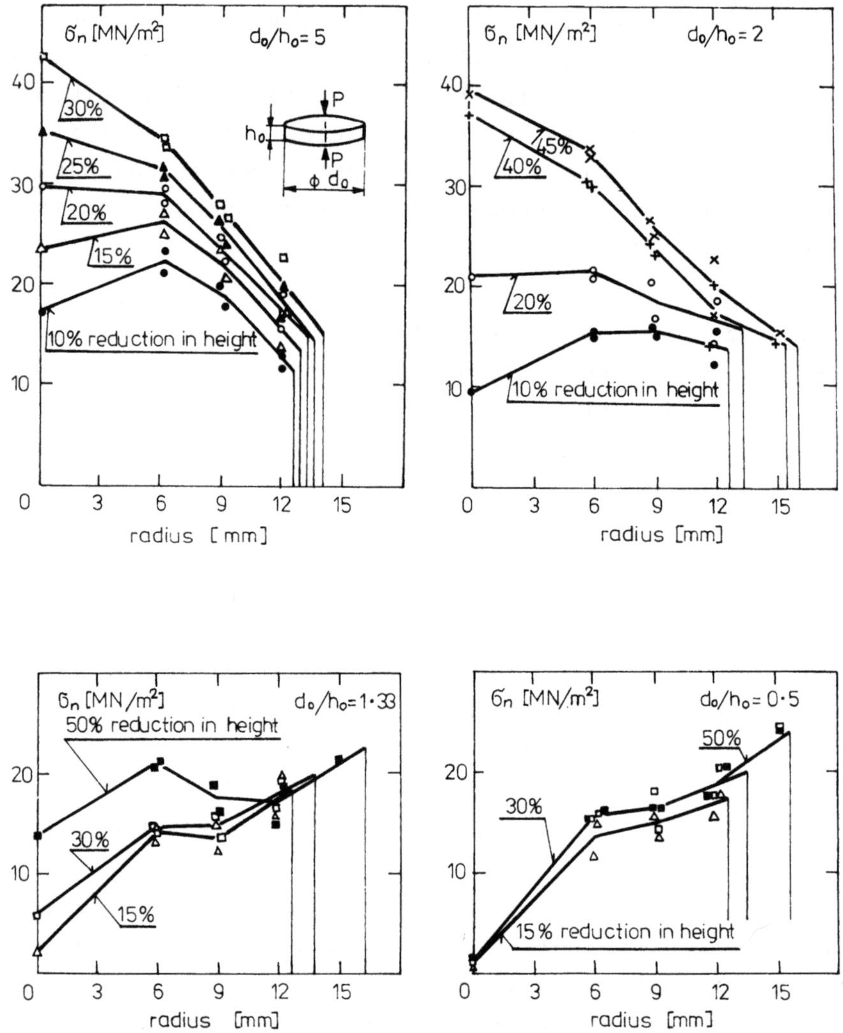

Fig.5. The measured normal stress distributions during compression of cylinders of different height without lubricant at the end-faces

are shown in Fig.5. It is seen that the normal stress distribution over the interface changes remarkably its character with varying current diameter-to-height /d/h/ ratio. Thus the first point is whether the theoretical and experimental normal stress diagrams are in a qualitative agreement.

Some slip-line field solutions of the problem of compression of a short cylinders have been given in [8]. One of them is shown in Fig.6. Region OAD is rigid, but

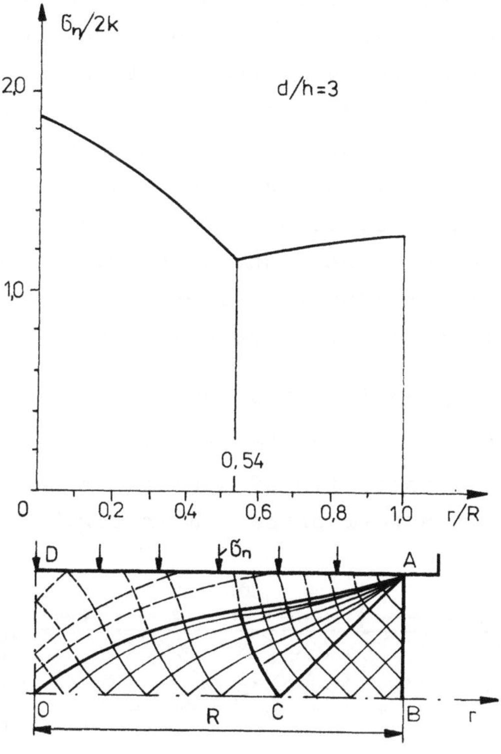

Fig.6. Slip-line solution and normal stress distribution for a short cylinder compressed between partially rough platens

the stress field based on the assumption that all the material is in the plastic state has been extended into it. The corresponding normal stress distribution over the interface is presented in the figure.

Since barreling is not taken into account in the solution, it is difficult to compare adequately the theoretical results with the experimental ones. However, some general remarks concerning a qualitative comparison can be made. The local minimum of the normal stresses on some radius of the interface is really observed when the current d/h ratio is about 3. Outside this radius both the theory and experiment indicate a slight increase of the normal stresses, but the drop in the stresses near the centre of the specimen observed for $d/h<3$ is not predicted by the theory. The agreement is better for shorter specimens when the normal stresses reach their maximum in the central part of the interface. In that cases qualitative differences between theory and experiment occur only near the external radius of the interface where the measured stresses continue to decrease in contrast to the theoretical diagram.

Plane strain compression of blocks

The classical slip-line field solution of the plane strain problem of the compression of a block between partially rough dies has the form presented in Fig.7 [9].

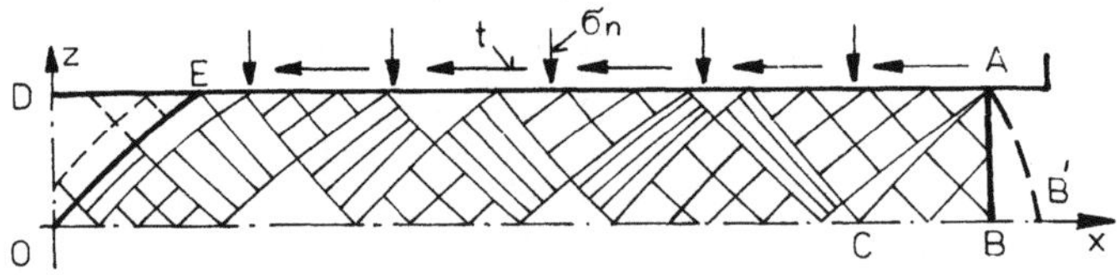

Fig.7. Slip-line solution for a block compressed between partially rough platens

The same solution is valid for any position AB′ of a free surface outside AB provided the block centre does not fall into a straight line region, what is not allowed through kinematical requirements.

Fig.8 presents some of the experimental normal stress distributions over the interface obtained for the block compression problem, and in Fig.9 two of the experimental

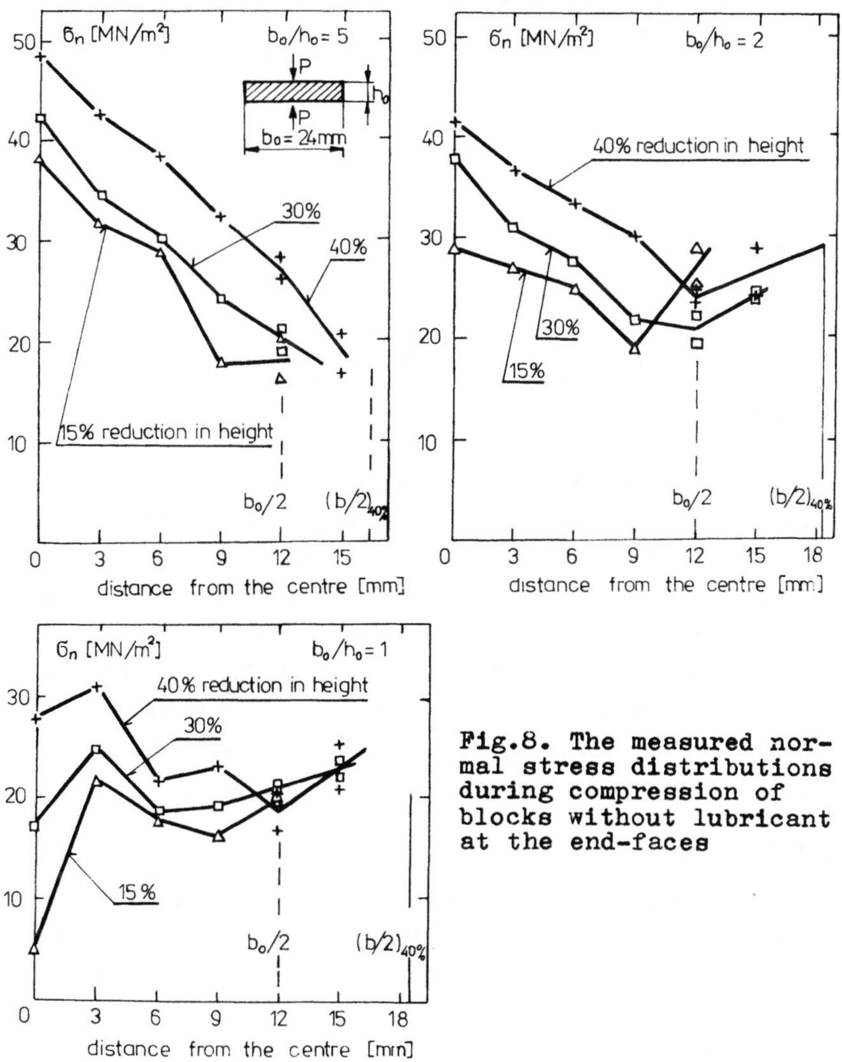

Fig.8. The measured normal stress distributions during compression of blocks without lubricant at the end-faces

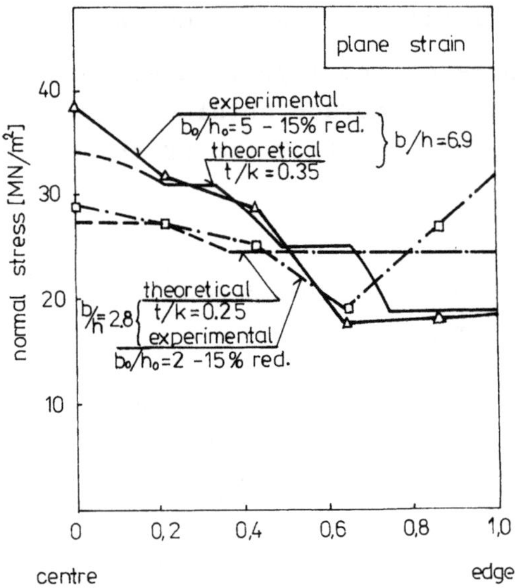

Fig.9. The comparison of the theoretical and experimental normal stress distributions for actual width-to-height ratio equal to 6.9 and 2.8

curves are compared with the theoretical ones. Generally speaking the theory predicts very well the stress distribution for widh-to-height ratios b/h greater than 6. For smaller values of b/h the experiment indicates that the normal stresses reach their minimum at some distance from the free surface, what is not explained by the solution presented. The sudden drop in normal stresses near the centre of the interface, observed for $b/h < 2$, is also not described by the rigid/plastic theory.

Compression of rings

The normal stress distributions measured during the compression of rings of two different heights are shown

Fig.10. The measured normal stress distributions during the compression of rings of two different heights

strongly on the value of diameter-to-height ratio.

The theoretical distribution of normal stresses calculated from an approximate solution has been given in [10]. It is qualitatively in a good agreement with the experimental data for short specimens /D_o/h_o=5/. For taller specimens, as it could be expected, this approximate solution does not provide satisfactory results.

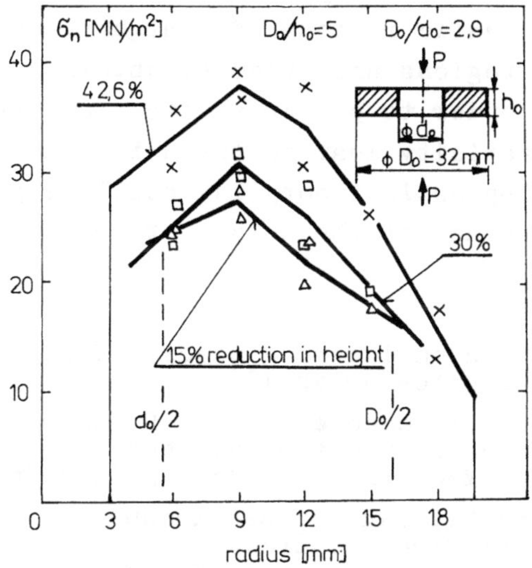

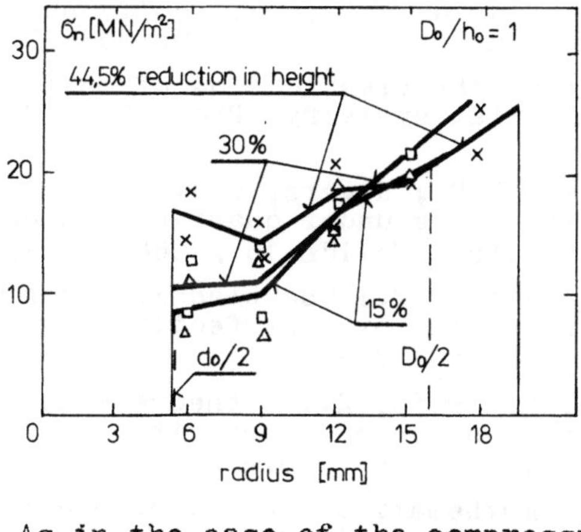

in Fig.10. As in the case of the compression of solid cylinders the character of a diagram obtained depends

Conclusions

The presented results show that the slip-line theory predicts the normal stress distribution on the tool-material interface quite well if the height of compressed specimens is sufficiently small, say if it is several times smaller than the diameter or the width of a specimen. In these cases theoretically rigid regions in which the stresses are not prescribed are small as compared with the plastic regions. For higher specimens, when such rigid regions are greater, obtained results suggest that in order to properly determine the stress distribution elasticity must be taken into account, though the rigid-plastic theory may still be good enough to predict the values of the total forces.

References

1 Unksow, E.P.; An engineering theory of plasticity. London: Bulterworth-Verlag 1961.

2 van Rooyen, G.T.; Backofen, W.A.: Distribution of interface stress in plane strain and axial symmetric compression. J.Mech.Phys.Solids 7 /1959/ 163-168.

3 Löwen, J.: Ein Beitrag zur Bestimmung des Reibungszustandes beim Gesenkschmieden. Ph.D. Thesis , Technischen Universität Hannover 1971.

4 Dietrich, L.: Contact stresses in the process of plastic flow of an axial-symmetrically shaped stamp. /In Polish/. Engng Trans. 18 /1970/ 353-370.

5 Shield, R.T.: On the plastic flow of metals under conditions of axial symmetry. Proc. Roy. Soc. A 233 /1955/ 267-286.

6 Suh, N.P.; Lee, R.S.; Rogers, C.R.: The yielding of truncated solid cones under quasi-static and dynamic loading. J.Mech.Phys.Solids 16 /1968/ 357-372.

7 Eason, G.; Shield, R.T.: The plastic indentation of a semi-infinite solid by a perfectly rough circular punch. ZAMP 11 /1960/ 33-42.

8 Kwaszczyńska, K.; Mróz, Z.: A theoretical analysis of plastic compression of short circular cylinders. Arch.Mech.Stos. 19 /1967/ 787-797.

9 Hill, R.: The mathematical theory of plasticity. Oxford: Clarendon Press 1950.

10 Hawkyard, J.B.; Johnson, W.: An analysis of the changes in geometry of a short hollow cylinder during axial compression. Int.J.Mech.Sci. 9 /1967/ 163-182.

Analysis of Metal Forming by the Rigid-Plastic Finite Element Method Based on Plasticity Theory for Porous Metals

S. SHIMA, K. MORI
Department of Mechanical Engineering,
Kyoto University, Sakyo-ku, Kyoto, JAPAN
and K. OSAKADA
Department of Mechanical Engineering,
Kobe University, Nada-ku, Kobe, JAPAN

Summary

Upsetting of a circular cylinder is analysed by the rigid-plastic finite element method based on the plasticity theory for porous metals. Deformation patterns and distributions of stresses and the equivalent strain are evaluated.
A method of examining the accuracy of the calculated stress distribution is proposed. The stress is thus shown to be statically admissible in good accuracy.

Introduction

The rigid-plastic FEM (finite element method) is considered to be useful for analysing metal forming processes. This method is basically the upper bound method; a velocity field in the deforming body is so determined that the energy due to plastic deformation and frictional loss in each deformation step is minimized. The concept of combining the upper bound approach with the FEM was first proposed by Hayes and Marcal [1]; the collapse load of a notched flat plate under uniaxial tension was evaluated. Using the same principle and with the use of the Lagrange multiplier, Lee et al. [2]∿[5] carried out extensive work, in which the stress was calculated by assuming the Lagrange multiplier to be equal to the mean stress. Price and Alexander [6] used the "penalty function" approach in minimizing the plastic energy to ensure incompressibility, and obtained the stress distribution by integrating the stress equilibrium equation from the boundary.

Since theory of plasticity for porous metals was proposed by a few researchers [7]∿[10], it has been applied to develop various theories and methods of anaysis for forming of sintered porous metals [11]∿[14]. Recently, it has been proved that the rigid-plastic FEM is quite useful for obtaining distributions of density, stresses and strain [15].

In this theory, when the relative density is assumed to be very high, say 99.5%, the theory would be applied to the approximate analysis of plastic deformation of pore-free metals. Since this theory has no need to satisfy volume constancy, the formulation for minimization of the energy is simplified, and further, the stress can be obtained directly from the strain rates. In this study, upsetting of a cylindrical billet is treated and the accuracy of the stress distribution is examined.

Plasticity Theory for Porous Metals

The basic equations of plasticity for porous metals [10] are given as follows for axi-symmetric case:

i) Yield criterion

$$\sigma_0 = \frac{1}{\rho^n}\Big[\frac{1}{2}\{(\sigma_r-\sigma_\theta)^2 + (\sigma_\theta-\sigma_z)^2 + (\sigma_z-\sigma_r)^2\} + 3\tau_{rz}^2 + (\sigma_m/f)^2\Big]^{1/2} \tag{1}$$

where σ_0 is the yield stress or the equivalent stress (σ_{eq}) for the matrix metal, ρ is the relative density of the porous metals, σ_m is the mean stress, and experimentally

$$\left.\begin{aligned} n &= 2.5 \\ f &= 1/2.49(1-\rho)^{0.514} \end{aligned}\right\} \tag{2}$$

ii) Stress-strain (rate) relation

$$\left.\begin{aligned}\dot{\varepsilon}_r &= \frac{3}{2}\,\frac{1}{\rho^{2n-1}}\,\frac{\dot{\varepsilon}_{eq}}{\sigma_{eq}}\{\sigma_r - (1 - 2/9f^2)\sigma_m\}\\ &\cdot\\ &\cdot\\ \dot{\gamma}_{rz} &= 3\,\frac{1}{\rho^{2n-1}}\,\frac{\dot{\varepsilon}_{eq}}{\sigma_{eq}}\,\tau_{rz}\end{aligned}\right\} \quad (3)$$

$$\dot{\varepsilon}_v = \dot{\varepsilon}_r + \dot{\varepsilon}_\theta + \dot{\varepsilon}_z = \frac{1}{\rho^{2n-1}}\,\frac{\dot{\varepsilon}_{eq}}{\sigma_{eq}}\,\frac{\sigma_m}{f^2} \quad (4)$$

where $\dot{\varepsilon}_{eq}$ is the equivalent strain rate referring to the matrix material and is given by

$$\dot{\varepsilon}_{eq} = \rho^{n-1}[(\tfrac{2}{9})\{(\dot{\varepsilon}_r-\dot{\varepsilon}_\theta)^2 + (\dot{\varepsilon}_\theta-\dot{\varepsilon}_z)^2 + (\dot{\varepsilon}_z-\dot{\varepsilon}_r)^2\} + (1/3)\dot{\gamma}_{rz}^2 + (f\dot{\varepsilon}_v)^2]^{1/2} \quad (5)$$

<u>Principle of the Analysis</u>

The upper bound theorem based on the above equations demonstrates that the rate of dissipation of plastic energy or power derived from any kinematically admissible velocity field is an upper bound on the power by the exact tractions [11],[13]. The exact velocity field may be approximated by a velocity field which minimizes the power $\dot{W}^*$. $\dot{W}^*$ is given by

$$\dot{W}^* = \int_V \rho\sigma_{eq}\,\dot{\varepsilon}_{eq}^*\,dV + \int_{S_T}\tau_f\Delta v^* dS - \int_{S_A} F_i v_i^*\,dS \geqq \int_{S_B} F_i v_i\,dS \quad (6)$$

where $\dot{\varepsilon}_{eq}^*$ is derived from a kinematically admissible velocity field, Δv^* is a velocity discontinuity such as tool-workpiece interfaces, S_T is a plane where the velocity discontinuity occurs, S_A is the surface where traction F_i is prescribed and S_B is the surface where velocity v_i is prescribed. The minimization of $\dot{W}^*$ is started from an appropriate initial velocity field. Note that, in this case, a kinematically admissible velocity field allows volume change in the material. However, it will be seen that the calculated volume change is very small when ρ is very near to unity, because the volumetric strain rate $\dot{\varepsilon}_v$ increases the power $\dot{W}^*$ to a very large extent.

Stress can be calculated from strain rates, e.g. from the first equation in Eq.(3) and Eq.(4), as follows:

$$\sigma_r = \rho^{2n-1}\sigma_{eq}\{\frac{2}{3}\,\dot{\varepsilon}_r + (f^2 - \frac{2}{9})\,\dot{\varepsilon}_v\}/\dot{\varepsilon}_{eq} \qquad (7)$$

Method of the Analysis

The problems treated here are upsetting of a cylindrical billet with completely rough platens, in which radial flow of the material at the interface is not allowed, and compression with a constant coefficient of friction at the tool-workpiece interfaces.

The height to diameter ratio of the cylinder is 1.0 (20mm × 20mm). The meridional plane of the cylinder is

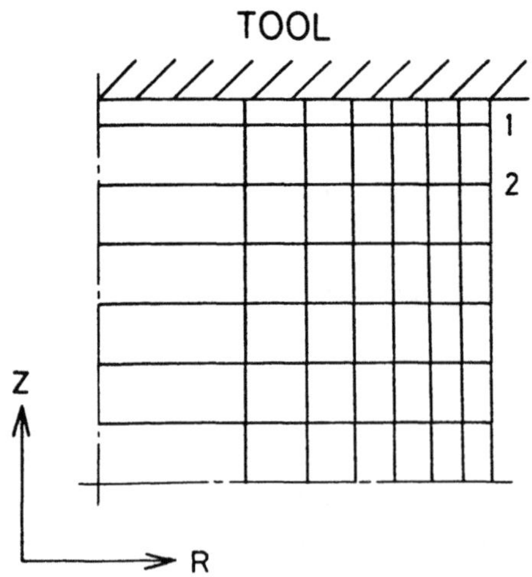

Fig.1. Finite element mesh used for analysis

divided into quadrilateral elements as shown in Fig.1. The method for calculating the strain-rates in each element from the nodal velocities is the same as the usual technique in the FEM [16]. Since the relative density ρ is very near to unity, the power, $\dot{W}_D^*$, required for plastic deformation is expressed by the following equation without violating the upper bound theorem.

$$\dot{W}_D^* = \sum_{i=1}^{m} (\rho\sigma_{eq}\,\dot{\varepsilon}_{eq}^*\,V)_i \cong \sum_{i=1}^{m} (\sigma_{eq}\,\dot{\varepsilon}_{eq}^*\,V)_i$$

$$= \dot{W}_D^*(u_1,\ v_1,\ \cdots\cdots,\ u_n,\ v_n) \qquad (8)$$

where m is the number of elements and n is the number of nodes.

The rate of energy loss due to friction at the tool-workpiece interfaces is of course zero in the case of upsetting with completely rough platens. If there is a relative slip at the tool-workpiece interfaces, the friction loss is given by

$$\dot{W}_F^* = \sum_{i=1}^{m'} (\tau_f\ \Delta v^*\ S)_i = \dot{W}_F^*(u_1,\ v_1,\ \cdots\cdot,u_n,v_n) \qquad (9)$$

where τ_f is the frictional stress, Δv^* the relative velocity at the interface, S the contact area and m' is the number of elements at the interface. τ_f was approximated by

$$\left.\begin{aligned} \tau_f &= \mu p_0 && \text{for } \mu p_0 < (1/\sqrt{3})\sigma_{eq} \\ \tau_f &= (1/\sqrt{3})\sigma_{eq} && \text{for } \mu p_0 \geqq (1/\sqrt{3})\sigma_{eq} \end{aligned}\right\} \qquad (10)$$

where μ is the coefficient of friction and p_0 is the tool pressure at the previous deformation step. This approximation was used to overcome the difficulty which is encountered in the upper bound approach, i.e. the Coulomb friction cannot be treated in the upper bound theorem.

The total power, $\dot{W}^*$, is

$$\dot{W}^* = \dot{W}_D^* + \dot{W}_F^* \qquad (11)$$

The initial velocity employed was

$$\left.\begin{aligned} u_0 &= C_1 r (C_2 - z) \\ v_0 &= C_1 (C_2 - z) + C_3 \end{aligned}\right\} \qquad (12)$$

This velocity field is derived from the assumption that the radial velocity changes linearly with r and z. The constants, C_1, C_2 and C_3, have been so chosen that the power is minimized. Then, minimization of $\dot{W}^*$ at the first step was carried out starting from the power, minimized as above, by the procedure which has already been presented elsewhere [15].
The reduction in height at each deformation step was 1.0∿1.2%. The calculation was carried out using a stress-strain curve expressed by

$$\sigma_{eq} = 76.3(\varepsilon_{eq} + 0.0397)^{0.21} \quad (kg/mm^2) \qquad (13)$$

This was obtained by simple compression of a medium carbon steel (∿0.25%C steel, JIS S25C).

Results of Analysis

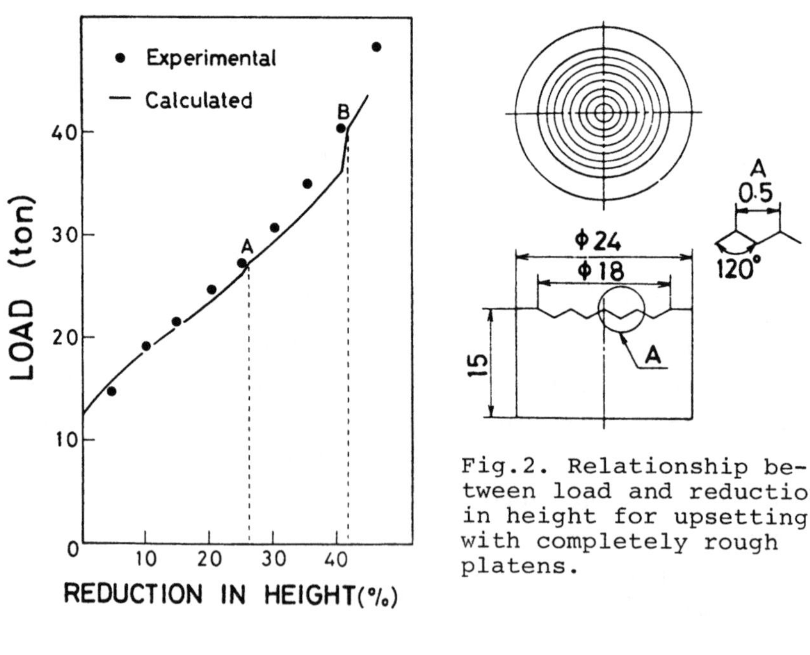

Fig.2. Relationship between load and reduction in height for upsetting with completely rough platens.

Fig. 2 shows the relationship between load and reduction in height for upsetting with completely rough platens. An experimental result for S25C steel is also shown for comparison. The experiment was carried out using platens with concentric grooves as shown in the figure. At point A(26%) and B(42%) folding occurred.

Fig.3 shows the relationship between load and reduction in height for compression with a constant coefficient of friction, $\mu=0.25$. For comparison, an experimental result for compression of S25C steel with dry friction with flat platens is also shown. It is seen in these figures that there is fairly good agreement between the calculated and the experimental results.

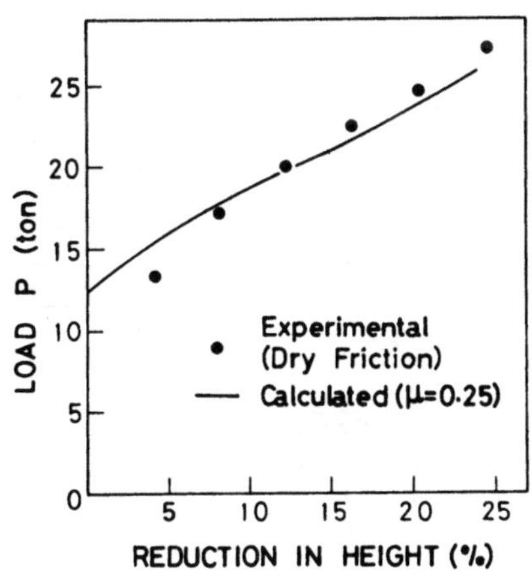

Fig.3. Relationship between load and reduction in height for compression.

Fig.4 shows deformation patterns at reduction in height of 26 and 45% for upsetting with completely rough platens. The volume change after 45% reduction was as small as 0.61%.

Fig.5 shows distributions of stress and equivalent strain at a reduction in height of 45% in compression with rough platens. The accuracy of the stress will be checked in the later section. The stresses at the equa-

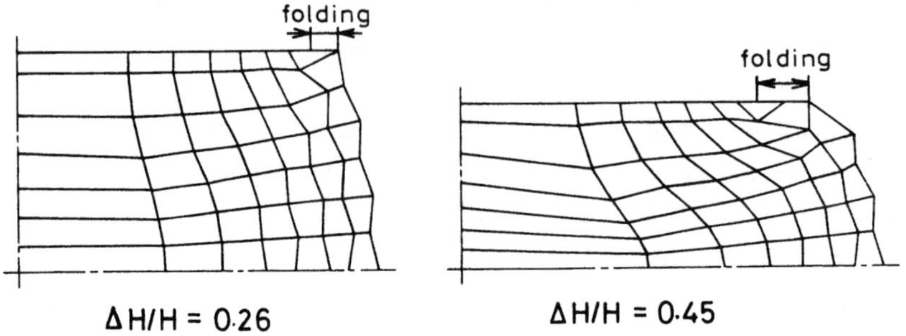

Fig.4. Deformations patterns in upsetting with completely rough platens.

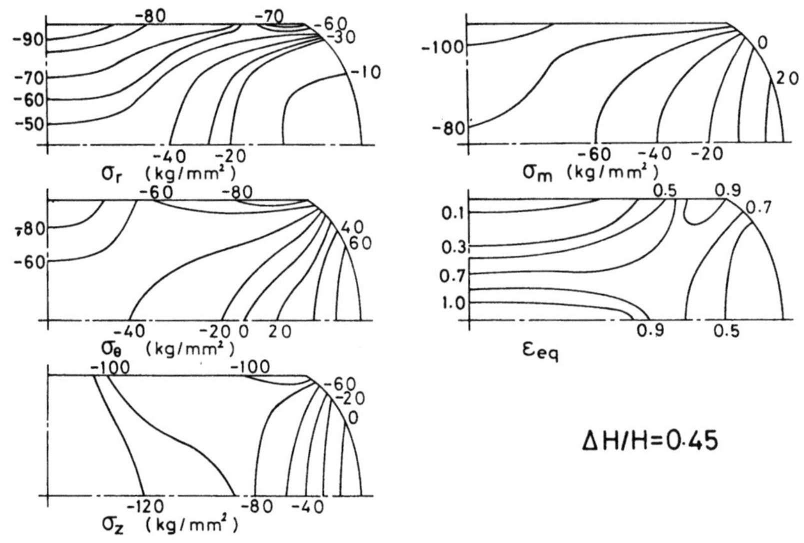

Fig.5. Distributions of stresses and the equivalent strain in upsetting with completely rough platens.

torial free surface are shown in Fig.6. Experimentally obtained stresses are also shown in the figure. The stresses were derived from the measured axial and hoop strain increments substituted in the Lèvy-Mises equations with the radial stress being zero; the axial strain increment was obtained from the change in the distance between two indentations which were so located that the equator lay in between, while the hoop strain increment was obtained from the change in the diameter of the equator. It is seen the calculated radial stress is almost zero and that the calculated axial and hoop stresses are respectively in good agreement with the experimental ones. Since compression with a constant coefficient of friction has provided similar results, these are not shown here.

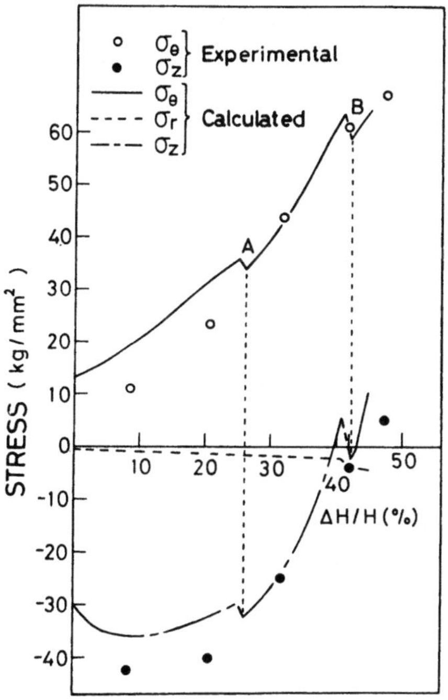

Fig.6. Variation of stresses at the equatorial free surface with reduction in height in upsetting with completely rough platens.

Examination of the Accuracy of the Stress Distributions

The accuracy of a stress field can be examined by checking whether or not the stress is statically admissible. A statically admissible stress field does not violate the yield condition and satisfies boundary conditons and static equilibrium conditions. In the present analysis, the yield condition is obviously satisfied. Regarding the boundary conditions, it has already been seen in the previous section that the stresses at the equator are in fairly good agreement with the experimental results. Therefore the static equilibrium should be checked.

In the ordinary elastic or elasto-plastic FEM, the stress is equilibrated at nodal points [16] and a similar idea is employed in this study. Fig.7 shows adjoining four elements, h-2, h-1, h and h+1, where σ_r, σ_θ, σ_z and τ_{rz} are stresses in element h, and $(P_{ir})_h$ and $(P_{iz})_h$ are respectively radial and axial nodal forces at node i. The forces acting at the nodes which belong to element h can be expressed by the stress using the usual method in the FEM as follows:

$$\begin{Bmatrix} (P_{ir})_h \\ (P_{iz})_h \\ (P_{jr})_h \\ (P_{jz})_h \\ (P_{mr})_h \\ (P_{mz})_h \\ (P_{kr})_h \\ (P_{kz})_h \end{Bmatrix} = V\frac{2}{J} \begin{bmatrix} B_i & D_i & 0 & C_i \\ 0 & 0 & C_i & B_i \\ B_j & D_j & 0 & C_j \\ 0 & 0 & C_j & B_j \\ B_m & D_m & 0 & C_m \\ 0 & 0 & C_m & B_m \\ B_k & D_k & 0 & C_k \\ 0 & 0 & C_k & B_k \end{bmatrix} \begin{Bmatrix} \sigma_r \\ \sigma_\theta \\ \sigma_z \\ \tau_{rz} \end{Bmatrix} \tag{14}$$

where V is the volume of element h; B, C, D and J are given by the coordinates of the nodes [14]. These forces are, in general, not in equilibrium. Therefore, the

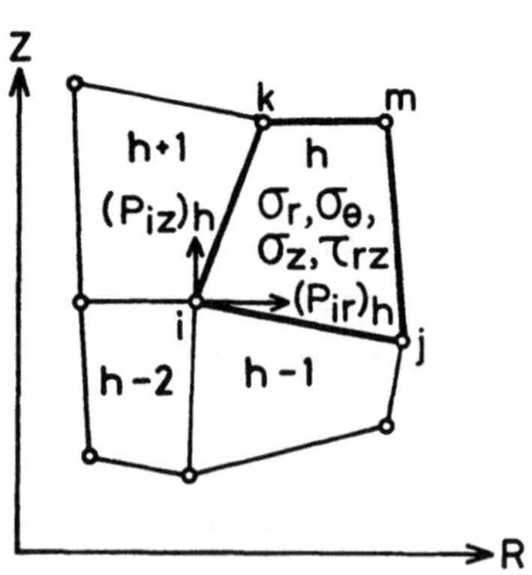

Fig.7. Stresses and nodal forces.

magnitude which is given by

$$\Delta P_i = [\{(P_{ir})_{h-2} + (P_{ir})_{h-1} + (P_{ir})_h + (P_{ir})_{h+1}\}^2 + \{(P_{iz})_{h-2} + (P_{iz})_{h-1} + (P_{iz})_h + (P_{iz})_{h+1}\}^2]^{1/2} \quad (15)$$

can be defined as the additional nodal forces required to bring the present velocity field into nodal equilibrium. Now, "excessive stress", $\Delta\sigma$, may be defined from Eq.(15) as

$$\Delta\sigma = \frac{1}{An_1} \sum_{i=1}^{n_1} \Delta P_i \quad (kg/mm^2) \quad (16)$$

where A is the total surface areas of all the elements divided by the number of the surfaces and n_1 is the number of the nodes which are not in contact with the surface of the die. If the "excessive stress", $\Delta\sigma$, is small enough compared with the flow stress, it can be said that the stress is statically admissible in good accuracy.

Fig.8 shows the variations of the "excessive stress" with reduction in height in the cases of compression between rough platens and compression with μ=0.25.

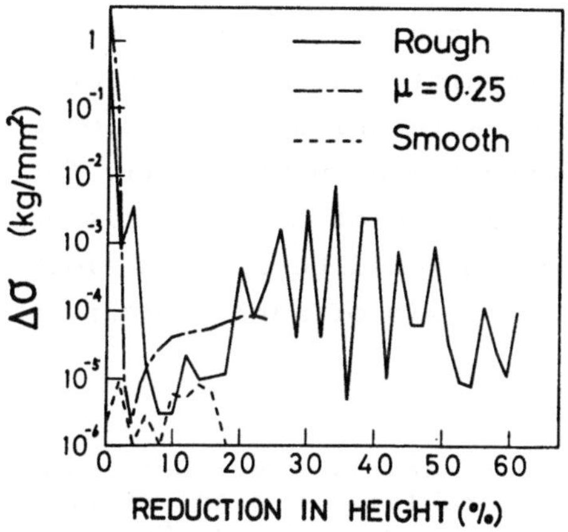

Fig.8. Variation of "excessive stress", $\Delta\sigma$, with reduction in height

For comparison, the result for frictionless compression is also shown. It is seen that in all cases the "excessive stress" is very small compared with the flow stress which is roughly 40kg/mm^2.

Discussion and Conclusions

From the view-point that the volumetric strain rate plays an important role in the analysis, the present approach is similar to the "penalty function" approach by Price and Alexander [6], but unlike their method, stress can be calculated directly from the constituitive equations. Further, since there is not the condition of volume constancy, the number of variables is smaller compared with that by Kobayashi et al. [2]∿[5].

In the present study, upsetting of a cylinder was analysed and stress distributions were obtained. The stress field was proved to be approximately statically admissible. In the analysis of upsetting of a cylinder, it is not very important what type of initial velocity should be chosen; the same results can be obtained regardless of the type of the initial velocity field. However, in other processes, such as extrusion or indentation, the initial velocity field has, in general, great effect on the obtained results. Therefore, an attempt should be made to develop a technique which provides a reasonable initial velocity field automatically for any forming process. If this kind of technique is established, the rigid-plastic FEM would be of much wider application in the field of analysis of metal forming processes.

References

1 Hayes, D.J.; Marcal, P.V.: Determination of upper bounds for problems in plane stress using finite element techniques. Int. J. Mech. Sci. 9 (1967) 245-251.

2 Lee, C.H.; Kobayashi, S.: New solutions to rigid-

plastic deformation problems using a matrix method. Tr. ASME, Ser.B 95(1973) 865-873.

3 Kobayashi, S.; Lee, C.H.; Shah, S.N.: Analysis of rigid-plastic deformation problems by the matrix method. J. Japan Soc. Tech. Plasticity 153 (1973) 770-778.

4 Kobayashi, S.; Lee, C.H.: Deformation Mechanics and workability in upsetting solid circular cylinders. 1st NAMWRC (1973) 185-204.

5 Shah, SN.; Lee, C.H.; Kobayashi, S.: Compression of tall, circular, solid cylinders between parallel flat dies. Int. Conf. Prod. Engg., Tokyo (1974) 295-300.

6 Price, J.W.H.; Alexander, J.M.: A study of isothermal forming or creep forming of a titanium alloy. 4th NAMWRC (1976) 46-53.

7 Kuhn, H.A.; Downey, C.L.: Deformation characteristics and plasticity theory of sintered powder materials. Int. J. Powder Met. 7 (1971) 15-20.

8 Green, R.J.: A plasticity theory for porous solids. Int. J. Mech. Sci. 14 (1972) 215-224.

9 Oyane, M.; Shima, S.; Kono, Y.: Theory of plasticity for porous metals. Bull. JSME 16 (1973) 1254-1262.

10 Shima, S.; Oyane, M.: Plasticity theory for porous metals. Int. J. Mech. Sci. 18 (1976) 285-291.

11 Oyane, M.; Tabata, T.: Slip-line field theory and upper-bound theory for porous materials. J. Japan Soc. Tech. Plasticity 15 (1974) 43-51.

12 Shima, S.; Inamoto, J.; Osakada, K.; Narutaki, R.: The finite element analysis of elastic-plastic deformation of porous metals. ibid. 16 (1975) 660-667.

13 Shima, S.; Tabata, T.; Oyane, M.: Upper bound theory for deformation of porous materials. Mem. Fac. Eng., Kyoto University 38 (1976) 117-137.

14 Shima, S.: Visio-plasticity analysis for porous metals. J. Japan Soc. Tech. Plasticity 18 (1977) 263-267.

15 Mori, K.; Shima, S.; Osakada, K.: Analysis of forming of porous metals using the rigid-plastic finite element method. Tr. JSME, to be published.

16 Ziekiewicz, O.C.; Cheung, Y.K.: The finite element method in structural and continuum mechanics.: McGraw-Hill 1967.

Mechanics of Creep in Metal Forming

N.N. MALININ

Moscow Higher Technical School named after N.E.Bauman, USSR

Introduction

In solving problems of metal forming processes the phenomenon of creep is not usually taken into account. The problems are solved by the chosen theory of plasticity, and the time factor is not taken into consideration. However, when metal undergoes deformations at high temperatures and stresses, the creep of the metal is essential even for short-time deformations, and therefore it should be taken into consideration. It is reasonable to use the flow or strain hardening theories of creep in calculations of metal forming as well as the main results of the experimental investigations of short-time creep in the case of large strains due to tension.

The results of experimental investigation of a short-time creep.

The investigation of short-time creep has been carried out at the materials testing laboratory of the Moscow Higher Technical School named after N.E.Bauman [1 ÷ 3] These tests showed that in some cases the first stage of a creep curve is very straight while in other cases these creep curves have curved initial parts. Fig.1 represents the first case (creep curves of an aluminium alloy at 450°C [1]) and Fig.2 - the second case (creep curves of boiler steel at 1190°C [2]). It is evident that the stresses at large strains cannot be considered constant during the test. Therefore, it is of great interest to compare the experimental creep curves with

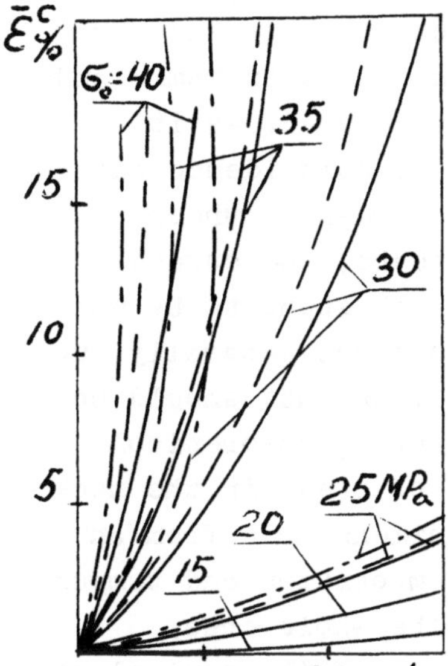

Fig. 1.

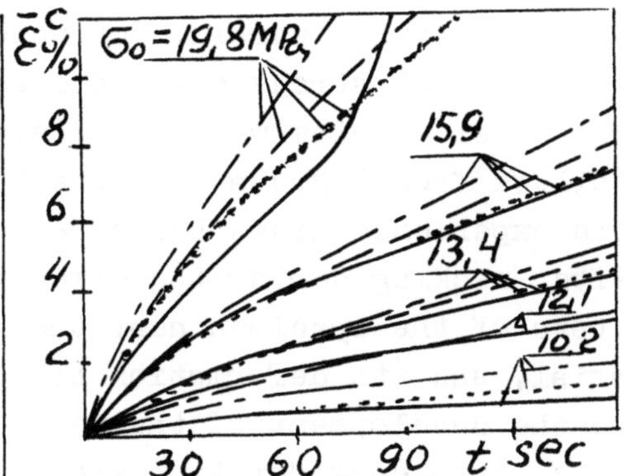

Fig. 2.

the theoretical ones according to various technical creep theories. Fig.1 and 2 [1, 2] show the theoretical creep curves according to the time-hardening theory (dash - dotted lines), the flow theory (dashed lines) and strain hardening theory (dotted lines). If the initial parts of the creep curves are linear (Fig.1), then the flow and strain hardening theories coincide, and dashed lines at Fig.1 correspond to the flow theory and to the strain hardening theory as well. The following dependencies were used when drawing the theoretical creep curves according to the time hardening, flow, and strain hardening theories respectively:

$$\bar{\varepsilon}^c=\sigma^n\Omega, \quad \xi^c=\sigma^n B, \quad \xi^c(\bar{\varepsilon}^c)^{\beta}=\alpha\sigma^{\nu},$$

where $\bar{\varepsilon}^c$ - the logarithmic creep strain, ξ^c - the creep strain rate, σ is the true stress, Ω and B are functions of time, and n, β, ν constants for a given material at the temperature of the test. The functions of time and the constants were determined by considering the initial parts of the creep curves up to 5% strain, for which the true and the conventional stresses as well as the logarithmic and the usual strains were not distinguished. As it follows from Fig.1 and 2, the best coincidence with the experience gives the strain hardening theory. After necking during the creep tests, the stress state of the specimen near the neck stops to be uniaxial, and its deformation takes place generally due to the development of the neck. This fact should be taken into consideration when drawing theoretical creep curves. The investigation of creep in the neck has been carried out [3] on specimens made from an aluminium alloy with the same composition and temperature as in the previous tests. The whole creep process was photographed by a camera. The equivalent logarithmic creep strain in the neck, which is equal to the logarithmic strain in the longitudinal direction (if we assume that the circumferential, and radial logarithmic strains are constant with respect to the radius, and equal) is calculated by the formula:

$$\bar{\varepsilon}_e = \bar{\varepsilon}_z = \ln \frac{F_o}{F}, \qquad (1)$$

where F_o and F are the initial, and the least cross-sectional area of the specimen, respectively. The equivalent stress was determined by Davidenkov's and Spiridonova's formula [4] :

$$\sigma_e = \frac{P}{\pi r_1^2 \left(1 + r_1/4\rho_1\right)} \qquad (2)$$

where P denotes the tensile force, r_1 - the radius of the least cross-section of the specimen, and ρ_1 - the radius of the curvature of the contour of the neck at the point of the least cross-section. These values were measured using the images of the neck

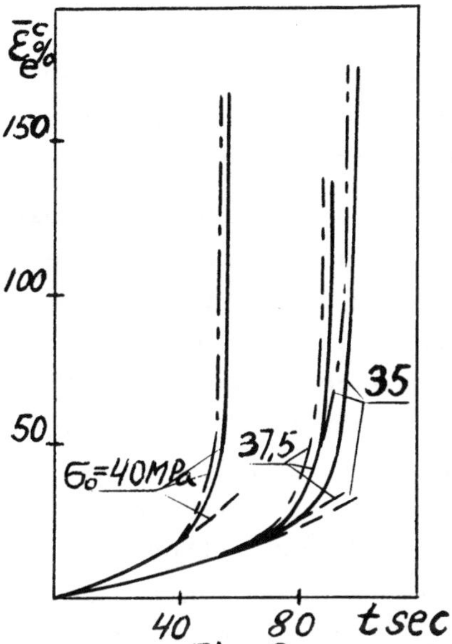

Fig.3

on the screen. Fig.3 shows the diagram of the dependence of the equivalent logarithmic strain on time. The experimental curves (solid lines) were drawn by using the formula (1), and the theoretical curves (dashed lines) - by the formula which follows from the flow and strain hardening theories, both of which coincide in this case:

$$\bar{\varepsilon}_e = K\int_0^t \sigma_e^n dt. \quad (3)$$

Formula (2) was used for determining the equivalent stress. The coincidence of the theoretical and experimental data is satisfactory. In the case when the phenomenon of short-time creep is essential, it would be influenced by the form of the stress-strain diagram [5,6] . For a material tested during short-time creep it is interesting to obtain the stress-strain diagrams for given laws of changing stresses and strains with time, and for the simplest cases when the rates of conventional stresses or usual strains are constant. Such tests have been carried out [7] using specimens made from an aluminium alloy with the same composition and temperature as it was mentioned above. Solid lines in Fig.4 denote the true stress-strain diagrams at different constant usual strain rates, and in Fig.5 at different constant conventional stress rates. Using the obtained experimental diagrams the curve of instant tension was drawn. For this purpose the logarithmic creep strain $\bar{\varepsilon}^c$ was subtracted from the full logarithmic strain $\bar{\varepsilon}$. The latter was calculated by equ.(3). All the obtained points are grouped near the straight line to which the modulus of elasticity $E = 6,4 \cdot 10^3$ MPa corresponds. Therefore plastic strains are practically absent in this particular case. The stress-strain diagram calculated by the formulae which are given in the article [5] are shown in Figs. 4 and 5 by means of dashed lines. After comparison of the diagrams we conclude that the calculated, and the experimental data coincide satisfactorily. This investigation gives us the possibility to consider our aluminium alloy as a nonlinear viscous material. The equation of state of such a material is given by the following formula:

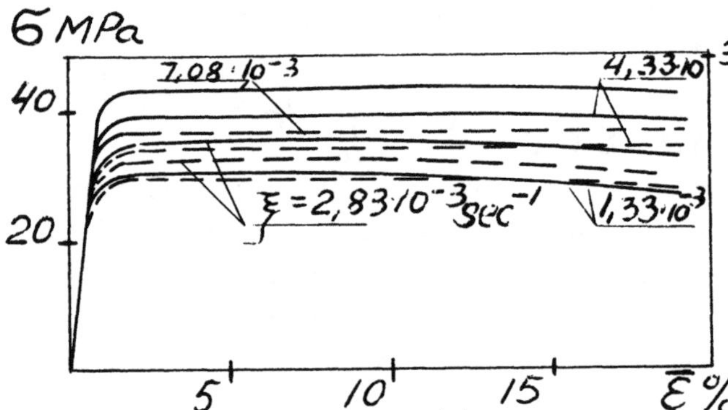

Fig. 4.

$$\xi_e = \xi_e^c = K \sigma_e^n .$$

One-dimensional problems

The solutions of such problems as upsetting, longitudinal rolling, extrusion, drawing, based on the plane section hypothesis as well as on the assumption of homogeneity of the stress state throughout the height of the workpiece are well known in literature [8 + 10] The former hypothesis leads to a contradiction with the law of the equality of shear stresses at the contact points of the tool and the workpiece, but together with the condition of incompressibility it leads directly to the establishment of the laws of strain change. The latter hypothesis reduces a two - dimensional problem to a one-dimensional one, and consequently simplifies its solution to a great extent. These two hypotheses allow to obtain simple formulae for metal forming loads, but, naturally, they do neither allow to study the stress and strain state at all points of the workpieces nor, consequently, to judge their strength while being worked. Such a statement of the problem was used for the solution

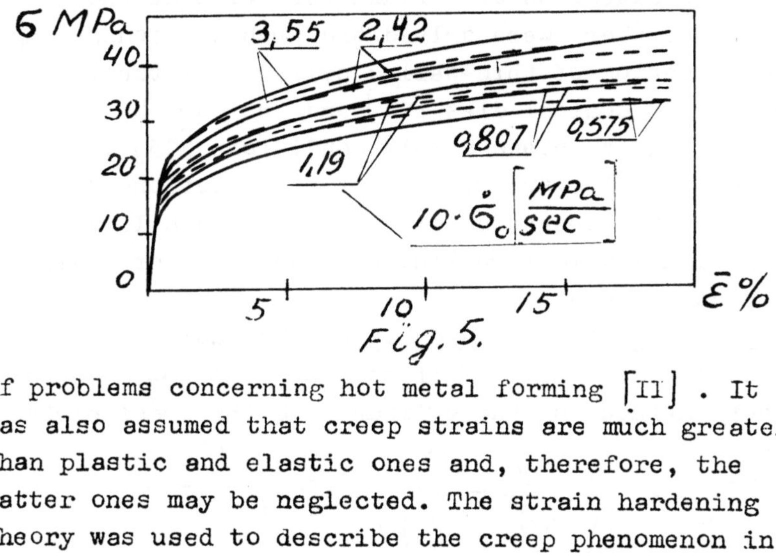

Fig. 5.

of problems concerning hot metal forming [11] . It was also assumed that creep strains are much greater than plastic and elastic ones and, therefore, the latter ones may be neglected. The strain hardening theory was used to describe the creep phenomenon in the following way:

$$\sigma_e = A\,\xi_e^{m_1}\left(\int \xi_e\, dt\right)^{m_2},$$

where σ_e is the equivalent stress, ξ_e the equivalent strain rate and $\int \xi_e dt$ - the Odqvist parameter. It is obvious that this theory permits to describe not only rate hardening, but also strain hardening. The intensity of friction forces on the contact surfaces between the tool and the workpiece was assumed to be proportional to the maximum shear stress. The latter assumption is a generalization of a commonly used one (in the case of a perfectly plastic solid), i.e. that the intensity of frictional forces is proportional to the yield stress of the material at pure shear. The application of the above-mentioned assumptions made it possible to solve in a closed form the following problems: the upsetting under plane-strain conditions, the upsetting of round solid and hollow cylinders, longitudinal sheet rolling under plane-strain conditions, the extrusion of a round bar. Without these assumptions there were solved the problems concerning the deformation of thin-walled cylindrical tubes within rigid conical dies [12] and the pure bending of a sheet under plane strain conditions [13]. As examples, we give the results of the solutions of the problem of the upsetting of a round solid cylinder, and the longitudinal rolling of a sheet under plain strain conditions.

<u>The upsetting of a round cylinder.</u> The contact pressure p and the intensity of frictional forces q on the contact surfaces of the cylinder and the press plates:

$$p = \frac{\sigma_e}{2}\left[2 + \frac{\chi}{h}\left(\frac{d}{2} - r\right)\right], \quad q = \chi\,\frac{\sigma_e}{2}.$$

The equivalent stress:

$$\sigma_e = A\left(\frac{v}{2h_0}\right)^{m_1}\left(\frac{h_0}{h}\right)^{m_1}\left(\ln\frac{h_0}{h}\right)^{m_2}.$$

The average conventional contact pressure:

$$p_{cp.} = \frac{4P}{\pi d_0^2} = A\left(\frac{v}{2h_0}\right)^{m_1}\varphi_2\left(\frac{h_0}{h}\right),$$

where

$$f_2\left(\frac{h_0}{h}\right)=\frac{1}{12}\left(\frac{h_0}{h}\right)^{m_1+1}\left(\ln\frac{h_0}{h}\right)^{m_2}\left[12+\gamma\frac{d_0}{h_0}\left(\frac{h_0}{h}\right)^{\frac{3}{2}}\right].$$

d_0 and d , h_0 and h are the diameters and heights of the cylinder in the undeformed and deformed states, respectively, and γ is an experimental coefficient.

<u>Longitudinal sheet rolling under plane strain conditions</u>. The formulae for longitudinal stresses at the entrance, and at the exit sections respectively:

$$\sigma_x=\sigma_{x1}+a\left[2J_1(\alpha_1)-2J_1(\alpha)-\gamma J_2(\alpha_1)+\gamma J_2(\alpha)\right],$$

$$\sigma_x=\sigma_{x2}-a\left[2J_1(\alpha)+\gamma J_2(\alpha)\right],$$

where

$$a=\frac{2^{1+2m_1+m_2}}{3^{\frac{1+m_1+m_2}{2}}}A\left(\frac{v h_1}{R^2}\right)^{m_1},$$

$$J_1(\alpha)=\int_0^{\alpha}\left[\ln\frac{h_1/R}{h_2/R+2(1-\cos\alpha)}\right]^{m_2}\frac{(tg\alpha)^{m_1}\sin\alpha\,d\alpha}{\left[\frac{h_2}{R}+2(1-\cos\alpha)\right]^{2m_1+1}},$$

$$J_2(\alpha)=\int_0^{\alpha}\left[\ln\frac{h_1/R}{h_2/R+2(1-\cos\alpha)}\right]^{m_2}\frac{(tg\alpha)^{m_1}\cos\alpha\,d\alpha}{\left[\frac{h_2}{R}+2(1-\cos\alpha)\right]^{2m_1+1}},$$

σ_{x1} ,σ_{x2} - are the stresses at the entrance and at the exit, v_1 is the velocity at the entrance, h_1 and h_2 denote the thicknesses of the workpiece at the entrance and at the exit, R is the radius of the rolls, α_1 - the contact angle, and α - the angular coordinate of some section $(0 \leq \alpha \leq \alpha_1)$. The angle α_c determining the position of a neutral section is found from the equation

$$a\left[2J_1(\alpha_1) - \chi J_2(\alpha_1) + 2\chi J_2(\alpha_c)\right] = \sigma_{x2} - \sigma_{x1}. \quad (4)$$

The contact pressure is evaluated by the following formula:

$$p = \frac{2}{\sqrt{3}}\sigma_e - \sigma_x ,$$

where the equivalent stress is equal to:

$$\sigma_e = \frac{\sqrt{3}}{2} a \left[\ln \frac{h_1/R}{\frac{h_2}{R} + 2(1-\cos\alpha)} \right]^{m_2} \frac{(\mathrm{tg}\alpha)^{m_1}}{\left[\frac{h_2}{R} + 2(1-\cos\alpha)\right]^{m_1}} .$$

The intensity of the frictional forces is determined from the formula: $q = \frac{\sigma_e}{\sqrt{3}}$.

Let us consider the example of the hot rolling of a sheet at the temperature 450°C. The material is the aluminium alloy described above. It is assumed that the contact angle is α = 20°, $\ell/h_{cp.} = 4$, that corresponds to h_2/R =0,0253 and h_1/h_2 = 5,74. The stresses at the entrance and at the exit equal zero: $\sigma_{x1} = \sigma_{x2} = 0$. The material which was considered above is a nonlinear viscous one. For this material holds at a given temperature

$$K = \frac{1}{A^{m_1}} = 63{,}4 \cdot 10^{-15} (\mathrm{MPa})^{-n} \sec^{-1}, \quad n = \frac{1}{m_1} = 7, \quad m_2 = 0.$$

After solving the equation (4) we obtain $\alpha_c = 4^\circ 19'$. Fig.6 shows the diagrams of contact pressures and the intensity of friction forces.

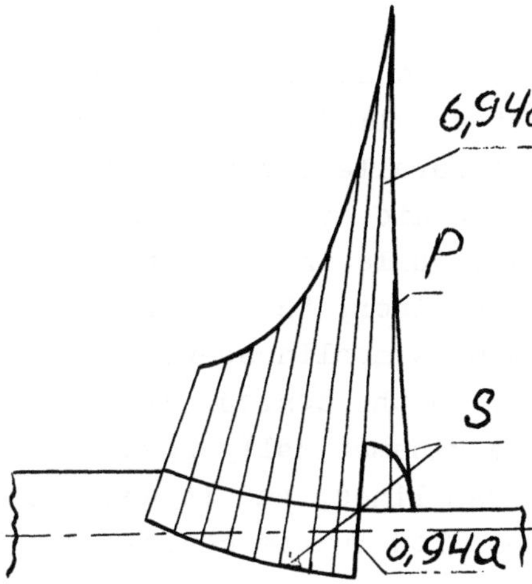

Fig. 6.

Two-dimensional problems

The solutions of two-dimensional problems concerning hot metal forming without using the above-mentioned hypotheses are very important because they permit to determine the states of stress and strain of the work-pieces. It gives the possibility to estimate their strength during processing. These solutions can be obtained by means of such well-known methods as the upperbound method, and the method of finite elements. Let us consider the application of these methods to the following problems: the extrusion of nonlinear viscous material in a conical die, and the upsetting of a solid circular cylinder when the flow theory is used.

Extrusion of a nonlinear-viscous material through a conical die. This problem was solved [14, 15] by the upper bound method on the basis of the principle of minimum total power of a system: $\delta(\mathcal{N}-\mathcal{L}) = 0$, where $\mathcal{N} = \int \sigma_e \xi_e \, dV$ is the power of the internal forces, and $\mathcal{L} = PV$ - the power of the external forces.

(P - extruding force, $\mathcal{V}$ - the velocity of the piston). The whole region of the deformed material was subdivided into some zones: the cylindrical container, the conical die, the length of parallel, the transition parts from the container to the die, and from the die to the length of parallel. While investigating the flow in the cylindrical container, the solution of the problem of a laminar flow of a nonlinear-viscous material in a cylinder was used. In the conical die and in the transitional parts the kinematically possible rate fields were given by stream lines, and it was assumed that there was a radial flow in the conical die between the transitional parts. The boundaries of the parts were assumed to be parabolic, while the equations of stream lines were given separately in each part in the form of polynomials. Changing the position of the internal boundaries, from the condition of minimum total power of the system was determined an actual rate field, and the position of the boundaries corresponding to it. The minimum of the functional was sought by using the method of the direct search. To check the results of the calculation there was carried out the experimental investigation of extruding an aluminium alloy at 4000°C in a die with the cone angle $2\alpha_1 = 90°$. The piston velocity is $\mathcal{V}$ = 0,008m/sec, the diameter and the height of the workpiece are $\mathcal{D}_W$ = 48 mm, H = 90mm, the diameter of the container amounts to $\mathcal{D} = 2R$ = 50 mm, and the extrusion ratio is equal to 4. The material constant at the test temperature is n = 3,44.

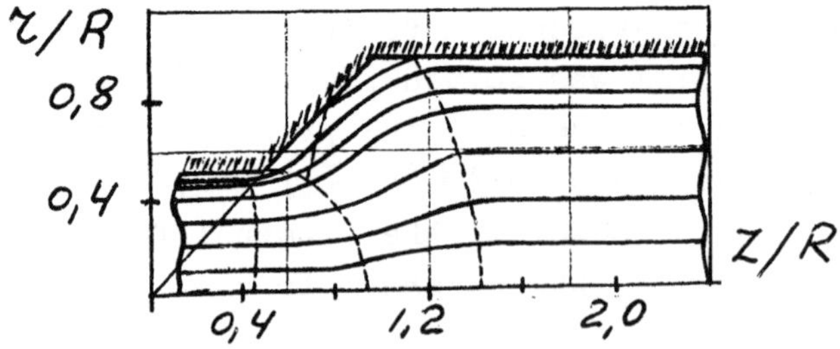

Fig.7.

Fig.7 shows the stream lines, and Fig.8 - the experimental (solid line) and the theoretical (dashed line) indicator diagrams. The difference between the theoretical and the experimental data is small.

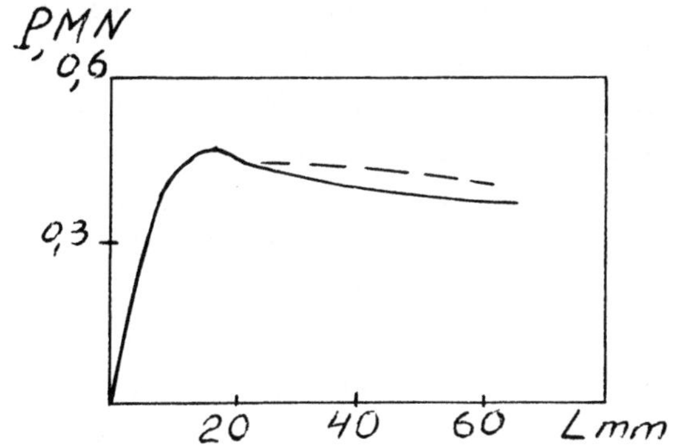

Fig. 8.

Upsetting of a circular cylinder. This problem was solved [16] by the method of finite elements in the form of a displacement method in accordance with the flow theory. According to this theory the dependence of the components of strain rates from the stresses are:

$$\xi_{ij} = \frac{3}{2}\frac{\xi_e}{\sigma_e}\left(\sigma_{ij} - \delta_{ij}\sigma_0\right), \qquad (5)$$

and

$$\xi_e = B\sigma_e^n, \qquad (6)$$

where σ_0 denotes the mean normal stress, δ_{ij} - the Kronecker delta, n - a constant, and B - a function of time. The latter value is determined experimentally for each material at a definite temperature. Using the dependency (6) we transform equation (5) into:

$$\xi_{ij} = \frac{3}{2}\xi_e^{1-\frac{1}{n}} B^{\frac{1}{n}}\left(\sigma_{ij} - \delta_{ij}\sigma_0\right). \qquad (7)$$

The obtained equation is analogous to the equation of state for nonlinear - elastic bodies, in which the strains are replaced by the strain rates, and $\xi^{\frac{1}{n}-1}B^{\frac{1}{n}}$ is used instead of the modulus of elasticity while Poisson's number is equal to one half. As it is known, the application of the method of finite elements in the form of the displacement method for this value of Poisson's ratio is impossible. Due to this reason, the latter value was taken to be close to 1/2, but not equal to it. It is obvious that the condition of incompressibility in this case is satisfied only approximately. When taking into account the adopted assumptions, the solution of equation (7) with respect to stresses permits to obtain one of the main relations of the method of finite elements in matrix form:

$$\{\sigma\} = \xi_i^{\frac{1}{n}-1} B^{-\frac{1}{n}} [\mathcal{D}_0]_{\mu<0,5} \{\varepsilon\}, \quad (8)$$

where the matrix $[\mathcal{D}_0]$ is analogous to that part of the elasticity matrix which alone depends on Poisson's number [17] . To formulate the algorythm it is necessary to have the relation connecting the rates of displacements of the nodes and the strain rates with respect to the volume of the finite element. Within the limits of a small step with respect to time, this relation has the following form:

$$\{\xi\} = [B]\{\dot{\delta}\},$$

where $\{\dot{\delta}\}$ is a vector-column of rates of nodal displacements and $[B]$ - the matrix determined by the approximation of the displacements with respect to the volume of a chosen finite element [17]. When applying Lagrange's principle of virtual displacements to each finite element which is in an equilibrium state, we obtain the stiffness matrix of the element $[K]$ in the following form:

$$[K] = \int_V \xi^{\frac{1}{n}-1} B^{-\frac{1}{n}} [B]^T [\mathcal{D}_c]_{\mu<0,5} [B] \, dV,$$

where dV denotes the element of the volume, and " T " means transposition of a matrix. Attributing the stress-strain state of the element to its centre of gravity, we finally obtain:

$$[K] = \xi_e^{\frac{1}{n}-1} B^{-\frac{1}{n}} [\bar{B}]^T [\mathcal{D}_o]_{\mu=0,5} [\bar{B}] V_e,$$

where $[\bar{B}]$ is the matrix determined for the centre of gravity of the element and V_e - is the volume of the finite element. Forming now the stiffness matrix of the whole body and setting up the vector-column of nodal forces, it is possible to obtain the system of nonlinear algebraic band-type equations for each step which are the equations of equilibrium. When solving these ones with respect to the rate of nodal displacements of the net of finite elements we obtain the solution of the problem. The subsequent definition of stresses from the relation (8), and strains which were accumulated up to a given moment of time, causes no difficulties. As an example, there was considered a process of hot axi-symmetric upsetting of a solid cylinder, the initial ratio of hight to the diameter being equal to 1. This cylinder was made from aluminium alloy [18] . The relative velocity of the plates was considered to be constant and equal to 0,167 mm/sec.

Sticking friction was assumed to hold between the plates and the adjacent material. The finite element was chosen in the form of a ring with a triangular cross section. To estimate the convergence of the solution there was made an experiment using the method suggested in [18] . Specimens with the sizes $h_o = \tau_o =$ 25mm were used. The results of the calculations allowed to ascertain that when dividing a workpiece into 384 axisymmetric rings, there was achieved a good coincidence between the theoretical and experimental grids. For a compression of 30% the latter ones are shown in Fig.9 by the lines 3 . Curves 1 and 2 denote the calculated grids for the values ν = 0,44 and 0,48 correspondingly. As it is seen from Fig.9 the accuracy of the solution grows with an increasing Poisson's number. To choose the coefficient greater than 0,48 is inexpedient as in this case, a more detailed finite - element discretization is necessary which leads to an essential increase of computer time. At the same time the discrepancy between the coordinates of the nets 2 and 3 does not exceed 5%. Fig.10 represents the experimental, and the calculated dependences of the load of deformation from time, for the two rates of upsetting for which different numbers of elements are used, and also for the above-mentioned approximate solution [18] . The same investigation has been carried out for the case of upsetting of a hollow cylinder [19].

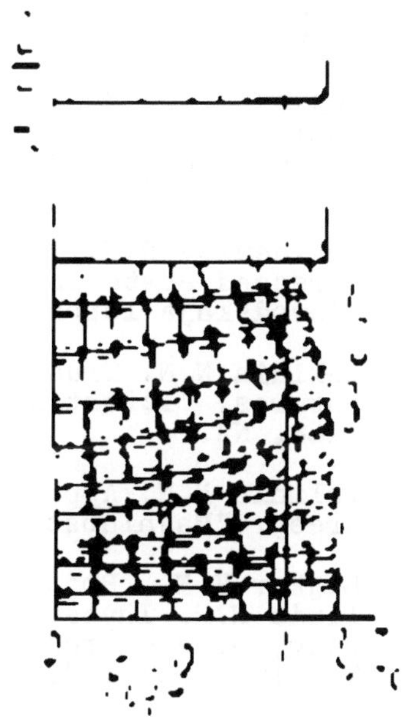

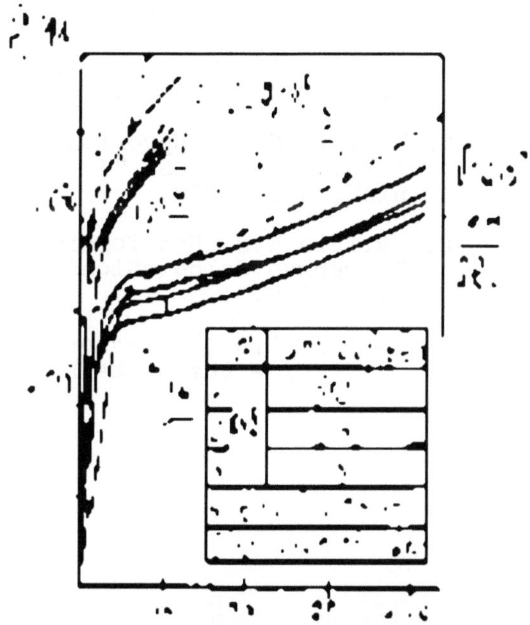

References

1. Bojkov V.N., Bojtsov Yu.I., Lazarenko E.S. Malinin N.N. Short-time creep of the D16 alloy at large strains. "Izvestija vuzov.Mashinostroenie", 1971, N4.

2. Bojkov V.N., Bojtsov Yu.I., Malinin N.N. Investigation of a short-time creep of a boiler steel. "Izvestija vuzov. Mashinostroenie", 1972, N9.

3. Bojkov V.N., Bojtsov Yu.I., Lazarenko E.S., Malinin N.N. Investigation of a short-time creep in a specimen's neck. "Izvestija vuzov. "Mashinostroenie", 1972, N10.

4. Davidenkov N.N., Spiridonova N.I. Analysis of stressed state in a neck of a tensiled specimen. "Zavodskaja laboratorija", 1945, N6.

5. Malinin N.N. True tensile diagrams at high temperatures. "Izvestija vuzov. Mashinostroenie", 1968, N1.

6. Malinin N.N. Applied theory of plasticity and creep. "Mashinostroenie", 1975.

7. Bojkov V.N., Bojtsov Yu.I., Lazarenko E.S., Malinin N.N. Investigation of a tension of the D16 alloy at elevated temperatures. "Izvestija vuzov. Mashinostroenie", 1971, N6.

8. Storozev M.V., Popov E.A. Theory of metal forming. "Mashinostroenie", 1977.

9. Tchelikov A.I., Grishkov A.I. Rolling theory. "Metallurgija", 1970.

10. Johnson W., Mellor P.B. Engineering plasticity. Van Nostrand Reinhold Company, 1973.

11. Malinin N.N. Approximate solutions of some metal forming problems. "Izvestija vuzov. Mashinostroenie", 1977, N12.

12. Malinin N.N. Hot forming of thin-walled tubes. "Strength Calculations", issue 19, Mashinostroenie, 1978.

13. Malinin N.N., Shirshov A.A. Heated sheet bending. "Izvestija vuzov. Mashinostroenie", 1978, N2.

14. Bobrovnikova N.N. Hot extruding of a nonlinear-viscous material. "Izvestija vuzov. Mashinostroenie", 1973, N5.

15. Bobrovnikova N.N. Some results of solving the problem of hot extruding of a nonlinear-viscous material. "Izvestija vuzov. Mashinostroenie", 1974, N1.

16. Malinin N.N., Romanov K.I. Solution of hot forming problems by the method of final elements. "Izvestija vuzov. Mashinostroenie", 1977, N8.

17. Zienkiewicz O.C. The finite element method in engineering science, Mc Graw Hill, 1971.

18. Romanov K.I. Experimental investigation of hot upsetting. "Izvestija vuzov. Mashinostroenie", 1977, N6.

19. Malinin N.N., Romanov K.I. Approximate solution of the problem of hot upsetting of a hollow cylinder. "Izvestija vuzov. Mashinostroenie", 1977, N7.

Modeling of the Tool-Workpiece Interface

JOHN A. SCHEY

Department of Mechanical Engineering
University of Waterloo, Ontario, Canada

Summary

From the response of lubricants to pressure, temperature, and strain rate, it is concluded that a coefficient of friction provides an adequate model of lubrication by polymers, boundary and E.P. agents, layer-lattice compounds, and organic liquids below their glass-transition temperature. In mixed lubrication, the fluid in the hydrodynamic pockets may be viscous or elastic. Soft metals can be described only by their shear strength, and the friction shear factor is sharply limited in its validity to unlubricated ideal rigid-plastic workpiece materials.

Introduction

The importance of friction between die and workpiece has long been recognized and studies of the effects of friction on forces, power requirements, and process limitations abound (1). Analysis initially described friction by a coefficient of friction

$$\mu = \frac{\tau_i}{p} \qquad (1)$$

where τ_i is the average shear strength of the interface, and p is the average interface pressure.

More recently, the frictional shear factor m gained popularity

$$\tau_i = m\,\tau_f \qquad (2)$$

where τ_f is the shear flow strength of the workpiece material and is often taken as $\sigma_f/\sqrt{3}$ (where σ_f is uniaxial flow strength). The relative virtues of these representations are debatable (2,3) but, for practical purposes, both can be satisfactory. The choice affects average pressures (and thus forces) only minimally. For well-lubricated processes the frictional contribution is small and overshadowed by uncertainties in flow stress, and both μ and m have tended to become adjustable factors for theory.

With the advent of numerical methods, stress and strain distributions may be calculated in much greater detail, and a realistic representation of the frictional boundary conditions is essential if these calculations are to be of value. Knowledge of the interface in metalworking is growing only slowly, but there has been a burst of activity on lubricant films in general. The purpose of this contribution is to evaluate lubricating mechanisms and identify acceptable methods of modeling them.

Lubricating Mechanisms

Lubricating mechanisms operate over very wide ranges: interface pressures vary from a fraction to high multiples of σ_f; sliding velocities range from zero to 50 m/s (with superimposed approach velocities of up to 20 m/s); bulk or inlet temperatures range from ambient to close to the melting point of the workpiece material, and are augmented by heat from the work of deformation and friction; the lubricant film may vary from vanishing to thick enough to separate all asperities; and the macro- and micro-geometries and compositions of the die and workpiece, coupled with the rheological and chemical characteristics of the lubricant, give rise to a vast array of interactions. For our purposes, a convenient framework is given by the type of lubricant used.

a. Dry Interfaces

These are encountered when no intentional lubrication is practiced. In the limit, the surfaces would be free of all contaminants, and adhesion would dominate the events. Adhesion between dissimilar metals and alloys is still only poorly understood, nevertheless, a rough estimate may be based on the mutual solubility at the interface temperature (4), allowing for the diffusion of minor elements to the surfaces (5). Sliding of the interface is likely to be arrested and true "sticking" may develop through solid-state welding.

The interface shear strength τ_i is equal to τ_f prevailing in the deformation zone only for an ideal rigid-plastic workpiece material (Fig. 1a), but is lower (or higher) for strain-hardening or strain-rate-sensitive materials (Fig. 1b), or for non-isothermal conditions (Fig. 1c). Average values of τ_i, m or μ are equally meaningless. The shear properties of the workpiece material need be known for various strains, strain rates, and temperatures; plastometer data (6) are seldom corrected for adiabatic heating or non-homogeneous deformation, and are more suited for calculating average pressures than stress and strain distributions.

Technical surfaces are never absolutely free of adsorbed films or reaction products (in particular, oxide films) which interfere with adhesion. However, the surface extension characteristic of bulk deformation exposes virgin surfaces which are more reactive than even a perfectly clean "old" surface. Thus, even contaminated surfaces exhibit the adhesion of the substrate, at least in surfaces undergoing large extensions or severe sliding (4).

Oxides grown on the substrate reduce adhesion and sometimes also τ_i. Unfortunately, all too little is known about their rheological properties. Their room temperature hardness is usually greater than that of the substrate, yet their shear strength must depend on temperature and hydrostatic pressure. Reoxidation of freshly exposed surfaces hinges on the availability and access of oxygen, and the whole interface situation is too uncertain for a priori modeling.

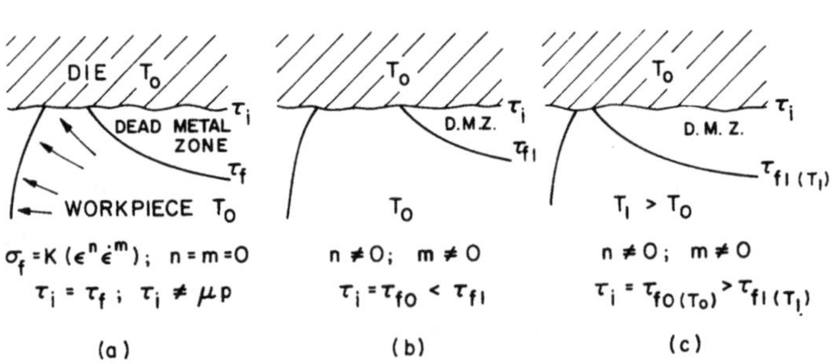

Fig. 1 Unlubricated interfaces

A similar, ever-present problem stems from the surface films inherited from prior processing or surface preparation. Organic residues from cleaning with surface-active fluids reduce adhesion (7) and friction (1). Residual films appear to have a pressure-dependent shear strength and, from measurements with pressure transducers built into the die surface (8,9), are best modeled by a coefficient of friction. This is confirmed from measured displacement distributions by the matrix method of analysis (10). Contrary conclusions for hot upsetting (9) may be due to chilling on the die faces.

A complication is that sticking is governed not only by adhesion but also by process geometry. Thus, sticking prevails in axial upsetting until the diameter-to-height ratio exceeds 3 (11,12); the sticking zone in rolling (13) increases with the h/L ratio (where h = mean workpiece thickness, L = projected length of contact).

Shearing of adhesive joints results in the formation of wear debris and/or metal transfer to the die surface. Contact then changes from workpiece vs. die to workpiece vs. workpiece material, except that the build up is heavily worked, contaminated and oxidized (14), and harder than the workpiece itself.

Asperity interaction is crucial even in elastic contact (15) and its importance cannot be overemphasized in metalworking (1,16). Thin oxide and contaminant films are particularly vulnerable to penetration by hard asperities, thus adhesion, τ_i and the sticking zone are much more affected by die than by workpiece roughness.

b. Soft Metal Films

With the advent of conversion coatings, soft metals have declined in importance but remain attractive from the theoretical point of view. If the film is continuous, a simple situation exists, with $\tau_i = \tau_s$ where τ_s is the shear flow strength of the soft metal. The best evidence indicates that τ_s is a function of temperature and strain rate, but not of hydrostatic pressure (17). If the material strain hardens, it changes its properties even during sliding, although not necessarily in any fixed relation to those of the workpiece, therefore, neither a constant μ nor m is useful. There is some doubt whether the pressure independence of shear strength also holds for thin films. For grossly deformed metal films and $p > 3$ GPa, the shear strength varied as (18)

$$\tau_s = \tau_o + \alpha' p \qquad (3)$$

where τ_o is shear strength at zero pressure and α' was 0.016 to 0.019 for Cu, Ag, and Au. The temperature dependence was an exponential function of T/T_m (the temperature of the film on the homologous temperature scale).

There appears no way (nor are there experimental data) to consider strain, strain rate, temperature, and pressure simultaneously. It is, nevertheless, clear that there is no direct relation between σ_f and τ_s, thus m has no application. The virtue of μ is similarly questionable, although a representation similar to Eq. (3) may be useful when $\dot{\gamma}$ (and strain-rate sensitivity) are low.

c. Polymeric Films

Many polymers deposited as thin films obey Eq. (3) (19). At high interface pressures τ_o becomes negligible and then

$$\tau_s = \alpha' p \qquad (4)$$

and α' is identical to μ. No relationship with m exists. For PTFE $\alpha' = 0.08$, for PE 0.10 to 0.14. Lower values are often found in metalworking, $\mu = 0.05$ is not uncommon for PTFE (3).

Temperature dependence follows an Eyring model of viscous flow

$$\tau_s = \tau_o' \exp(-Q/RT) \qquad (5)$$

where τ_o' is a material constant, and R the gas constant. The activation energy Q is quite similar for several polymers, suggesting that shearing involves sliding of parallel-oriented chains.

Complications arise from several sources. The above relationships apply only to polymers above their glass-transition temperature T_g. Below T_g sliding may be accommodated (as, for example, with PMMA) by film fracture or by sliding at the polymer/metal interface (19). In metalworking, some polymer is usually transferred to the die surface but this appears insufficient to prevent pick-up and scoring if the polymer film loses continuity.

d. Layer-Lattice Compounds

The problem is often that of forming continuous films before die pick-up could occur. By the time the film is well worked, a transfer film forms on the contacting surface and shear takes place between the two films, at least with MoS_2. Shear strain rate thus does not enter as a variable except as it affects temperature. The steady-state τ_s is pressure-sensitive, Eq. (4) and approximates a constant μ, typically 0.05 for MoS_2 (3). The shear strength of a partially worked film (20) also depends on whether the film is formed mechanically (as on Al, Ti) or through reactions (as on Cu, Fe), and on oxygen access and temperature.

e. E.P. and Boundary Lubricants

Both groups reduce τ_i through the formation of thin films. Little is known about the pressure dependence of E.P. layers (Cl or S compounds), although by analogy to MoS_2 one would expect it. Reactions are dependent on time, temperature, and substrate composition, and can take place under metalworking conditions only if there is repeated contact. This limits their action to the die; contact times are likely to be too short for their formation on the workpiece surface.

Boundary films form almost instantaneously and have a most beneficial effect in reducing τ_i. The shear strength of stearic acid and its soaps is pressure- and temperature-dependent (19) like that of polymer films, with a similar Q in Eq. (5), suggesting that methylene groups slide on each other. Not surprisingly, a constant μ of around 0.05 is often found (3). Above the melting point of the polar compound (or its soap) the liquid behaves like a viscous fluid, except that the polar compound is also attached to the surfaces as an orientated layer of polymer-like properties. The situation becomes a special case of mixed lubrication. Pressure raises the melting point; many boundary agents (e.g. oleic acid) liquid at room temperature and atmospheric pressure solidify at high pressures, and then presumably exhibit the shear behavior of stearic acid.

Both E.P. and boundary additives rely on chemical reactions for developing their full potential and the compositions of the die and workpiece surfaces are critical (21).

There is a growing body of literature dealing with the mathematical modeling of boundary and E.P. lubrication (22). Successes such as are relate to wear rather than friction. In particular, there is no quantitative prediction available for the formation of a "friction polymer": the freshly formed surface acts as a catalyst in decomposing the lubricant, the fragments of which recombine to form a low-molecular-weight polymer. This polymer exhibits, presumably, the characteristic polymer shear behavior. The cumulative time necessary for friction polymer formation may never be reached in the laboratory and this may explain, at least in part, why laboratory simulation of metalworking lubrication is so often unsuccessful.

The entire field of chemical reactions on surfaces, called mechanochemistry (23) or tribochemistry (24) is developing rapidly.

f. Fluid Films

Organic fluids (mineral oils and synthetics) are important on their own and as carriers of additives. It has long been known that their viscosity η is a function of pressure, approximately as

$$\eta = \eta_o \exp(\alpha p) \qquad (6)$$

where η_o is dynamic viscosity at atmospheric pressure. They solidify at some critical pressure (in sliding/rolling contacts at 0.7 GPa) and then exhibit a pressure-dependent shear strength, Eq. (4), with $\alpha' = 0.07$ for a number of fluids (25). Sliding takes place at the interface between two films attached to the metal surfaces and then temperature in the shear plane remains the only other significant variable.

All this makes good sense if these fluids are regarded as polymers with T_g above room temperature. A consideration of specific volume changes shows that T_g must move higher with higher pressures and faster cooling rates; this promotes transition to the glassy state under high interface pressures and short engagement times, such as prevail in EHD contacts (26). Direct observation with infrared spectroscopy (27) puts T_g as low as 0.1 GPa at low shear rates and above 1 GPa for high shear rates (and temperatures around 140°C). While lubricant models for high pressures are far from settled, a model (35) based on elastic-solid behavior appears to have broad validity.

The regime of solid-like behavior is extended by the adsorption of oriented molecular layers. The transition to boundary lubrication, at least on the tips of asperities, is then a natural one.

These concepts have not been considered yet in metalworking but they are likely to restrict the applicability of plastohydrodynamic analyses. Experimental verification of full fluid-film lubrication from measured pressures is difficult because the frictional component is negligible compared to the deformation term. Only rolling allows a direct check; in rolling with negative forward slip, traction in the lubricant film is solely responsible for strip transport. Hydrodynamic lubrication is found limited to very small reductions (29), and some other mechanism must be active under practical conditions even with the best lubricant. Hydrodynamic lubrication may exist in wire drawing with a soap, although a pressure-dependent τ_i model may be more realistic.

g. <u>Surface-Roughness Effects</u>

With the exception of some sheet-metalworking (and glass-lubricated hot working) operations, the average fluid film thickness is almost always less than the combined surface roughness of die and workpiece. The possible harmful effects have already been mentioned, it now remains to balance the picture by pointing out the possible beneficial effects of roughness, particularly on the workpiece. A rougher surface helps to build a film of greater average thickness (1,29,30). In the course of deformation this film breaks up into small hydrodynamic or hydrostatic pockets (31, 32) separated by large areas protected only by boundary films. In the course of further deformation, lubricant is squeezed out from these pockets which thus serve as microscopic reservoirs. Even the die surface may benefit from a roughness of controlled magnitude and orientation (33).

In most practical situations, lubrication is of the mixed type, with a β fraction of the surface area in boundary contact. An average μ_{mix} may then be defined (34) as

$$\mu_{mix} = \beta\, \mu_B + (1 - \beta)\, \mu_H \qquad (7)$$

where μ_B applies to boundary contact and μ_H to the hydrodynamic pockets. A first attempt at modeling the interface in rolling under realistic conditions correctly predicts the general trends (35).

h. Melting Solids

Glasses are perhaps the only fluids displaying Newtonian, pressure-independent viscosity. Melting off from an excess supply is critical in hot extrusion with large extrusion ratios (36) whereas the viscosity at the interface temperature is critical for pre-applied films (37). With the latter, graphitic lubricants applied to the die protect in case of lubricant breakdown, but may also change the shear properties of the glass.

Shear Properties of Interface

It is obvious that much work needs to be done yet, nevertheless, a pattern of surprising simplicity emerges: when the shear strength of the interface (for room temperature and at a constant shear strain rate) is plotted against interface pressure (Fig. 2), only three basic forms of response are evident:

1. Metals alone are insensitive to interface pressure.
2. Polymers, boundary layers, soaps, and layer lattice compounds share an often linear response to pressure (fats, waxes and E.P. reaction films are likely to respond similarly).
3. Fluids (mineral and synthetic oils) exhibit a steeply (exponentially) rising shear strength until at some critical pressure they become polymer-like. This pressure can be as low as 0.3 GPa but as high as 2 GPa for low-viscosity oils used in high-speed rolling and drawing.

Modeling of Interface

The concept of the frictional shear factor m requires, Eq. (2) that τ_i be in a fixed relation to τ_f and independent of interface (hydrostatic) pressure. This is satisfied only for unlubricated surfaces and then only for an ideal rigid-plastic workpiece material (Fig. 1a).

The concept of the coefficient of friction μ requires, Eq. (1) that τ_i be linearly proportional to interface pressure. This is approximately true (Fig. 3) of many practical lubricant classes, therefore, a constant coefficient of friction offers a quite adequate model of many solid organic, organo-metallic, and layer lattice films. It also describes an oil/additive system at high enough interface pressures (extreme right of Fig. 3) to solidify the oil. At lower pressures (below 0.3 to 2.0 GPa, depending on η

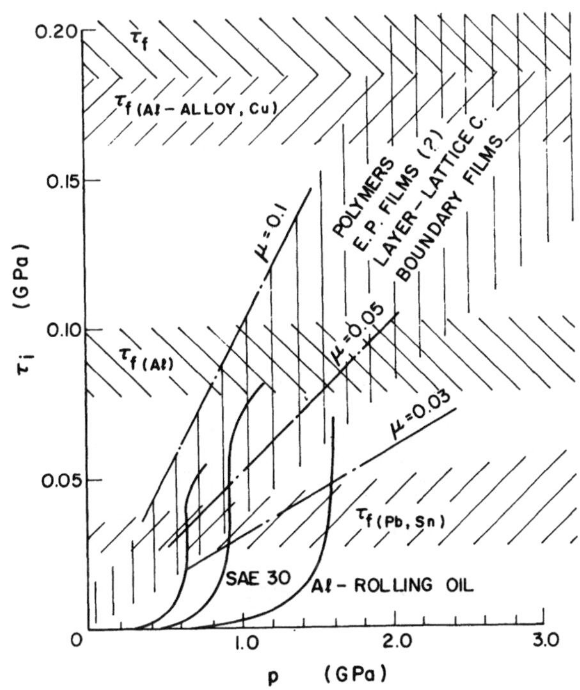

Fig. 2 Interface shear strength as a function of pressure, $\dot{\gamma} = 1/s$

and α of the oil), true mixed lubrication exists, with an oil of viscous character trapped in hydrodynamic pockets. Polar additives form a partially-oriented film of high apparent viscosity and possibly solid-like rheology, and a pressure-sensitive τ_i (and constant μ) may still be a reasonable approximation. Only for metals (and truly unlubricated surfaces) is μ totally misleading; at high interface pressures sticking can prevail at quite low μ values (right of Fig. 4).

The ultimate accuracy of calculation is assured, of course, if the true strength τ_i of the interface is known. For predictive calculations the relationships shown in Figs. 2 and 3 should give adequate guidance. A locally varying coefficient of friction is justified only for detailed calculations (38). Moderate increases in temperature can be taken into account generally by moving towards a lower τ_i with solids and a lower η with fluids, although a larger temperature rise will usually change the lubricating mechanism itself and may also lead to lubricant breakdown.

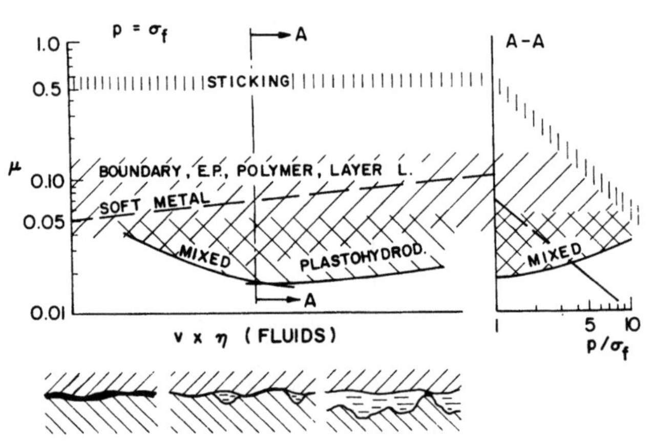

Fig. 3 Coefficient of friction for various lubricating mechanisms

References

1 Schey, J.A., Ed.: Metal Deformation Processes: Friction and Lubrication. New York: Dekker 1970.

2 Male, A.T.; DePierre, V.; Saul, G.: ASLE Trans. 16 (1973) 177-184.

3 Schey, J.A., in: Proc. 4th North-Am. Metalworking Res. Conf. Dearborn: SME 1976, 108-114.

4 Newnham, J.A.; Schey, J.A.: J. Lubric. Techn. 91 (1969) 351-359.

5 Ferrante, J.; Buckley, D.H.: ASLE Trans. 15 (1972) 18-24.

6 Altan, T.; Boulger, F.W.: J. Eng. Ind. 95 (1973) 1009-1019.

7 Hartweck, W.: Stahl v. Eisen 97 (1977) 638-639.

8 Cole, I.M.; Sansome, D.H.: Proc. 9th Int. MTDR Conf. Oxford: Pergamon 1968, 271-286.

9 Bühler, H.; Löwen, J.: Arch. Eisenhüttenw. 43 (1972) 603-607.

10 Oh, S.I.; Kobayashi, S.; Thomsen, E.G.; Proc. 3rd NAMRC. Dearborn: SME 1975, 159-167.

11 Unksov, E.P.: An Engineering Theory of Plasticity. London: Butterworths 1961.

12 Wallace, P.W.; Schey, J.A.; in: Proc. 8th Int. MTDR Conf. Oxford: Pergamon 1967, 1361-1380.

13 Tarnovskii, I. Ya.; Pozdeyev, A.A.; Lyashkov, V.B.: Deformation of Metals During Rolling. Oxford: Pergamon 1965.

14 Smith, A.M.; Barlow, E.; Amor, M.P.; Davies, N.C.: ASLE Trans. 21 (1978) 226-230.

15 Bowden, F.P.; Tabor, D.: The Friction and Lubrication of Solids, Parts I and II. Oxford: Clarendon 1954 and 1964.

16 Kloos, K.H. in: Mechanische Umformtechnik (O. Kienzle, Ed.). Berlin: Springer 1968, 293-342.

17 Pugh, H. Ll. D., Ed.: Mechanical Behavior of Materials Under Pressure. Amsterdam: Elsevier 1970.

18 Riecker, R.E.; Towle, L.C.: J. Appl. Physics 38 (1967) 5189-5194.

19 Amazu, J.K.A.; Briscoe, B.J.; Tabor, D.: ASLE Trans. 20 (1977) 354-358.

20 Reid, J.V.; Schey, J.A. in: Proc. 2nd Int. Conf. Solid Lubrication. Des Plaines, Ill.: ASLE 1978.

21 Schey, J.A.; Newnham, J.A.: Lubric. Engrg. 26 (1970) 129-137.

22 Beerbower, A.: ASLE Trans. 14 (1973) 90-104.

23 Rowe, C.N.; Murphy, W.R.: in Proc. Tribology Workshop (Atlanta). Washington: NSF 1974, 327-401.

24 Thiessen, P.A.; Meyer, K.; Heinicke, G.: Grundlagen der Tribochemie. Berlin: Akademie Verlag 1967.

25 Johnson, K.L.; Cameron, R.: Proc. Instn. Mech. Engrs. 182 (1967) Pt. I, 307-319.

26 Alsaad, M.; Blair, S.; Sanborn, D.M.; Winer, W.O.: Internal Report, Georgia Inst. Techn. Atlanta 1977.

27 Lauer, J.L.; Peterkin, M.E.: ASLE Trans. 21 (1978) 250-256.

28 Tevaarwerk, J.L.: in Proc. Int. Conf. Fundamentals of Tribology. Cambridge, Mass.: MIT Press, 1978.

29 Reid, J.V.; Schey, J.A.: ASLE Trans. 21 (1978) 191-200.

30 Mizuno, T.; Matsubara, K.; Kimura, H.: Bull. Japan Soc. Mech. Eng. 12 (1969) 359-367.

31 Ratnagar, D.D.; Cheng, H.S.; Schey, J.A.: J. Lubric. Techn. 96 (1974) 591-594.

32 Wilson, W.R.D., ibid, 99 (1977) 10-14.

33 Schey, J.A.: Proc. Eurotrib 77, Vol. 1, 67/1-67/4.

34 Vogelpohl, G.; Öl Kohle 12 (1936) 943-947.

35 Tsao, Y.H.; Sargent, L.B.: ASLE Trans. 20 (1977) 55-63.

36 Baqué, P.; Pantin, J.; Jacob, G.: J. Lubric. Techn. 97 (1975) 18-24.

37 Kulkarni, K.M.; Schey, J.A.; Wallace, P.W.; DePierre, V.: J. Inst. Metals 100 (1972) 33-39.

38 Bartram, R.; Lung, M.; Mahrenholtz, O.: Arch. Eisenhüttenw. 41 (1970) 969-973.

Drawing of Fiber Reinforced Tubes

J.F. Bratt

Faculty of Mechanical Engineering
The Norwegian Institute of Technology

Abstract

The paper presents an analytical study of bounds to the plastic drawing stresses in tubes reinforced by high strength fibers. It is shown that the drawing stresses vary with the fiber orientation,and this is practically independent of the ratio of the compressive to the tensile yield stresses of the matrix material. An equation predicting the thickness is discussed.

Notation

σ_t, σ_L	Normal stresses in the tangential and longitudinal directions
σ_{Le}, σ_{Li}	Normal stresses at exit and entrance
σ_o, τ_o	Yield normal stress, yield shear stress
p, α	Die pressure, die angle
ρ	Ratio of yield compressive to yield tensile stresses
ε_t, ε_L	Strain rates
r_i, r_e	Tube radii at entrance and exit
v_i, v_e	Velocities at entrance and exit
h_i, h_e	Thicknesses at entrance and exit
w	Power of deformation
θ	Average fiber orientation angle
k	Gradient coeffisient
u	Meridional velocity
R	Radius from cone apex
N_{ij}	Membrane force

Introduction

Bounds to drawing forces in conventional metal tubes are based on well-known flow mechanisms, the analysis of which is presented in texts by, among others, Johnson and Mellor [1] and Avitzur [2]. Fiber reinforced tubes see reference [3], have not been analyzed extensively in the above manner.

The general yield criterion for anisotropic materials has been discussed by Hill [4], Calcote [5] and McLaughlin [6]. The present paper deals with the prediction of drawing forces in tubes made of a matrix material with a strength-differential, reinforced with fibers of high yield stresses. It is assumed that yielding takes place without appreciable fibre breaking, debonding or buckling, and that the fiber weight fraction is small. The analysis establishes theoretically the influence of the fiber orientation on the drawing forces and diameter reductions.

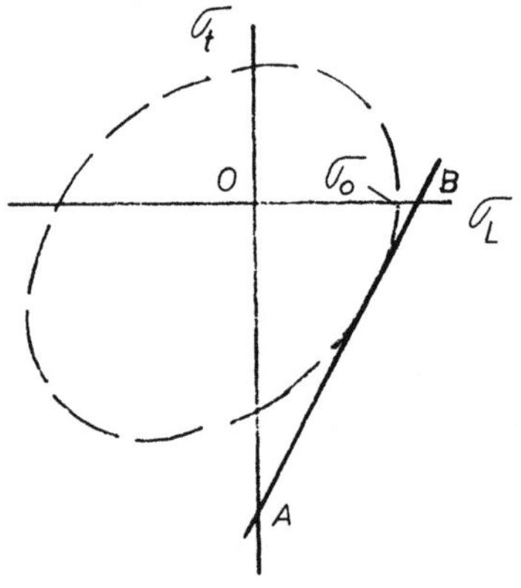

Fig. 1. Yield condition for fiber reinforced material with fibers oriented at $\pm\ \theta$ to the tube axis.

The yield condition

A questionable assumption in classical yield criteria is that yielding is independent of the sign of the stress. Tube drawing is confined to the fourth quadrant in the stress plane, ideally, and it is necessary to determine whether the matrix tensile and compressive yield stresses may be far apart. Many authors point out the dependence of the yielding on the sign of the stresses, see f.ex. references [7], [8], [9] and [14].

In the following presentation numerical values of ρ = 1.25 and 1.50 will be used for the tube matrix material. According to data by Stassi [10] this ratio increases with temperature.

Examples of reinforcing fibers are steel and tungsten, see ref. [6]. The fibers remain symmetrically oriented at fixed, average angels $\pm\ \theta$ during the drawing process. The yield condition may be expressed in terms of the average yield stresses of the constituents:

$$\sigma_{ij} = \sigma_{ij}{}^{m} + \frac{N_{ij}{}^{f}}{h} \tag{1}$$

in which m and f refer to matrix and fiber respectively. McLaughlin [6] suggests that the terms on the right side should be added at points on the matrix and fiber yield surfaces, where surface normal vectors are proportional to the strain rate vector, ε_{ij}. The above condition is shown in figure 1 for states of plane stress. Reports on the yielding of such materials exist, see [11], but experiments concerning the yield surface in the 4th quadrant of the stress plane are lacking.

With the above limitations the biaxial yield condition for the matrix material may be written

$$\sigma_L{}^2 + \sigma_t{}^2 - \sigma_L\sigma_t + (\sigma_L+\sigma_t)(\rho-1)\sigma_o = \rho\sigma_o \tag{2}$$

Similar expressions have been suggested by several authors, see f.ex. [12]. When the two-family fiber texture yield condition is addes, the following yield criterion results

$$\sigma_t - k\sigma_L = \frac{\sigma_o}{2k-1}((\rho-1)(2k^2+1-3k) - 2q \cdot$$

$$\cdot\ (k^2-k+1)) \tag{3}$$

where $k = \tan^2\theta$, $q = ((\rho-1)^2 - \frac{S}{3(k^2-k+1)})^{\frac{1}{2}}$

and $S = ((\rho^2+1)(2+k-k^2) - \rho(5-2k+2k^2))$

The assumed state of plane stress follows from the conditions that $h \ll r$ and that die pressure is small compared with the lateral and tangential stress components.

Lower bound solution

The average strain rates along AB in figure 1 are given by the relation

$$\frac{\varepsilon_L}{\varepsilon_t} = -k \tag{4}$$

Looking at radial equilibrium of the forces on an element the contact pressure is given by

$$p = -\frac{h}{r}\sigma_t \cos\alpha \tag{5}$$

In order to estimate the drawing stress the equilibrium of the forces acting in the longitudinal direction is considered, [1]:

$$\frac{d}{dr}(\sigma_L hr) - \sigma_t h + \frac{\mu pr}{\sin\alpha} = 0 \tag{6}$$

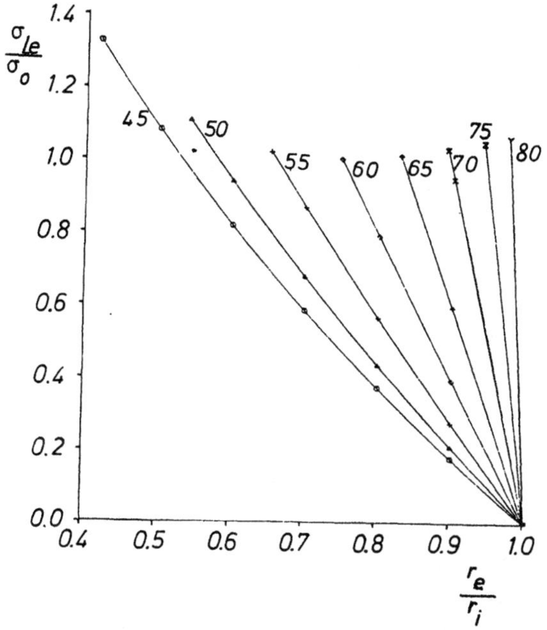

Fig. 2. Drawing stress as a function of radius reduction ratio, $\alpha = 9^{o}$, $\mu = 0.048$ and $\rho = 1.25$. Lower bound.

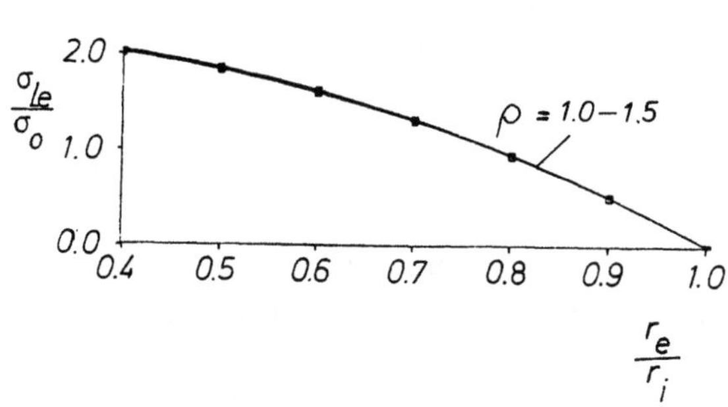

Fig. 3. Drawing stress versus radius reduction ratio. ρ varies between 1.0 and 1.5. $\alpha = 6^{o}$, $\mu = 0.085$ and $\theta = 60^{o}$.

Substituting σ_t from equation (3) in (6) and taking into account the boundary conditions that when

$$r = r_i,\ \sigma_L = \sigma_{Li} = 0 \text{ and } r = r_e,\ \sigma_L = \sigma_{Le} \qquad (7)$$

equation (6) becomes after integration

$$\frac{\sigma_{Le}}{\sigma_o} = \frac{1}{2k-1}\ (2q(k^2-k+1) - (\rho-1)(2k^2+1-3k)) \cdot$$
$$\cdot\ (1+B)\ \frac{1}{Bk}\ (1-(\frac{r_e}{r_i})^{Bk}) \qquad (8)$$

where $B = \mu\cot\alpha$. Equation (8) predicts a lower bound to the drawing stress. It is programmed and plotted in figure 2. In these graphs σ_{Le}/σ_o is drawn versus r_e/r_i for a number of fiber orientations. In figure 2 $\alpha = 9^o$, $\mu = 0.048$, $\rho = 1.25$ respectively. μ and α are averages from measured data presented in [13]. The graphs are terminated at the maximum permissible radius reduction ratios. The left side of equation (8) at which this occurs is given by point B in figure 1.

In figure 3 is shown equation (8) plotted with $\theta = 60^o$, $\alpha = 6^o$, $\mu = 0.085$ and with ρ varying between 1.0 and 1.5.

Substituting the value of σ_{Le} at point B in figure 1 in equation (8) the following relation results:

$$\frac{r_i}{r_e} = (1 + \frac{\mu}{\tan\alpha})^{\frac{\tan\alpha}{k\mu}} \qquad (9)$$

Equation (9) is shown graphically in figure 4 for average combinations of μ and α given by the following table:

Curve No:	1	2	3	4	5	6	7	8
α:	6°	9°	6°	9°	7°	8°	7°	8°
μ:	.085	.080	.043	.048	.070	.075	.042	.045

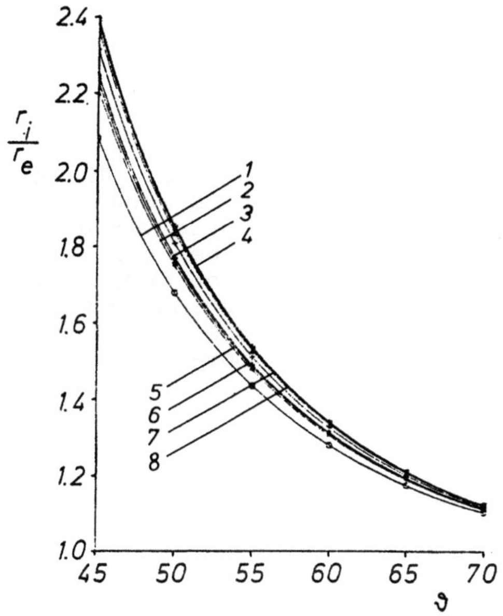

Fig. 4. Radius reduction ratio versus fiber orientation angle for various combinations of α and μ. Lower bound.

From equation (4) and the incompressibility condition the tube thickness variation follows,

$$h = h_i \left(\frac{r}{r_i}\right)^{k-1} \tag{10}$$

The relation between the tube thickness and the maximum reduction in radius can be written

$$\left(\frac{h_e}{h_i}\right)_{max} = \left(\frac{r_e}{r_i}\right)_{max}^{k-1} \tag{11}$$

Figure 5 shows the inverse of equation (11) for the same combination of α and μ as in figure 4.

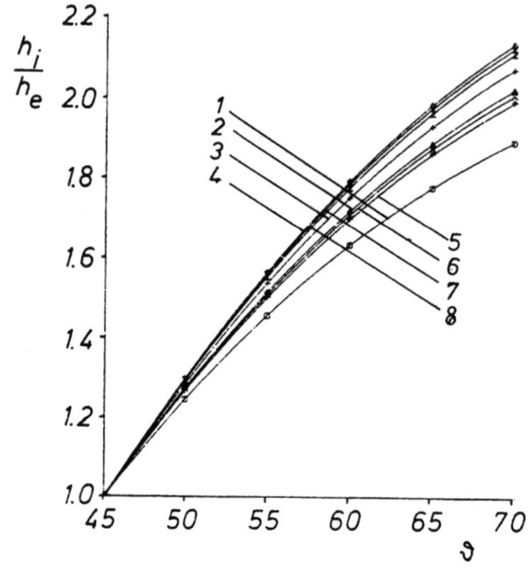

Fig. 5. Thickness reduction ratio versus fiber orientation angle for various combinations of α and μ. Lower bound.

Upper bound solution

The power involved in the drawing operation are regarded to be the sum of those due to velocity discontinuity, deformation and friction, [2]. At the entrance and exit of the die the velocity discontinuity losses are

$$W_{vd} = \int_{Si} \tau_o v_{si} dA + \int_{Se} \tau_o v_{se} dA \qquad (12)$$

In pure shear $\sigma_L = -\sigma_t = \tau_o$, which is substituted in equation (3). The resulting expression for τ_o is assumed in equation (12). After integration the equation becomes

$$W_{vd} = 4\pi r_e h_e v_e \sigma_o Q \sin\alpha \qquad (13)$$

where $Q = \frac{1}{(2k-1)(1+k)} \left((\rho-1)(2k^2+1-3k) - 2q \cdot (k^2-k+1)\right)$

The conditions of incompressibility and continuous flow result in

$$v_i = v_e\left(\frac{r_e}{r_i}\right)\left(\frac{h_e}{h_i}\right) \tag{14}$$

which has been used in deriving (13). The power of deformation is given by

$$W_d = \int_V (\sigma_L \varepsilon_L + \sigma_t \varepsilon_t)\, dV \tag{15}$$

It is seen that due to rotational symmetry

$$\varepsilon_L = \frac{\partial u}{\partial R} \quad \text{and} \quad \varepsilon_t = \frac{u}{R} \tag{16}$$

Substituting equation (11) in equation (14) and rearranging, the expression for u is

$$u = -\, v_e\left(\frac{R_e}{R}\right)^k \cos\alpha \tag{17}$$

which when substituted in equation (16) yields

$$\varepsilon_L = k v_e \frac{R_e^{\,k}}{R^{k+1}} \cos\alpha \quad \text{and} \quad \varepsilon_t = -v_e \frac{R_e^{\,k}}{R^{k+1}} \cos\alpha \tag{18}$$

Replacing dV and the strain rates in equation (15) and integrating results in

$$W_d = 2\pi r_e h_e v_e \sigma_o \cos\alpha \ln\left(\frac{r_i}{r_e}\right) Q(1+k) \tag{19}$$

Assuming a mean value of σ_t during the drawing the friction power becomes

$$W_f = 2\pi r_e h_e v_e \frac{\mu}{\tan\alpha} \cos\alpha \left(1-\frac{k}{2}(1+B)\frac{1}{Bk}\right.$$

$$\left.\left(\frac{r_e}{r_i}\right)^{Bk}\right)\right) \cdot \sigma_o Q(1+k) \tag{20}$$

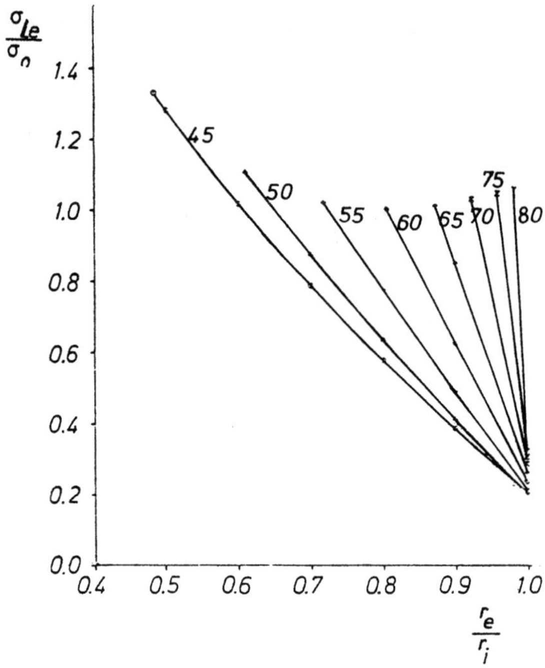

Fig. 6. Drawing stress as a function of radius reduction ratio. $\alpha = 9^o$, $\mu = 0.084$ and $\rho = 1.25$. Upper bound.

$$\frac{\sigma_{Le}}{\sigma_o} = (\ln(\frac{r_i}{r_e})\ \cos\alpha + \frac{2}{1+k}\ \sin\alpha + \ln(\frac{r_i}{r_e})\frac{\mu\cos\alpha}{\tan\alpha}\ \cdot$$

$$\cdot\ (1 - \frac{k}{2}\ (1+B)\ \frac{1}{Bk}\ (1-(\frac{r_e}{r_i})^{Bk}\ ))P \qquad (21)$$

where $P = \frac{1}{1-2k}((\rho-1)(1-3k+2k^2) - 2q(1-k+k^2))$

The maximum reduction in radius is given by equation (21) with the left side equal to kP, i.e.:

$$\ln\left(\frac{r_i}{r_e}\right)\cos\alpha + \frac{2}{1+k}\sin\alpha + \ln\left(\frac{r_i}{r_e}\right)\frac{\mu\cos\alpha}{\tan\alpha}\cdot$$

$$\cdot\left(1 - \frac{k}{2}(1+B)\frac{1}{Bk}\left(1-\left(\frac{r_i}{r_e}\right)^{-Bk}\right)\right) - \frac{1}{k} = 0 \qquad (22)$$

Equation (21) is programmed and σ_{Le}/σ_o is plotted versus reduction ratios in figure 6 for the same fiber orientations as those shown in figure 2 for the lower bound solution. Likewise equation (22) is programmed and plotted in figure 7 for the same sets of α and μ as given in the table for the lower bound solution.

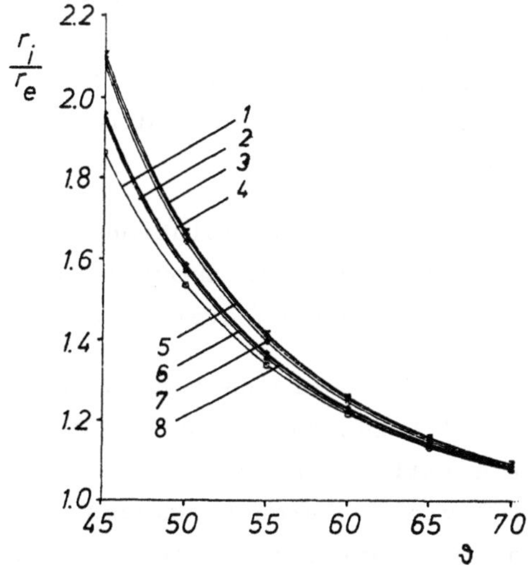

Fig. 7. Radius reduction ratio versus fiber orientation angle for various combinations of α and μ. Upper bound.

Discussion and conclusion

The fiber grid pattern is distorted during drawing; hence the angle should be used in the reduction calculation is the mean value of θ_i and θ_e. F.ex. a tube with $\theta_i = 60^o$ will end up with $\theta_e = 40^o$ at 20% reduction. The orientation to be used in the graphs, then, is 50°. To show the effect of the varying fiber angle on the drawing stress, an incremental theory should be used.

Fiber buckling at low densities may occur in weak matricies due to increased unsupported fiber lengths. Fiber yielding is predicted by considering equilibrium of a ring element in a matrix of moderate yield strength. Tubes without geometry imperfections, of steel reinforced aluminium f.ex., may yield completely provided the fiber yield stress is less than 6.5 times that of aluminium, the ratio of the fiber radius to the tube radius is less than 1/10, and the fiber orientation is tangential. Possible buckling is less likely to happen in the last stage of the drawing since the states of stress and the fiber angles here are more favourable. If a floating plug is used, buckling in the radial direction will be prevented altogether.

A comparison of the upper and lower bound solutions shows that the difference between the bounds is about 10%; the difference increases with increasing average angle θ.

Equation (10) shows that thickness strains vanish when $k = 1$. When $k = 2$ ($\theta = 54.74^{\circ}$) the thickness variation equals the radius reduction ratio. In the region $\theta = 70^{\circ}$ and above this equation yields large thickness strains. However, the maximum radius reduction is limited in this region suggesting large strains.

The maximum radius reduction ratio varies from $r_i/r_e = 2.20$ at $\theta = 45^{\circ}$ to 1.15 at $\theta = 70^{\circ}$. The upper and lower bounds differ by about 10%.

The foregoing leads to the following conclusions:

1. The drawing stress lies between the lower bound solution, equation (8) and the upper bound solution, equation (21). The drawing stress is not significantly influenced by ρ. Bounds to the maximum radius reduction as a function of average fiber orientation angle may be expressed by equation (9) and (22).

2. In the region investigated it appears that equation (11) predicts thickness strains.

References

[1] Johnson, W. and Mellor, P.B.,"Plasticity for mechanical Engineers", D.Van Nostrand Co. Ltd., London, 1970.

[2] Avitzur, B., "Metal forming: Processes and analysis", McGraw-Hill Book Co., New York, 1968.

[3] Chawla, K.K., "A report on the international conference on composite materials 1975". J. Engg. Materials and Technology, Oct. 1975.

[4] Hill, R., "The mathematical theory of plasticity", Oxford University Press, London, 1950.

[5] Calcote, L.R., "The analysis of laminated composite structures", Van Nostrand Reinhold Co., New York, 1969.

[6] McLaughlin, P.V., Jr., "Plastic limit behaviour and failure of filament reinforced materials", Int. J. Solids Struct., Vol. 8, 1972.

[7] Drucker, D.C., "Plasticity theory, strength-differential (sd) phenomenon, and volume expansion in metals and plastics", Metallurgical Transactions, Vol. 4, No. 3, 1973.

[8] Hu, L.W. and Pae, K.D., "Inclusion of the hydrostatic stress component in formulation of the yield condition", J. Franklin Institute, Vol. 275, 1963.

[9] Bridgman, P.W., "Studies in large plastic flow and fracture", McGraw-Hill Book Co., New York, 1952.

[10] D'alia, F. Stassi, "Flow and fracture of materials", Instituto Di Technologie Meccaniche, Palermo 1966.

[11] Jackson, P.W. and Cratchley, D., "The effect of fibre orientation on the tensile strength of fibre-reinforced metals", J. Mech. Phys. Solids 14, 1966.

[12] Bratt, J.F. and Kanan, Ö., "Determination of the yield condition in the third quadrant of the stress plane", JAM, Trans. ASME, March, 1966.

[13] Evans, W. and Avitzur, B., "Measurement of friction in drawing, extrusion and rolling", J. Lubrication Tech., Trans. ASME, series F, Vol. 90, Jan. 1968.

[14] Tselikov, A., "Stress and strain in metal rolling", Mir Publishers, Moscow, 1967.

Stresses and Strains in Deformation of Clad Metals

O. PAWELSKI and W. RASP
Max-Planck-Institut, Düsseldorf, Germany

A series of remarkable wire fractures in a Japanese wire drawing plant brought up the idea of analyzing the plane strain deformation of clad metals[1]. When ordinary patented steel wires have been drawn no problems occured.But when the same wires were coated with a thicker copper layer it was extremely difficult to deform this composite material. Obviously, the effect of coating was a change of the stress state in the steel core in such a way that the danger of cracking was increased. Fig. 1 gives an example of a wire which had been broken immediately after the exit from the die.

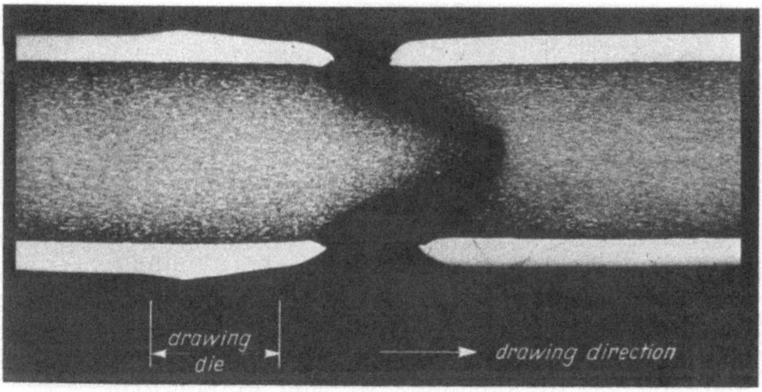

Fig. 1: Fracture in drawing of a copper-clad steel wire

This experience gave the motivation for a more detailed plastomechanical analysis of stresses and strains in the plastic deformation of clad metals by wire drawing and cylinder upsetting[2].

Basic Analysis

At first, the state of stress is considered acting on both sides of the plane interface where two materials A and B with different yield stresses are in contact, Fig. 2. Considering the equilibrium equations the condition of plane strain is assumed. When crossing the interface in the y-direction it is evident that the partial derivatives in this direction of the normal stress σ_y and the shear stress τ_{xy} must be finite and that these both stresses must be continuous. But as the derivative $\partial\sigma_x/\partial y$ is not included in the equations of equilibrium it can increase to infinity. Therefore σ_x may be discontinuous.

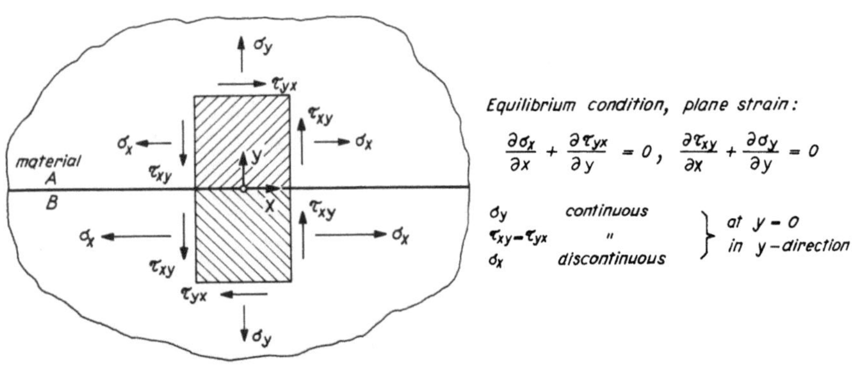

Fig. 2: Stresses on volume elements at the interface of materials A and B with different yield stresses

The amount of this discontinuity depends on the state of stress and the yield stresses of the contacting materials. As the whole composite must be in the plastic range the yield criteria for both components A and B have to be fulfilled. This may be demonstrated by a graphic representation of the Mohr´s stress circles in Fig. 3. The two circles A and B must intersect each other in a point representing the common stresses σ_y and τ_{xy}. Then for the jump of the stress σ_x a strong and a weak solution is obtained.

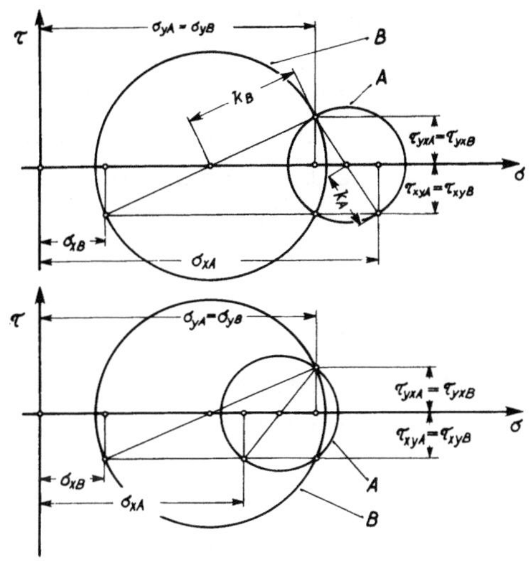

Fig. 3: Mohr's stress circles for plane strain deformation of materials A and B

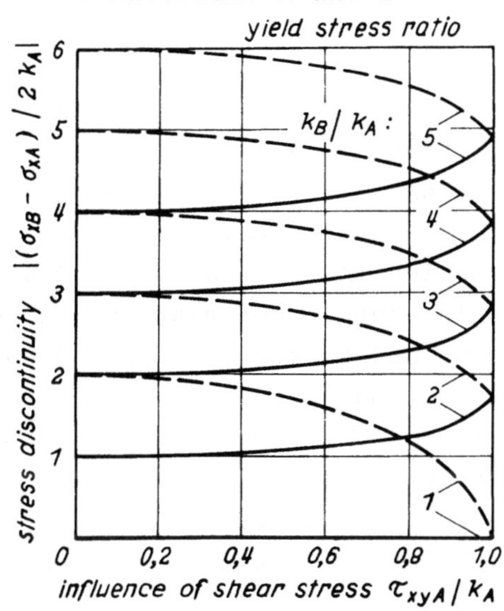

Fig 4: The amount of the stress discontinuity

Fig. 4 shows the amount of the stress discontinuity dependent on the shear stress τ_{xy} and the shear yield stresses k_A and k_B. In the limit case of a vanishing shear stress τ_{xy} the stress discontinuity results in the sum or in the difference of the two yield stresses. With increasing shear stress the difference between the both values for the stress discontinuity decreases. When τ_{xy} reaches the shear yield stress k_A of the softer component there is only one solution for the jump in the stress left.

Since the bigger circle B in Fig. 3 can be placed in a mirror-inverted position as related to the two intersection points it follows that generally there are four possible stress states in the plastic deformation of a two metal-composite. In spite of that there remain only two values for the stress discontinuity. Which of these stress states has to be chosen depends on the specific process conditions.

Wire Drawing

Now the special problem of wire drawing is discussed. As indicated in Fig. 5 there is a core of material B and a sleeve of material A which are drawn together through a drawing die with the semiangle α . As in the case of plane strain deformation there is also a stress discontinuity in axisymmetric deformation. Whereas the vertical pressures p_A and p_B must be continuous the horizontal stresses q_A and q_B are different on both sides of the interface. If the problem is treated with the elementary slab method the stresses p and q are taken as principal stresses. Then it follows that p_A equals p_B. The horizontal stresses q_A and q_B are constant across their individual sections. They differ from each other in the difference of the yield stresses k_f of the materials A and B. It is

$$q_A - q_B = k_{fA} - k_{fB} .$$

The unknown functions which are functions of the longitudinal coordinate x alone, are p_A, q_A, q_B and the interface shear stress τ . For their evaluation there are two equilibrium equations in the x-direction and two yield criteria for both the materials. For the yield condition the criterion of Tresca is chosen. Further Coulomb-friction at the die surface with μ as friction coefficient is assumed.

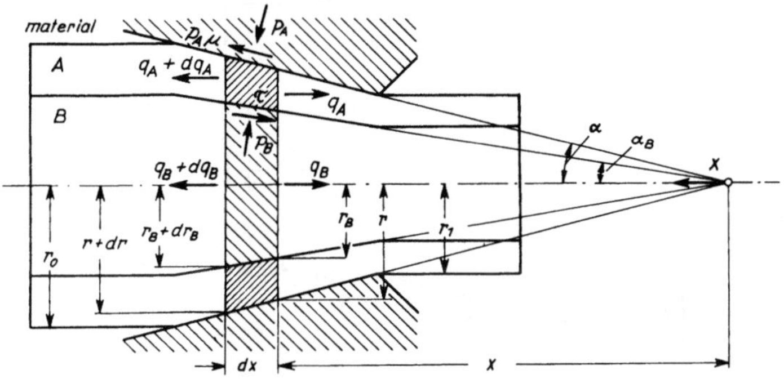

Fig. 5: Stresses in drawing of wire out of sandwiched metals

It is not worthwhile to discuss the solution of these equations in detail. The main result is that the known formulae for the drawing force like the Siebel or Sachs equation can be used further if an average value for the yield stress is used, calculated by a linear mixture rule from the individual yield stresses k_{fA} and k_{fB}. With the volume fraction

$$\rho^2 = (r_B / r)^2$$

it is

$$\bar{k}_f = k_{fA} + \rho^2 (k_{fB} - k_{fA}) .$$

ϱ^2 = 0 represents the case of 100 % A and ϱ^2 = 1 represents the case of 100 % B.

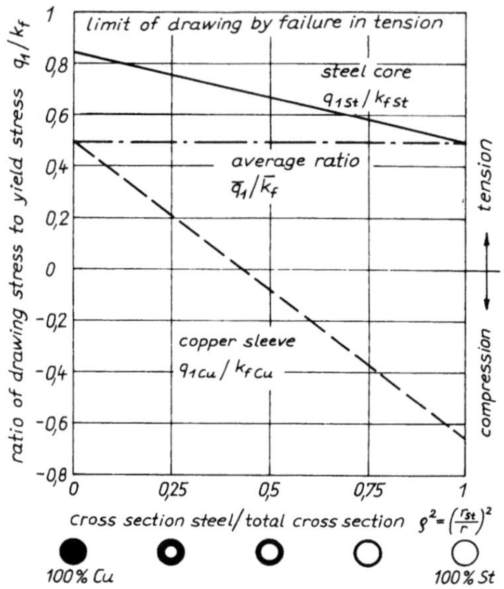

Fig. 6: Ratio of drawing stress to yield stress

Fig. 6 gives an example of the calculated horizontal stresses at the die exit in drawing of a copper-clad steel wire. In order to compare them with the case of pure tension they are related to the corresponding yield stress. Whereas according to the linear mixture rule the average ratio is constant for all volume fractions - in the example about 0.5 - the ratio for the steel core increases with the thickness of the copper sleeve up to 0.85. On the other hand in the copper sleeve the ratio is always smaller than the average ratio. The important conclusion is that the harder part of the composite wire is loaded with a higher tension and therefore more endangered to fracture than in drawing the uncoated metals. This is the reason why the Japanese had such big difficulties in drawing steel wires with thicker layers of copper. Therefore the theory of plasticity is not only of academic interest, but sometimes useful for the industry, too.

Own experiments in the laboratory have shown that the linear mixture rule for the drawing stresses is in fairly good agreement with measured values. Fig. 7 gives an example for the drawing of a nonferrous alloy AlMgSi0,5 clad with mild steel St 37. The reversed case with steel at the inside could not be realized in our experiments because the steel core remained rigid and the nonferrous sleeve was drawn like a tube on a mandrel.

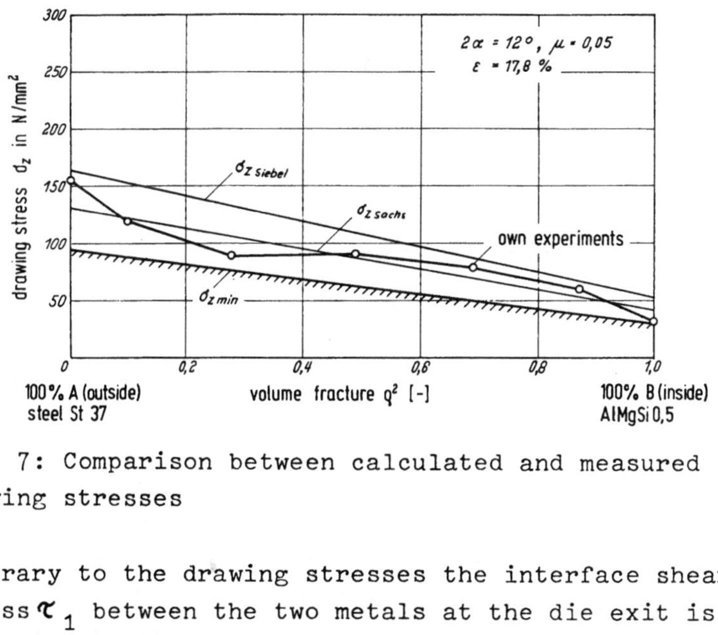

Fig. 7: Comparison between calculated and measured drawing stresses

Contrary to the drawing stresses the interface shear stress τ_1 between the two metals at the die exit is not a linear function of the volume fraction. As shown in Fig. 8 it passes through an extremum. In the case of a hard core and a soft sleeve τ_1 even changes its sign as compared with the frictional shear stress τ_{R1} at the die surface. With the harder part at the outside the maximum value of the interface shear stress may exceed the frictional shear stress by a factor of about 2.

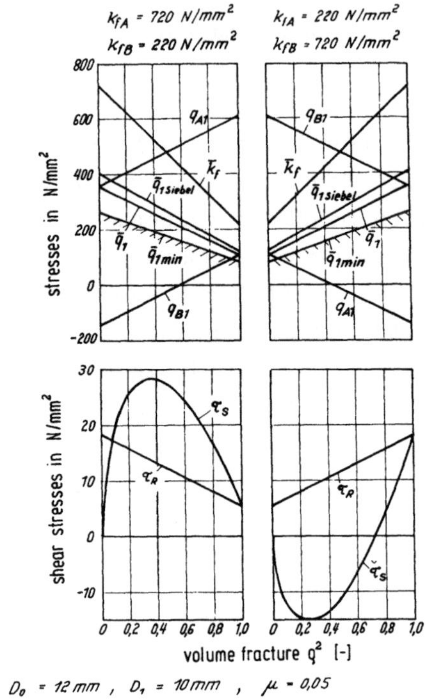

Fig. 8: Stress distribution at the exit of the drawing die

Upsetting of Cylinders

The second metal working operation presented in this paper is the upsetting of composite cylinders. As shown in Fig. 9 for a special example the stress discontinuity now influences the pressure distribution across the diameter. The top graph is valid for a hard core and the lower graph for a soft core. The jump in the stress equals again the difference in the yield stresses.

By integration of the pressure distribution the average tool pressure p_m is calculated. As plotted in Fig. 9 for missing friction, i.e. for $\mu=0$, p_m is again a linear function of the volume fracture ϱ^2. For other values of μ slightly curved lines are obtained. But it can be stated that the deviation from the linear mixture rule is negligible even up to a friction coefficient of 0.5.

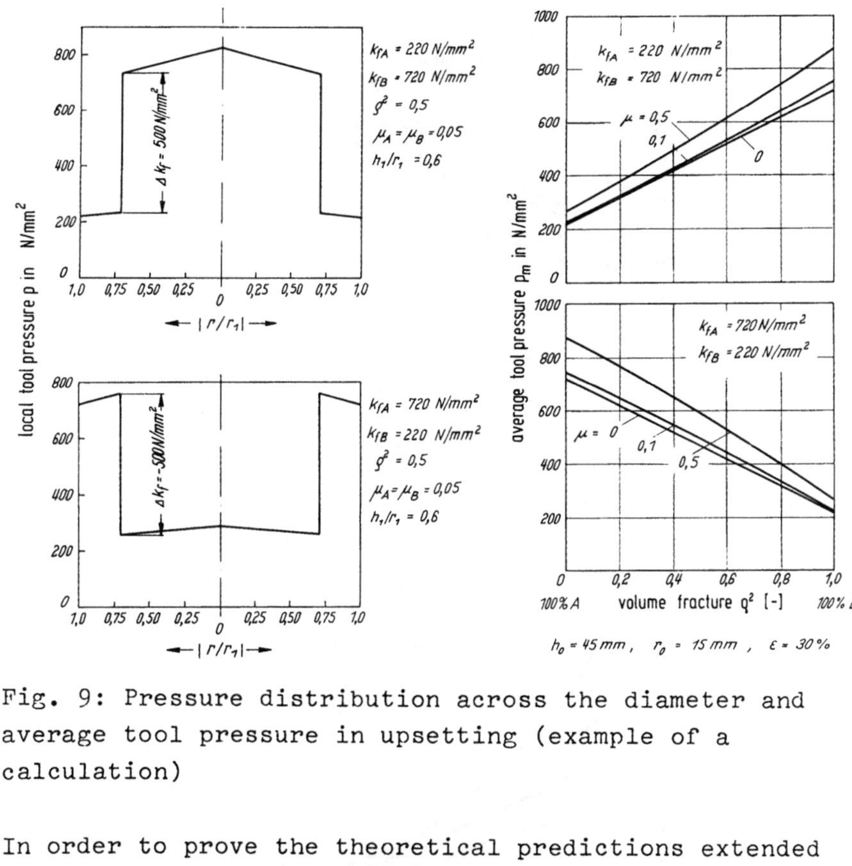

Fig. 9: Pressure distribution across the diameter and average tool pressure in upsetting (example of a calculation)

In order to prove the theoretical predictions extended laboratory tests were carried out. For a first series the test cylinders were made from steel St 37 in the core and an aluminium-alloy AlMgSi0,5 in the sleeve. In a second series the combination of materials was inverted. The volume fracture ranged from 0 to 1. The natural strain was chosen between 0.05 and 1. Rape seed oil was used as a lubricant so that the friction coefficient may be estimated at about 0.05.

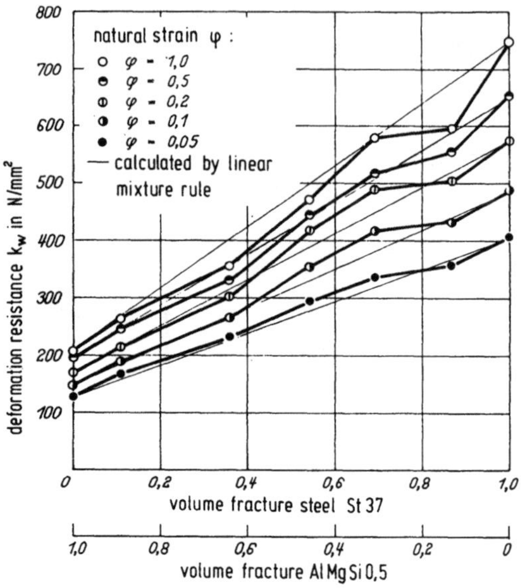

Fig. 10: Examination of the linear mixture rule for upsetting, core: steel St 37, sleeve: AlMgSi0,5

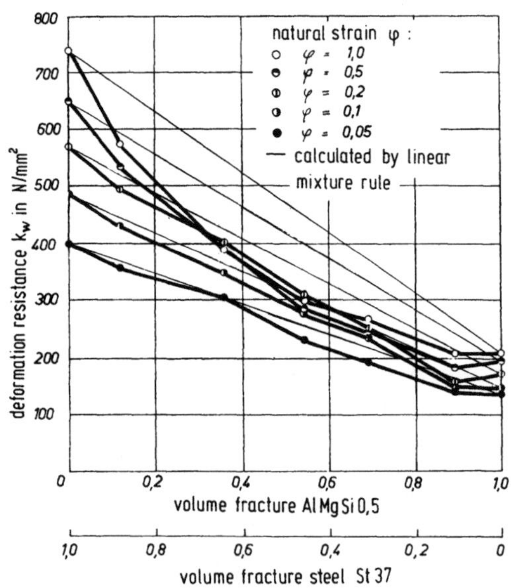

Fig. 11: Examination of the linear mixture rule for upsetting, core: AlMgSi0,5 , sleeve: steel St 37

Fig. 10 shows the results of the first test series with steel cores. The deformation resistance k_w or the average tool pressure p_m increases linearly with the volume fraction. Only for the larger deformations the measured lines tend to bend a little bit downwards. Apart from that, the linear mixture rule is again confirmed. The deformation pattern and with it the stress distributions are completely changed when the softer aluminium-alloy is put in the core.Now the deformation resistance of a composite cylindrical specimen is much lower than calculated by the linear mixture rule, Fig. 11. The deviation can be up to about 30%.

For an explanation it shall be looked at the shape of the specimens after 70 % upsetting. Fig. 12 gives some photographs of longitudinal sections of the test cylinders with steel cores. The harder cylinder at the inside shows the typical barrelling which is known from the upsetting of a homogeneous specimen. The softer sleeve adapts itself to this type of material flow. Only for smaller wall thicknesses a buckling of the sleeve or a formation of a cavity is observed. On the whole the deformation pattern is quite similar to that of upsetting a homogeneous specimen. Accordingly it is understandable that the deformation stresses follow the linear mixture rule.

The flow of material in the opposite case of steel at the outside is also shown in Fig. 12. Now the deformation pattern is determined by buckling, bulging, and folding processes acting at the steel sleeve. Obviously, the softer part in the inside of the composite is forced to follow the shape given by the outer sleeve. Thus the core does no longer take the shape of a barrel. It seems that there is a flow of material in the core which may be compared with a radial extrusion process.

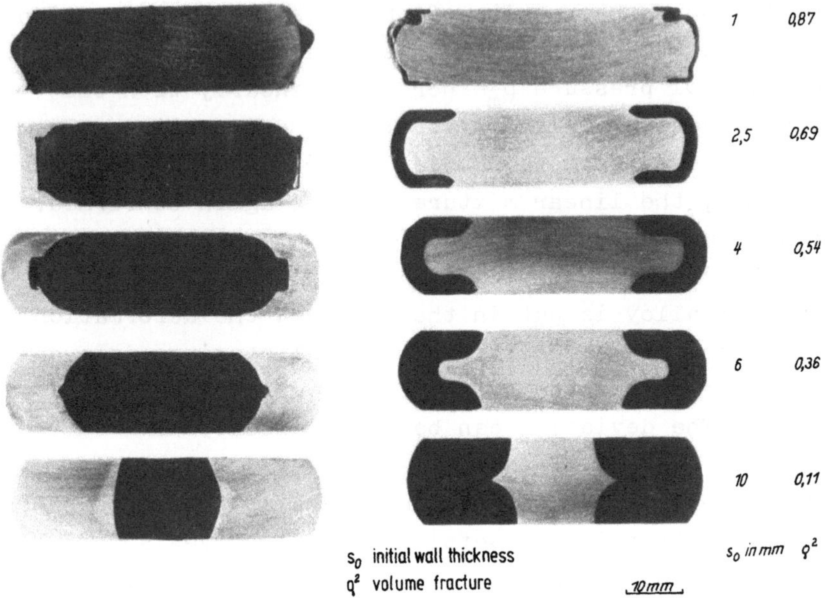

core: steel St 37 core: AlMgSi0,5
sleeve: AlMgSi0,5 sleeve: steel St 37

Fig. 12: Longitudinal sections of clad metal after 70% upsetting

It can be assumed that the observed instable deformation of the steel sleeve needs less energy than the stable upsetting process. This would give an explanation for the negative deviation of the measured tool pressures from the linear mixture rule.

Therefore it was tried to apply the theories of buckling and bulging and finally the theory of yield hinge. The formation of so-called yield hinges is explained in Fig. 13 for the case of folding a tube under compression. The plastic work consists of two parts. The first part is necessary for the plastic bending in the hinges. The second part is due to the plastic stretching of the portions of the wall between the single hinges. Then the critical compression stress related to the yield stress of the tube material is only a function of the ratio inner radius/outer radius.

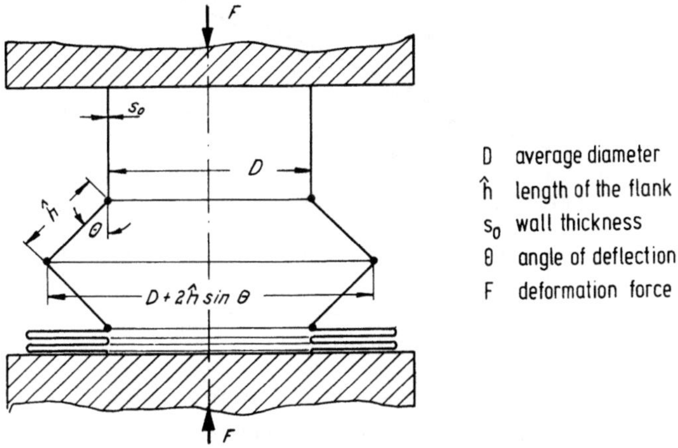

Fig. 13: Theoretical assumptions for folding of a tube under compression (after J.M.Alexander)

To simplify matters it was assumed that in the composite specimen there should be no interaction between the inner and the outer part, so that there is the possibility of a free folding of the sleeve. Furthermore the portion of the tool pressure due to the deformation of the core was calculated for the case of homogeneous upsetting with friction.

In the case of instability, the average tool pressure acting at the composite cylinder was then calculated by using the critical pressure for the outer part and the upsetting pressure for the core. In Fig. 14 the results are simultaneously plotted for the case of St 37 at the inside and AlMgSi0,5 at the outside. In agreement with the experimental observations the theories of buckling, bulging and the theory of yield hinge lead to values not much lower than those predicted by the linear mixture rule.

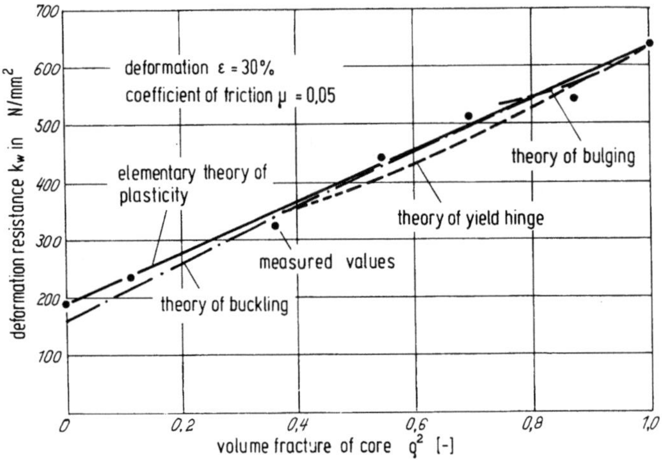

Fig. 14: Deformation resistance in upsetting of clad metals measured and calculated by different theories, core: steel St 37 , sleeve: AlMgSi0,5

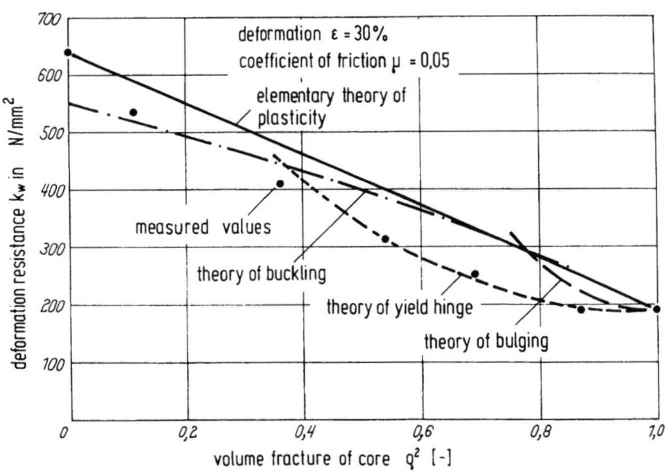

Fig. 15: Deformation resistance in upsetting of clad metals measured and calculated by different theories, core: AlMgSi0,5 , sleeve: steel St 37

But when arranging the steel at the outside of the specimen the critical values for instable deformation by buckling, bulging, and especially by formation of yield hinges are significantly lower than the deformation resistance in normal upsetting, Fig. 15. As it may be seen by comparison of the theoretical curves with the experimental points the agreement is very satisfactory.

Conclusions

The investigation has shown that in metal working operations following the conditions of the elementary slab theory it is possible to calculate the stresses in deformation of clad metals by applying a linear mixture rule for the yield stresses. Furthermore an important conclusion is that the state of stress in the harder part of the composite is superimposed by additional tension stresses depending on the geometry of the composite workpiece and mainly on the difference in the yield stresses of the individual materials. The consequence is an increase of the danger of cracking. At least, one has to take into account that the deformability is lower than in the deformation of homogeneous workpieces.

In special cases of more or less unstable or inhomogeneous deformation of clad metals in which the elementary theoretical procedure fails, the methods of advanced theory of plasticity give a good insight into the deformation stresses and the flow of material.

References

1. Weinstein,A.S., and O.Pawelski: Plane Strain Drawing of Sandwiched Metals. 8th M.T.D.R.Conference, University of Manchester 1967.

2. Rasp,W.: Spannungen und Formänderungen beim Umformen metallischer Schichtverbundwerkstoffe. Dr.-Ing. Thesis, TH Aachen 1977.

Joint Examination Project of Validity of Various Numerical Methods for the Analysis of Metal Forming Processes –

Report Given and Comments Made at the Round Table Discussion of the Symposium

H. KUDO and S. MATSUBARA

Department of Mechanical Engineering

of Yokohama National University,

Japan

Summary

Upsetting process of a cylindrical workpiece under unified conditions is analyzed by fourteen research groups using their own numerical methods and computational procedures. The results are collected and compared, and were found to show considerable discrepancies. Possible reasons for them and further steps to be taken to improve the results are discussed.

Foreword

The present Project is a comparative calibration test of various numerical methods of analysis of large plastic deformation problems. An upsetting of a cylindrical workpiece having a specified set of constitutive equations between parallel and perfectly rough dies down to a reduction in height of 60% was the common objective to be analyzed. The numerical method and computational procedure to determine the upsetting load, equatorial stresses and side profile of workpiece were left in the hands of the participators. Fourteen results have been collected and compared during the Symposium in Tutzing.

Considerable discrepancies among the results were observed and the significance of the present Project as well as the interpretation of the results were discussed in a Session 'Round Table Discussion'. It was decided, at least among the participators of the Project, to put it forward.

Aim and Background of the Project

The present Project was originally set up by the present reporters as an international cooperative work of Scientific and Technical Committee 'Forming' of the International Institution for Production Engineering Research (CIRP) in 1976.

Numerical methods are now being widely used in the analysis of metal forming processes. Since those different methods hitherto published have been applied to different forming problems with different process parameters for which no exact solutions were available, comparison of their advantages and disadvantages viewed from reliability and computational time cannot be made. It is also unbearable for a single office to make this comparison.

The aims of the Project were, therefore,
(i) to enable individual participators to estimate their own methods for further development,
(ii) to offer to prospective users of numerical methods a key to the selection, and
(iii) to develop cooperatively, if possible, more reliable and economical methods of analysis of metal forming problems.

By May 1977, seven results were submitted, five from Japan, one from F.R.G. and one from the Netherlands. They showed considerable scatters and some unconvincing points. The results were compiled and summarized in an Interim Report, August, 1977 for distribution to the participators. Since then there has been no opportunity to get together to discuss the results.

As the present reporters expected that the Tutzing Symposium would be a good opportunity to put the Project forward, revised letters of invitation to the Project were, with the concurrence of Prof. H. Lippmann, the Chairman of the Scientific Committee of the Symposium, sent to prospective participators in the meeting who seemed to the present reporters interested in the numerical methods of analysis. The letters were addressed to, other than the members of the Scientific Committee and Local Organizing Committee, 28 people belonging to 14 different countries.

Common Objective and Its Process Parameters to be Analyzed

The upsetting of a cylindrical billet having an initial diameter of 20 mm and height of 30 mm between perfectly rough and rigid parallel platens was the common objective to be treated.

The mechanical properties and constitutive equations of the billet material are specified as shown in Table 1. Each participator could use either the Prandtl-Reuss Material or the Lévy-Mises material depending on his own numerical method. But he was recommended to submit his results for both the strain-hardening and non-hardening materials, the yield stress curves being given in Fig. 1. The

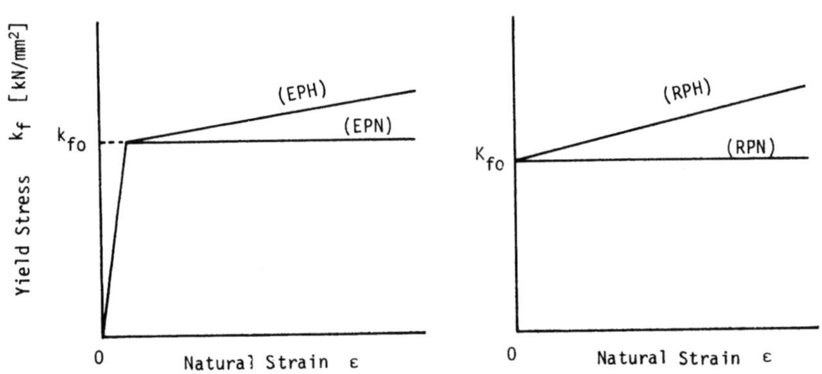

Fig. 1. Yield Stress Curves of the Model Workpiece Material

flow stress curve for the hardening material was a modified flow stress curve of annealed 0.45% carbon steel by linearization.

The interfacial condition 'sticking' was defined by 'no radial

Table 1 Specified Mechanical Properties and Constitutive Equations of Workpiece Material

Yield and Strain-Hardening Criteria Due to	Initial Yield Stress k_{fo} [kN/mm²]	Flow Rule Due to	Young's Modulus E [kN/mm²]	Poisson's Ratio ν	Rate of Strain-Hardening $dk_f/d\varepsilon_{pl}$ [kN/mm²]	Material Designation
von Mises and Hill $\bar{\sigma} = f(\int d\bar{\varepsilon}_{pl})$	0.700	Prandtl-Reuss $d\varepsilon_{ij}^{pl} = \sigma_{ij}'d\lambda$	200	0.300	0.300	[EPH]
					0.000	[EPN]
		Lévy-Mises $d\varepsilon_{ij} = \sigma_{ij}'d\lambda$	infinity	0.499 −0.500	0.300	[RPH]
					0.000	[RPN]

displacement', i.e., $du_r = 0$, over the workpiece-die interface.

Calculation Method and Procedure

The calculation method and procedure were left in the hands of the participators. It was suggested that they would carry out their calculation with different numbers and shapes of discretized elements and different deformation increments to find out the optimum procedure with regard to the computational accuracy and computing time.

It was also suggested to the participators to use, at least, the finite element idealization shown in Fig. 2 for reference. Another

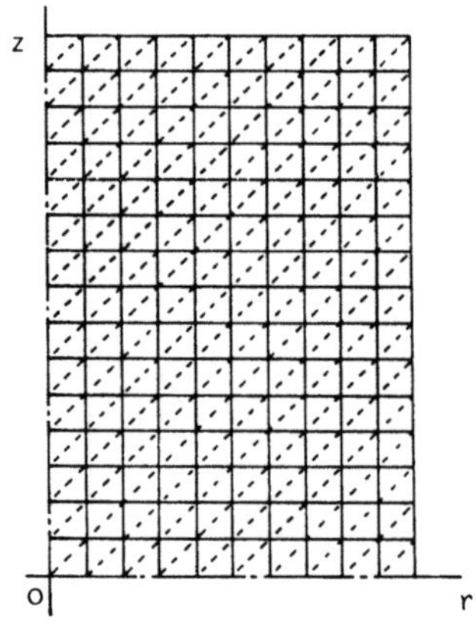

Fig. 2. Recommended Finite Element Idealization as a Reference

suggestion given to the participators was that they would check their algorism and programme by performing calculation for the friction-free upsetting, i.e., $\tau_{rz} = 0$ over the interface, and by comparing the results with those of the known exact solution that shows the homogeneous deformation and stresses.

Results to be Output and Compared

The participators were requested to submit the results on the items given in Table 2, and they were also requested to state the items shown in Table 3.

Table 2 Results to be Submitted and Compared

1.	Extents of yielded area in workpiece at reductions in height r of 0.2, 0.5 and 1%.
2.	Bulge profiles of workpiece side at r = 10, 20, 30, 40, 50 and 60%.
3.	Relationship between current maximum and minimum diameters of workpiece, d_{max}/d_o and d_{min}/d_o, and r up to 60%.
4.	Relationship between upsetting load F and r up to 60%.
5.	Relationship between equatorial circumferential and axial stresses, σ_θ and σ_z, and r up to 60%.

Table 3 Items to be Stated Regarding Numerical Method, Computational Procedure and Computer

Numerical Method	1.	Numerical method adopted.
	2.	Whether the stiffness matrix is based on the large strain-large displacement theory.
	3.	Whether the rotation of stress axis is considered.
Computational Procedure	4.	Number and shape of the discretized element.
	5.	The displacement or interpolation function.
	6.	Deformation increment taken per computing step.
	7.	Procedure of determining upsetting load.
	8.	Procedure of determining equatorial stresses.
Computer	9.	Manufacturer and type of computer used.
	10.	Computational time required for one cycle of addition, subtraction, multiplication and division and for one deformation increment.

Participators in the Project and Their Method and Procedures

The names and places of the participators in the Project and the methods, procedures and computers used by them are listed in Tables 4 and 5.

Table 4 is for those participators who analyzed upsetting process of the elastic-plastic material using the finite element method. The asterisked participator indicates one who submitted his results on the occasion of the previous operations. All participators

Table 4 List of Names of and Calculation Procedures Elastic-Plastic Finite Element Method Based Axis Rotation

Name & Place of Participator	Shape & Number of Element		Deformation Increment(%)	Load Determined from
ISEKI, H., Inst. Technol., Tokyo	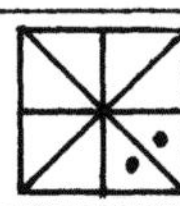	300	0.001 −0.16	normal force over interface
MCMEEKING, R.M. & AL-KHATTAT, I.M., Univ. Stanford	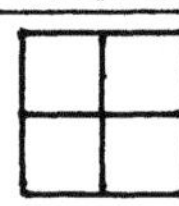	150	0.003 −1.667	ibid.
TAKAHASHI, H.*, Univ. Yamagata	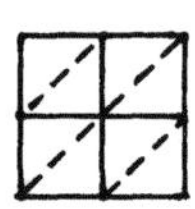	300 & 150	1	ibid.
YAMADA, Y., HIRAKAWA, T., & WIFI, A.S., Univ. Tokyo	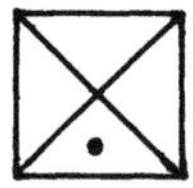	48 or 308	(0.03−0.166) (<2% red.) −0.25	ibid.
SKLAD, M., Politech. Warszawska	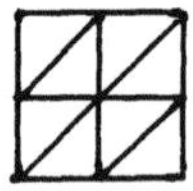	300	−2	ibid.

except for the last adopted the finite strain and finite-displacement theory taking the rotation of stress axis with respect to material element into account.

The data for those participators who treated the rigid-plastic material [RP] by the finite element-upper bound method based on the minimum energy principle are summarized in Table 5. Again the asterisked name indicates the participator in the previous operations. In both the elastic-plastic and rigid-plastic treatments,

and Computers Used by Participators Who Adopted the
Mostly on Finite Strain Theory and Including Stress

Equat. Stresses Determined from	Manufacturer and Type of Computer	Computing Time: add./subt./mult. /div. (μsec) & def.incre.(sec)	Remarks
average of the two triangles	HITAC 8700/8800	0.2/0.2/0.2 /1.1 & 6	friction-free test O.K.; no folding up to 52%
averaging and extrapolation	IBM 370/168	0.3/0.3/0.61 /2.65 & 15.2	friction-free test O.K.; no folding up to 60%
	NEC, ACOS 700	0.5/0.5/1 /2 & 30	equilibrium of nodal forces instead of force increments considered; triangular element for deviatoric stresses and square element for mean normal stress; folding between 30 and 40%
value at the solid circle	HITAC 8700/8800	0.2/0.2/0.2 /1.1 & 20	friction-free test O.K.; no folding up to 30%
	ELWRO, ODRA 1305	16/16/32 /66 & 1200	no finite-strain & stress rotation considered; max. reduction 10%

Table 5 List of Names of and Calculation Procedures and the Finite Element-Upper Bound Approach Method Dissipation for Rigid-Plastic Material

Name & Place of Participator	Shape & Number of Element	Deformation Increment [%]	Load Determined from
ANDRESEN, K. & STECK, E., Tech. Univ. Braunschweig	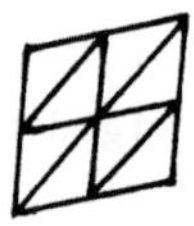 60	5	total rate of energy consumption
KLIE, W., Tech. Univ. Hannover	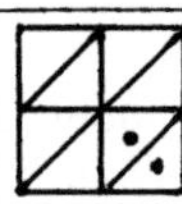 300	0.2—0.5	σ_z over interface
KOBAYASHI, S., Univ. California	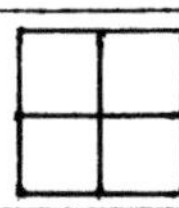 60	1	ibid.
ERLMANN, K. & MARECZEK, G., Tech. Univ. Hannover	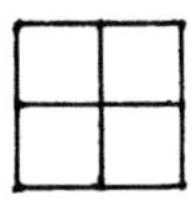 150	2	ibid-
NAKAJIMA, H., WATANABE, K. et al., Nippon Steel Corp.	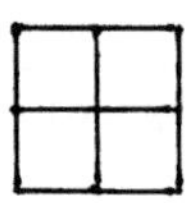 49	1 or 5	ibid
REBHOLZ, M. & ROLL, K., Univ. Stuttgart	weighed residual method based on least square	66	σ_z over interface
ROLL, K., Univ. Stuttgart	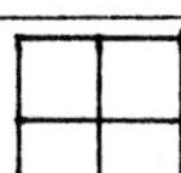 40 or 70	3.3	σ_z over $z = 0$
SHIMA, S. Univ. Kyoto & OSAKADA, K., Univ. Kobe	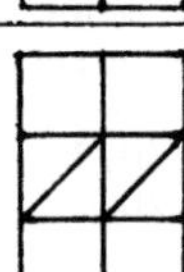 80 & 740	1	σ_z over interface as well as total rate of energy consumption
ZIJSLING, D.H., Tech. Univ. Twente*	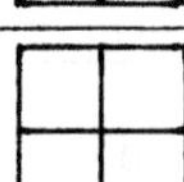 150	2.5—2.0	σ_z over interface

Computers Used by Participators Who Adopted Mostly
Based on the Principle of Minimum Rate of Energy

Equat. Stresses Determined from	Manufacturer & Type of Computer	Computing Time: add./subt./mult. /div. (μsec) & def.incre.(sec)	Remarks
?	ICL 1906S	-/-/- /- & 70	friction-free test O.K.; folding at 30% reduction
average of 2 triangles	CYBER 76	0.11/0.11 /0.14/0.55 & 3.04	no folding at 60% reduction
?	CDC 7600	0.11/0.11 /0.14/0.55 & 5.8	friction-free test O.K.; folding at 35%
?	CYBER	0.11/0.11 /0.14/0.55 & 3.04	punch diameter of d_o; folding presumably between 30 and 40%
?	HITACHI M-170S	0.54/0.54 /4.62/11.42 & 77.2	folding at 40%; no marked effect of deformation increment except load
?	CONT. DATA 6600 & CYBER 175	0.4/0.4/1.0 /2.9 & 20	solved iteratively
extrapolation and averaging	ibid	0.4/0.4/1.0 /2.9 & Ca 50—80	no folding
?	FACOM M-190	0.16/-/-/- & 14	porous material having an initial density of 99.5%; friction-free test O.K.; folding between 40 and 50%
?	DEC-10 (KL)		folding at 38%

the displacement or interpolation function was a linear or a four-point quadratic bilinear function with respect to r and z depending on the triangular or rectangular elements adopted respectively.

Comparison of Collected Results

A) Extent of Yielded Area

Fig. 3 illustrates the extents of yielded area at reductions of ca. 0.2, 0.5 and 1%. There were no significant differences in the extent among all the results obtained for the [EPH], [EPN], [RPH] and [RPN] materials.

B) Side Profile and Deformation

Fig. 4(a) and (b) summarize and compare the side profiles of bulged workpiece at the reductions of 30 and 60% for the [EPH] and [RPH] materials. The name of the presenter of each curve will be identified by his initials attached. The results for [EPN] and [RPN] are similar to Fig. 4, though only one result for the [EPN] material has been submitted. The thick solid lines marked by ff denote the results for the friction-free upsetting, which have been drawn simply from the condition of volume constancy.

There are some irregular discrepancies among the profiles. Many of the [RPH] results show that folding of the side surface onto the die surface occurs at a reduction in height lying between 30 and 40%, while some exhibit no folding even at the 60% reduction.

The same observation can be also made in Fig. 5, in which the ratios of the current maximum and minimum diameters to the initial diameter of the workpiece are graphed in relation to the reduction. It is obvious that in friction-free upsetting, both d_{max}/d_o and d_{min}/d_o follow the single curve denoted by ff, while in upsetting with sticking friction d_{min}/d_o remains unity as long as folding does not occur.

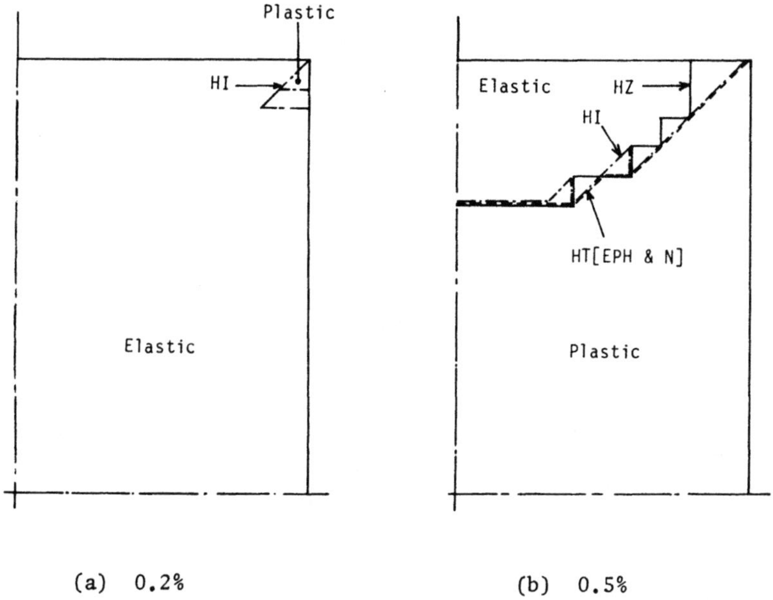

Fig. 3. Predicted Extent of Yielded Area a

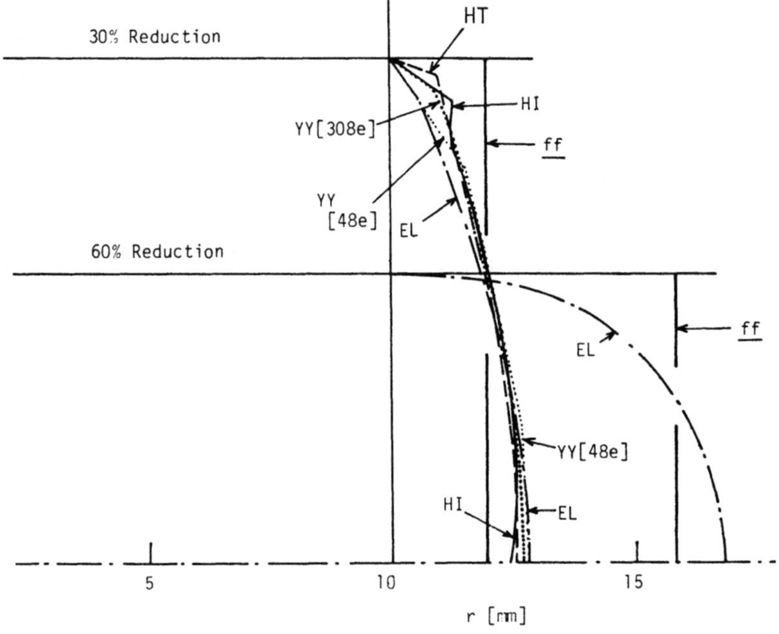

(a) [EPH]

Fig. 4. Comparison of Predicted Profiles of Side Surface at Reductions of 30 and 40%

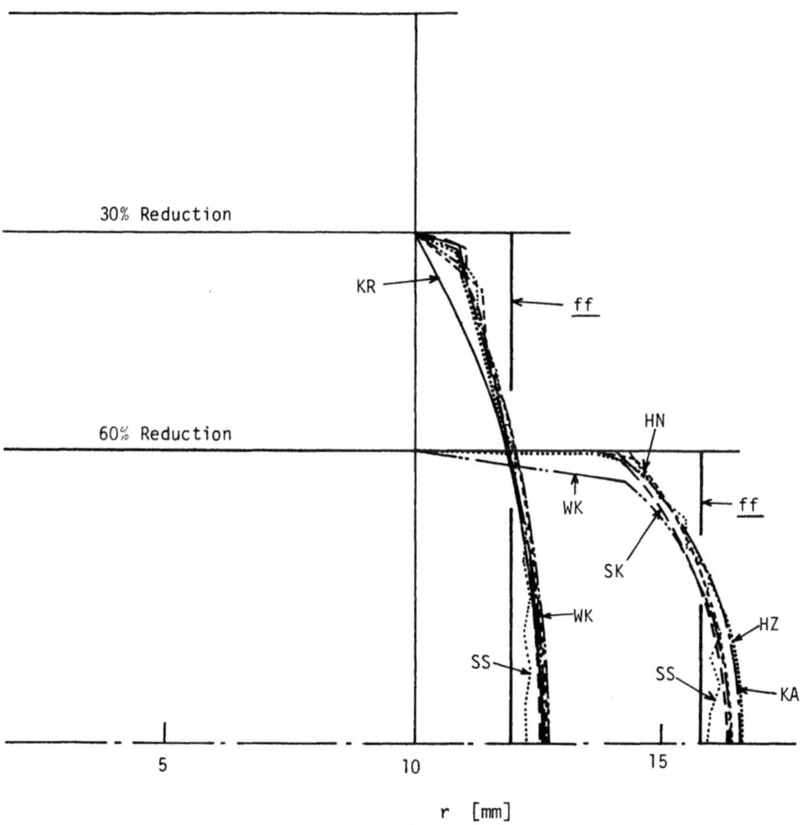

30% Reduction
60% Reduction
KR
ff
HN
WK
SK
WK
SS
HZ
SS
KA
5
10
15
r [mm]

(b) [RPH]

Fig. 4.

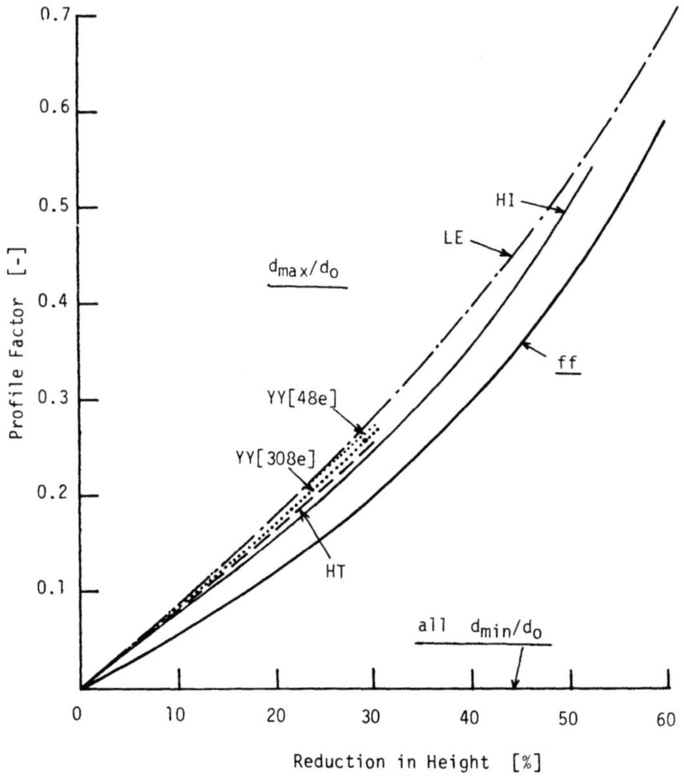

(a) [EPH]

Fig. 5. Comparison of Predicted Bulge Factors, i.e., the Ratios of Maximum and Minimum Diameters, d_{max} and d_{min} to Initial Diameter d_o of Workpiece at Reductions of 30 and 40%

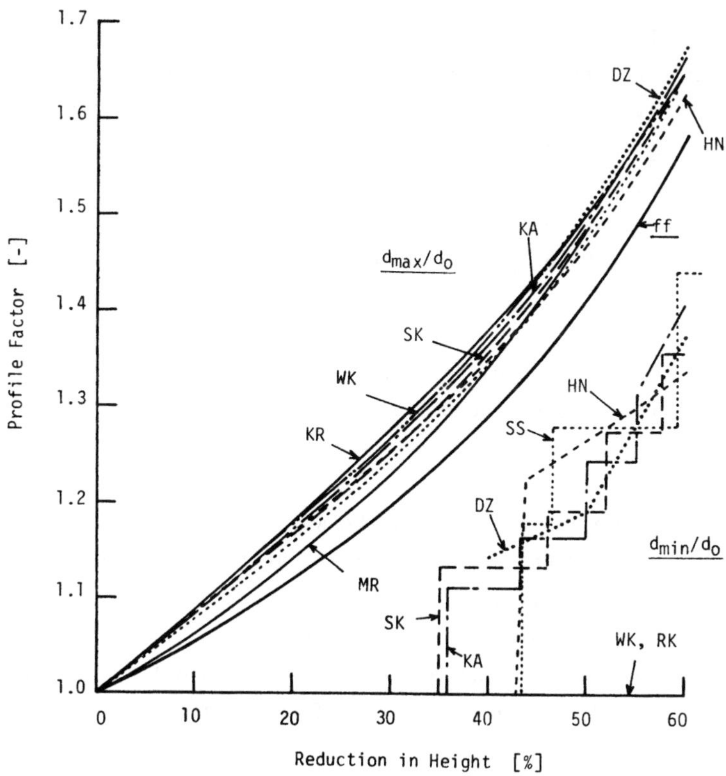
Profile Factor [-]
1.7
1.6
1.5
1.4
1.3
1.2
1.1
1.0
0
10
20
30
40
50
60
Reduction in Height [%]
dmax/do
dmin/do
DZ
HN
KA
ff
SK
WK
KR
MR
HN
SS
DZ
SK
KA
WK, RK

(b) [RPH]

Fig. 5.

Although presentation of the predicted deformation pattern over the diametral section of workpiece was not requested in the present operation of the Project, it will be interesting to see what it is like. Fig. 6 reproduces the one obtained by Zijsling for the reduction of 50% on the occasion of the first operation.

C) Upsetting Load

The predicted upsetting load-reduction curves are put together and compared in Fig. 7(a), (b) and (c) for [EPH], [RPH] and [RPN] respectively. Again considerable discrepancies are found among the results obtained by different participators. Some curves fall even below the friction-free curve ff. The drops in the curves due to HN were stated to have appeared when an element folded onto the die surface.

There are no significant differences in the results obtained for [EPH] and [RPH] by using the different methods.

D) Equatorial Stresses

As will be seen in Tables 3 and 4, the way of determining the stresses is declared only by some participators. The present reporters assume, however, that with triangular elements σ_θ and σ_z in the two triangles which lie closest to the equator have been averaged, while with rectangular elements σ_θ' and σ_z' on the equator have been added to the average σ_m of the element.

Fig. 8(a), (b) and (c) show the predicted equatorial stresses versus reduction curves for [EPH], [RPH] and [RPN]. In friction-free upsetting, it is obvious that σ_z equals the yield stress while σ_θ remains zero.

In the present case, in which the material sticks to the die surface, σ_θ appears and grows as upsetting progresses, while σ_z tends to increase algebraically until it becomes tensile. Although all the results indicate the same tendencies, the discrepancies among them are most marked.

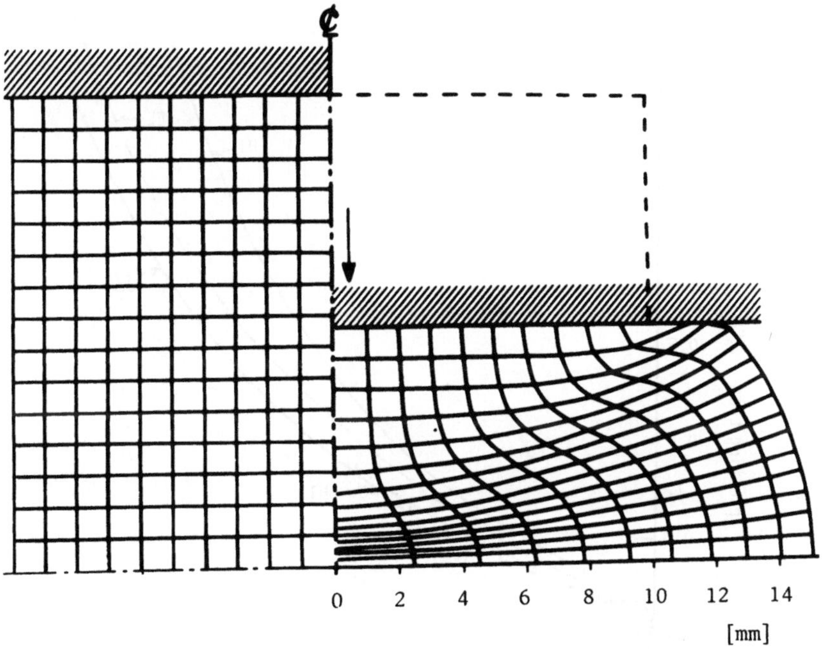

Fig. 6. Examples of Distorted Grid on Diametral Section of Workpiece Upset Down to 50% of Initial Height (after HZ)

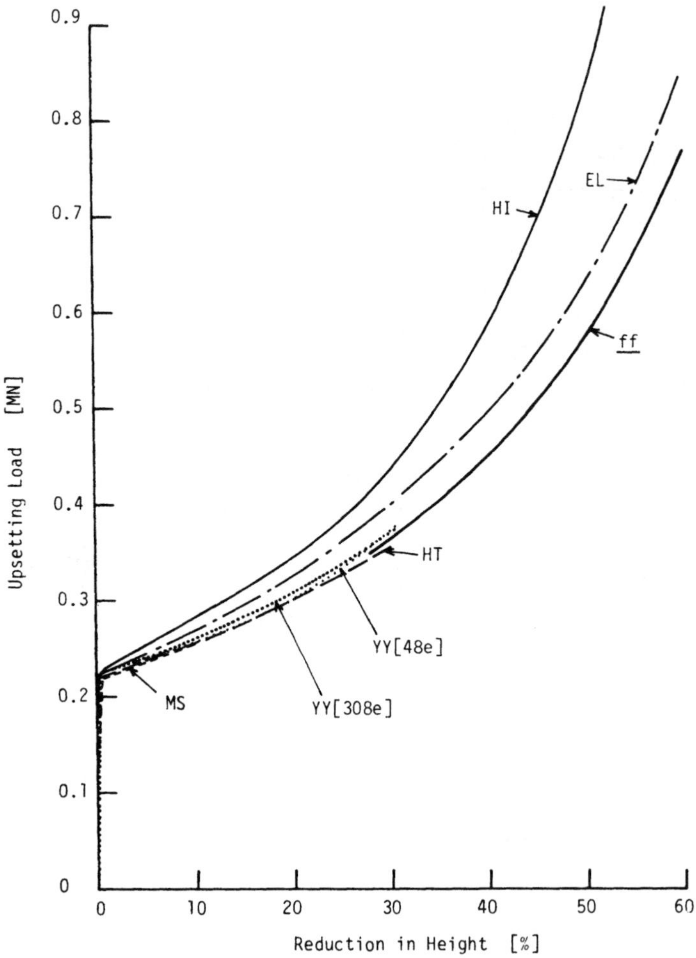

(a) [EPH]

Fig. 7. Comparison of Predicted Load - Reduction Curves

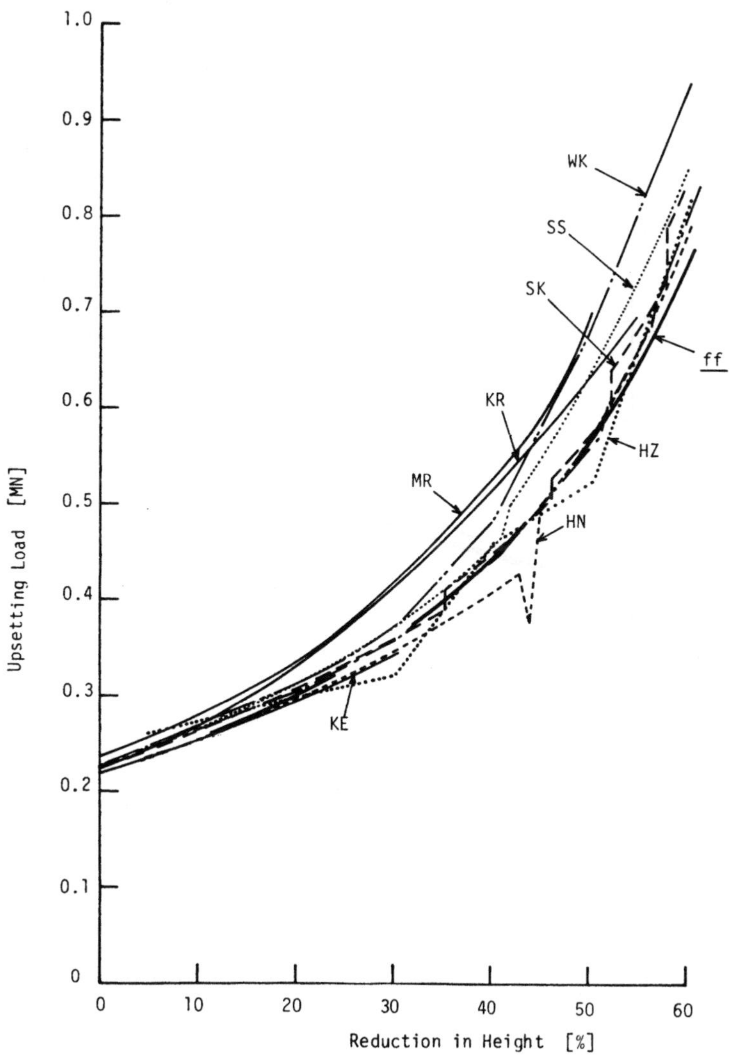
1.0
0.9
0.8
0.7
0.6
0.5
0.4
0.3
0.2
0.1
0
Upsetting Load [MN]
0
10
20
30
40
50
60
Reduction in Height [%]
WK
SS
SK
ff
KR
HZ
MR
HN
KE

(b) [RPH]

Fig. 7.

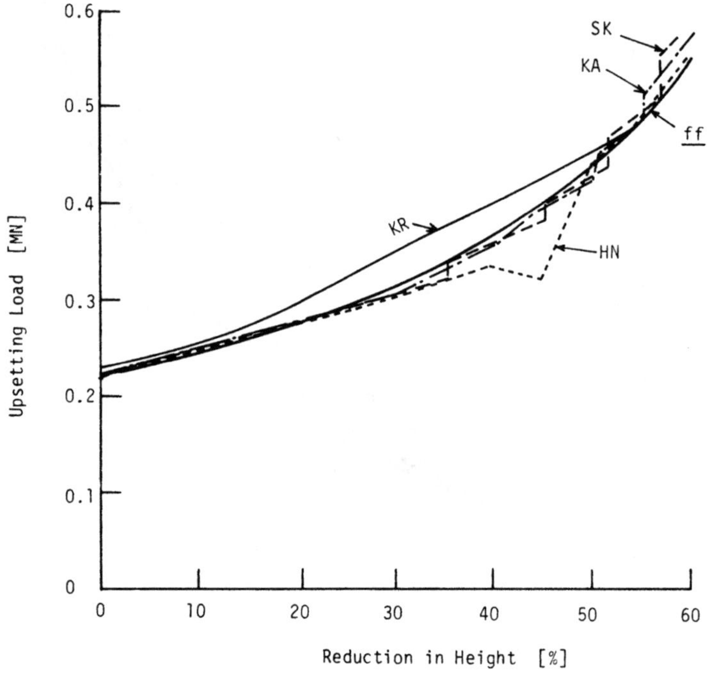
0.6
0.5
0.4
0.3
0.2
0.1
0
Upsetting Load [MN]
0
10
20
30
40
50
60
Reduction in Height [%]
SK
KA
ff
KR
HN

(c) [RPN]

Fig. 7.

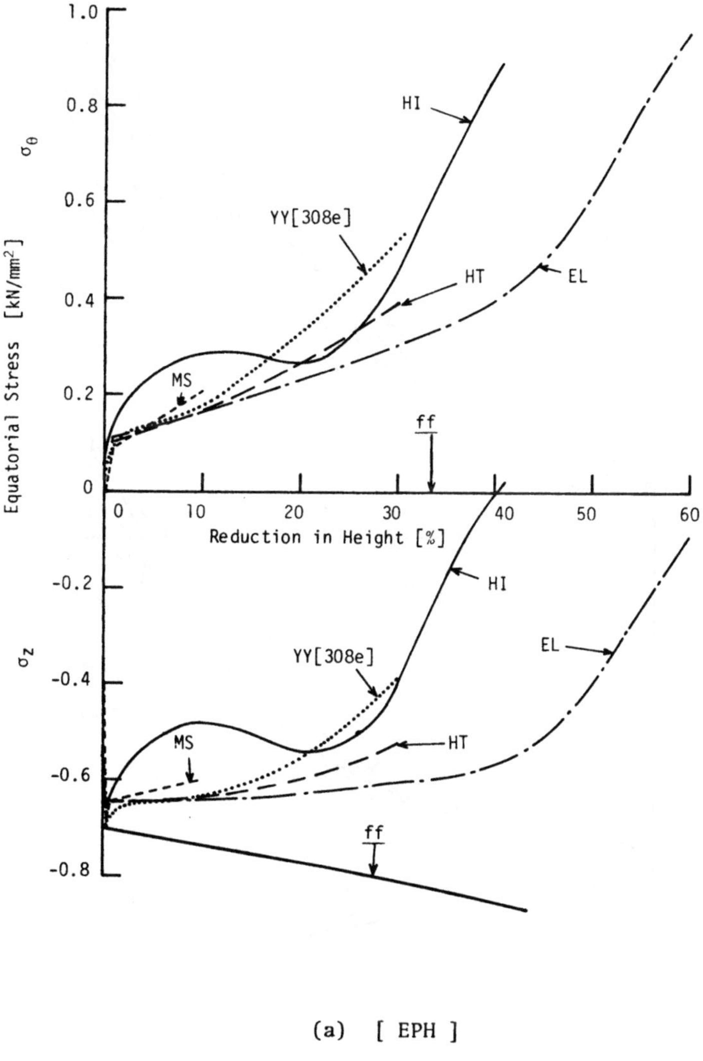

(a) [EPH]

Fig. 8. Comparison of Predicted Equatorial Stresses, σ_θ and σ_z, versus Reduction Curves

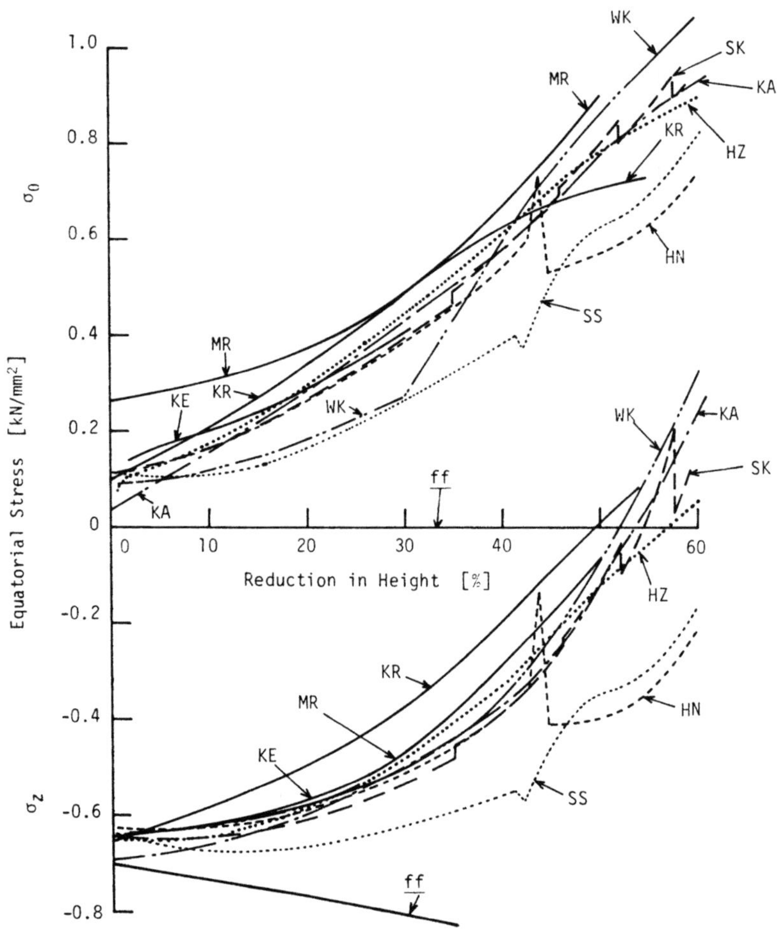

1.0
0.8
0.6
0.4
0.2
0
-0.2
-0.4
-0.6
-0.8
σ_0
σ_z
Equatorial Stress [kN/mm²]
Reduction in Height [%]
0
10
20
30
40
60
WK
MR
SK
KA
KR
HZ
HN
SS
KE
ff

(b) [RPH]

Fig. 8.

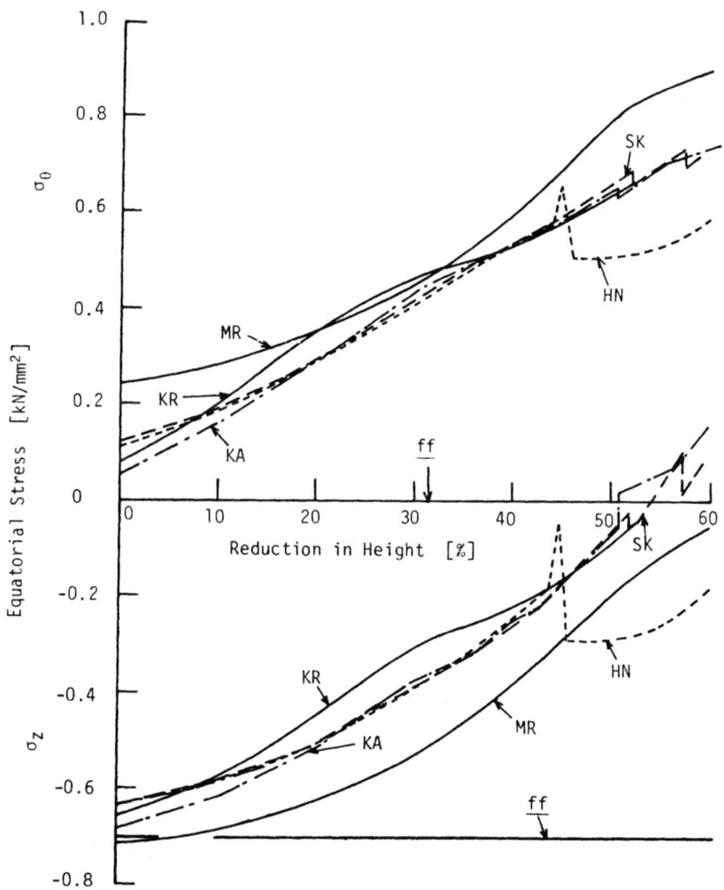
1.0
0.8
0.6
0.4
0.2
0
-0.2
-0.4
-0.6
-0.8
σ0
σz
Equatorial Stress [kN/mm²]
Reduction in Height [%]
0
10
20
30
40
50
60
MR
KR
KA
SK
HN
ff

(c) [RPN]

Fig. 8.

Discussion and Conclusion

It was appreciated at the Symposium by a number of attendants including non-participators in the Project that the present attempt was worthwhile since it would promote refining and improving numerical methods of analysis of metal forming processes and also would help those who were interested in the methods in understanding the state of art.

The differences in the number and shape of the finite elements were unanimously recognized as one of the most significant factors that were responsible for the discrepancies observed in the collected results. It was agreed that if finer elements had been used around the edge of the workpiece earlier folding would have been predicted in accordance with the experimental observation.*

Selection of the deformation increment was thought to be another important factor. Tables 3 and 4 show that larger increments were chosen in the rigid-plastic analyses based on the upper bound theorem than those used in the elastic-plastic finite element methods, though both gave the similar results.

A case was demonstrated by Yamada in which changing the increment from 1 to 0.5 and 0.25% resulted in noticiable reduction in the upsetting load.

It was also pointed out that the discrepancies could be attributable to differences in many other factors such as the way of determining the equivalent strain, iteration criterion, way of maintaining incompressibility, etc. which were not specified in the present Project. Nevertheless, all the participators were so optimistic that they felt that they would reach satisfactory ends in the near future, if the Project would be continued with better orientation

---

* This will be confirmed if the side profiles obtained by YY et al. with 48 and 308 elements are compared, see the curves marked by YY(48e) and YY(308e) in Fig. 4(a).

and information exchange. One of the participators who had conducted some benchmark tests in other fields of application of the finite element method stated that the present discrepancies were not surprising as the result of the first round and that even mere recomputation carried out by individuals would greatly reduce the discrepancies.*

During the Round Table Discussion, there were also some criticisms and suggestions from the floor. It was pointed out that the condition of sticking was ambiguous or physically impossible and that the present Project was meaningless since computational details had not been specified beforehand. It was true that some attendants felt that the numerical methods introduced there were no more than an academic play. It was also suggested that the computed results should be compared with results of a carefully controlled experiment.

Finally, it was agreed among the participators that they would continue the present Project to achieve the anticipated end and that Prof. E.H. Lee and his collaborators at the University of Stanford would serve hereafter as the planner and compiler.

Achnowledgement – The present reporters thank all the participators in the Project who submitted their results. They also acknowledge Prof. H. Lippmann for providing us with the opportunity of putting forward the Project and Miss Y. Harada of their Department for preparing the typescript.

---

* According to an agreement made at the Round Table Discussion, some of the results were revised after the Symposium and were sent to the present reporters. The present report includes those new results; as a matter of fact, they are more consistent with the others than were the previous ones.

MIX
Papier aus verantwortungsvollen Quellen
Paper from responsible sources
FSC® C105338

If you have any concerns about our products,
you can contact us on
ProductSafety@springernature.com

In case Publisher is established outside the EU,
the EU authorized representative is:
Springer Nature Customer Service Center GmbH
Europaplatz 3, 69115 Heidelberg, Germany

Printed by Libri Plureos GmbH
in Hamburg, Germany